**Earth Observation for Land
and Emergency Monitoring**

Earth Observation for Land and Emergency Monitoring

Edited by Heiko Balzter

National Centre for Earth Observation
University of Leicester
Centre for Landscape and Climate Research
Department of Geography
Leicester, UK

This edition first published 2017
© 2017 John Wiley & Sons Ltd

Registered Office
John Wiley & Sons Ltd, The Atrium, Southern Gate, Chichester, West Sussex, PO19 8SQ, UK

Editorial Office
The Atrium, Southern Gate, Chichester, West Sussex, PO19 8SQ, UK

For details of our global editorial offices, customer services, and more information about Wiley products visit us at www.wiley.com.

Wiley also publishes its books in a variety of electronic formats and by print-on-demand. Some content that appears in standard print versions of this book may not be available in other formats.

Library of Congress Cataloging-in-Publication data applied for

ISBN: 9781118793794 (hardback)

Cover Image: (Background) © Matt Champlin/Gettyimages;(Satellite) © Andrey Armyagov/Shutterstock
Cover Design: Wiley

Set in 10/12pt Warnock by SPi Global, Pondicherry, India
Printed and bound in Malaysia by Vivar Printing Sdn Bhd

10 9 8 7 6 5 4 3 2 1

Contents

List of Contributors

Heiko Balzter
National Centre for Earth Observation
University of Leicester
Centre for Landscape and Climate Research
Department of Geography
Leicester, UK

P. Rodriguez-Veiga
National Centre for Earth Observation
University of Leicester
Centre for Landscape and Climate Research
Department of Geography
Leicester, UK

S. Saatchi
NASA Jet Propulsion Laboratory
Pasadena
California, USA

J. Wheeler
University of Leicester
Centre for Landscape and Climate Research
Department of Geography
Leicester, UK

K. Tansey
University of Leicester
Centre for Landscape and Climate Research
Department of Geography
Leicester, UK

M.A. Stelmaszczuk-Górska
Friedrich-Schiller-University Jena
Department of Earth Observation
Jena, Germany

C.J. Thiel
Friedrich-Schiller-University Jena
Department of Earth Observation
Jena, Germany

C.C. Schmullius
Friedrich-Schiller-University
Department of Earth Observation
Jena, Germany

N.J. Tate
University of Leicester
Department of Geography
Leicester, UK

B.F. Spies
Airbus Defence and Space
Geo-Intelligence
Meridian Business Park
Leicester, UK

A. Lamb
Airbus Defence and Space
Geo-Intelligence
Meridian Business Park
Leicester, UK

S. Shrestha
University of Leicester
Centre for Landscape and Climate
Research
Department of Geography
Leicester, UK

C. Smith
University of Leicester
Centre for Landscape and Climate
Research
Department of Geography
Leicester, UK

Z. Bochenek
Remote Sensing Centre
Institute of Geodesy and Cartography
Warsaw, Poland

M. Chernetskiy
Friedrich-Schiller-University
Institute of Geography, Department for
Earth Observation
Jena, Germany

N. Gobron
EC Joint Research Centre
Ispra (VA), Italy

J. Gómez-Dans
University College London (UCL)
London, UK

P. Lewis
University College London (UCL)
London, UK

S.J. van Beijma
Airbus Defence and Space
Geo-Intelligence
Europa House
Southwood Crescent
Farnborough, UK

A. Comber
Chair in Spatial Data Analytics
School of Geography
University of Leeds
Leeds, UK

D. Stratoulias
Balaton Limnological Institute
Centre for Ecological Research
Hungarian Academy of Sciences
Tihany, Hungary

I. Keramitsoglou
Institute for Astronomy, Astrophysics,
Space Applications and Remote Sensing
National Observatory of Athens
Pendeli
Athens, Greece

P. Burai
Envirosense Hungary Kft.
Debrecen, Hungary

L. Csaba
Envirosense Hungary Kft.
Debrecen, Hungary

A. Zlinszky
Balaton Limnological Institute
Centre for Ecological Research
Hungarian Academy of Sciences
Tihany, Hungary

V.R. Tóth
Balaton Limnological Institute
Centre for Ecological Research
Hungarian Academy of Sciences
Tihany, Hungary

S.C.J. Palmer
Balaton Limnological Institute
Centre for Ecological Research
Hungarian Academy of Sciences
Tihany, Hungary

V. Nicolás-Perea
University of Leicester
Centre for Landscape and Climate
Research
Department of Geography
Leicester, UK

P. Kourkouli
GAMMA Remote Sensing AG
Gümligen
Switzerland

U. Wegmüller
GAMMA Remote Sensing AG
Gümligen
Switzerland

A. Wiesmann
GAMMA Remote Sensing AG
Gümligen
Switzerland

J. Papke
GAMMA Remote Sensing AG
Gümligen
Switzerland

T. Strozzi
GAMMA Remote Sensing AG
Gümligen
Switzerland

C. Fourie
German Aerospace Center (DLR)
German Remote Sensing Data Center
(DFD)
Oberpfaffenhofen, Germany

E. Schoepfer
German Aerospace Center (DLR)
German Remote Sensing Data Center (DFD)
Oberpfaffenhofen, Germany

L. Moser
German Aerospace Center (DLR)
German Remote Sensing Data Centre
(DFD)
Oberpfaffenhofen, Germany

A. Schmitt
German Aerospace Center (DLR)
German Remote Sensing Data Center
(DFD)
Oberpfaffenhofen, Germany

S. Voigt
German Aerospace Center (DLR)
German Remote Sensing Data Center
(DFD)
Oberpfaffenhofen
Germany

M. Ofwono
Institute of Geodesy and Cartography
Warsaw, Poland

K. Dabrowska-Zielinska
Institute of Geodesy and Cartography
Warsaw, Poland

J. Kaduk
University of Leicester
Centre for Landscape and Climate
Research
Department of Geography
Leicester, UK

Preface

The material presented in this book was produced within the European Centre of Excellence in Earth Observation Research Training, GIONET. GIONET was funded by the European Commission's Marie Curie Programme as an Initial Training Network, Grant Agreement number PITN-GA-2010-264509.

I was the scientist-in-charge of this four-year project. The 14 individual early-career researchers employed by GIONET were based in seven partner organizations, and I want to acknowledge the scientific team leaders involved in this work: Andreas Wiesmann and Urs Wegmuller at Gamma Remote Sensing Ltd, Switzerland; Chris Schmullius from Friedrich-Schiller-University Jena in Germany; Alistair Lamb at Airbus Defence and Space in the UK; Katarzyna Dabrowska-Zielinska, Stan Lewinski, Martyna Gatkowska, Zbigniew Bochenek and Agata Hoscilo at the Institute of Geodesy and Cartography in Poland; Viktor Toth at the Balaton Limnological Institute Ecological Research Centre of the Hungarian Academy of Sciences in Hungary and Stefan Voigt and Elisabeth Schöpfer at the German Aerospace Center.

Many associated partners also contributed to GIONET by participating in meetings, summer schools and by hosting researchers during secondments, and I want to acknowledge in particular Etienne Bartholomé and Andreas Brink from the Joint Research Centre of the European Commission in Ispra, Geoff Smith from SpectoNatura in the UK, Ralph Humberg and Andreas Kühnen from Trimble in Germany, Nick Fernando and Gareth Crisford from ITT-VIS in the UK, and the partners who joined us during the course of the project, not least Pablo Rosso from Blackbridge in Germany, Zofia Banaszak from Danko Plantbreeding in Poland, Csaba Lenart from Envirosense in Hungary, Leif Eriksson from Chalmers University in Sweden, Iphigenia Keramitsoglou from the National Observatory in Athens, Greece, Pietro Teatini from University of Padova in Italy, Peter Hunter from University of Stirling in the UK and Chris Schmullius in her role for Earth Observation Services, Germany.

The PhD supervisory team at the University of Leicester is also gratefully acknowledged, and includes my dear colleague and former landlord the late Pete Fisher, Kevin Tansey, Claire Smith, John Remedios, Nick Tate and Lex Comber.

Particular thanks go to Virginia Nicolas-Perea who supported the GIONET project as project manager and PhD advisor.

And finally, the main chapter authors are acknowledged for their efforts as incumbents of the 14 GIONET researcher positions, namely Pedro Rodriguez Veiga, James Wheeler, Sybrand Van Beijma, Matthew Ofwono, Shailesh Shrestha, Christoff Fourie, Linda Moser, Penelope Kourkouli, Jessika Papke, Stephanie Palmer, Martyna Stelmaszczuk-Gorska, Maxim Chernetskiy, Dimitris Stratoulias and Bernard Spies.

1

Earth Observation for Land and Emergency Monitoring Core Services

Heiko Balzter

National Centre for Earth Observation, University of Leicester, Centre for Landscape and Climate Research, Department of Geography, Leicester, UK

The Copernicus programme is Europe's flagship operational Earth Observation programme. It is comprised of a number of core services aiming at specific user groups. Many policies and international initiatives rely on Earth Observation to deliver information services. The downstream development of new satellite applications is a rapidly growing global market and creates jobs, economic growth and prosperity for societies.

The Copernicus initiative is delivering many core monitoring services of the oceans, the land surface, air quality, climate change and the polar ice sheets. It is a laudable programme in its aspiration to provide operational long-term observations of critical parameters from space. However, it has limitations and there is room for developing new and improved innovative information services around the existing Copernicus service portfolio.

The initial operations of the European Copernicus programme from 2011 to 2014 have delivered a comprehensive range of satellite applications in support of sustainable forestry. The Geoland-2 project has established the operational Copernicus land monitoring core service, which is now implemented with a global, European, local and in-situ component. Global data products to support sustainable agriculture and forestry include surface albedo, fractionally absorbed PAR (FAPAR), fraction of PAR absorbed by vegetation for photosynthesis processes, Leaf Area Index (LAI), Top of Canopy spectral reflectance, Fractional cover (Fcover), Normalized Difference Vegetation Index (NDVI), Vegetation Condition Index (VCI), Vegetation Productivity Indicator (VPI), Dry Matter Productivity (DMP), burnt area, active fires, land surface temperature, soil moisture, areas of water bodies, water level (lakes and rivers), and vegetation phenology at 1 km resolution.

At the European scale, the vector-based CORINE land cover (reference year 1990) is being updated (last produced in 2000 and 2006, and currently being updated to 2012). It consists of 44 land cover classes and uses a Minimum Mapping Unit (MMU) of 25 ha area, or a minimum width of 100 m for linear landscape structures. Land cover changes are mapped with an MMU of 5 ha by visual interpretation of high-resolution satellite imagery. CORINE has a wide range of applications, underpinning the European

Communities policies in the domains of environment, but also agriculture, transport, spatial planning etc. The High-Resolution-Layers (HRL) at 100 m spatial resolution include two forestry data products: tree cover density and forest type. In GIO-land an additional two forest products are being produced for the European Commission's Joint Research Centre (JRC): tree cover presence/absence, and dominant leaf type at 25 m spatial resolution. The tree cover density dataset maps the level of tree cover density in a range from 0–100%, has no MMU (minimum number of pixels to form a patch) and a minimum mapping width of 20 m. The forest type products in their original 20 m resolution version consists of the dominant leaf type (MMU of 0.5 ha, 10% tree cover density threshold applied), and a support layer showing trees under agricultural use and in urban contexts (derived from CORINE and imperviousness 2009 data). For the final 100 m product trees under agricultural use and urban context from the support layer are removed.

This book introduces the reader to the outcomes from four years of research in support of the Copernicus Land Monitoring Core Service and the Emergency Monitoring Core Service.

The research was funded by the Marie Curie PEOPLE programme in Framework Programme 7, as an Initial Training Network. The GIONET project established a European Centre of Excellence in Earth Observation Research Training in 2011, when Copernicus was called "GMES" (Global Monitoring for Environment and Security), and just entered into its GMES Initial Operations phase (GIO).

GIONET trained 14 PhD researchers in academia, industry, and research centres in advanced remote sensing skills, accompanied by interpersonal, entrepreneurship and management skills. Seven organizations from five European countries employed the researchers and were supported by a large group of associated partners.

This book is structured into thematic chapters, covering Forest Monitoring (Part I), Land Cover and Land Cover Change Monitoring (Part II), Coastal Zone and Freshwater Monitoring (Part III), Land Deformation Mapping and Humanitarian Crisis Response Strategies (Part IV) and Earth Observation for Climate Adaptation (Part V). A Conclusions chapter summarizes the main findings presented in the book.

The UN initiative "Reducing Emissions from Deforestation and Forest Degradation" (REDD+) provides a strong user pull for forest information from space. In Part I on forest monitoring a concept for global forest biomass mapping is presented, making use of geographically varying forest allometric models, spaceborne profiling LiDAR (ICESAT-GLAS) and Synthetic Aperture Radar (SAR). Synergies between multi-temporal and multi-frequency interferometric radar and optical satellite data for biomass mapping and change detection are discussed and a SAR mapping application to the Congo Basin presented.

Conceived in 1985 as the CORINE programme, land cover monitoring is the most operational element of the Copernicus programme. The methodology remains largely unchanged. Part II on land cover and land cover change monitoring presents approaches that go beyond the current implementation of largely optical/near-infrared based land cover monitoring methods. Classification methods with multi-frequency, multi-temporal SAR data over semi-arid and forested African landscapes are explained and contrasted against the capabilities of optical-near-infrared high-resolution satellite images. A methodological framework for multi-scale remote sensing concludes this chapter.

The European Water Framework Directive requires monitoring of the ecological status and water quality of all major water bodies and their habitats. Earth Observation is only beginning to influence this application area. In Part III on coastal zone and freshwater monitoring, a study of salt marsh habitats in Wales is presented. Salt marshes are regarded as effective buffers against sea level rise and can be mapped with multi-sensor data to support Integrated Coastal Zone Management. Freshwater applications focus on the ecology of emergent and submerged macrophytes in Lake Balaton, Hungary, using airborne hyperspectral and LiDAR remote sensing to map the extent of reed dieback syndrome, and satellite remote sensing to map and monitor optically active water quality parameters, such as chlorophyll-a as a proxy for phytoplankton biomass and blooms.

The recent past was characterized by many humanitarian crises and natural disasters. Part IV of this book describes the use of radar interferometry for land deformation mapping applications and demonstrates the use of machine learning algorithms in the context of humanitarian crisis response strategies. After a short review on radar interferometry, a new hybrid method using Differential SAR Interferometry/Persistent Scatterer Interferometry for ground-motion monitoring from spaceborneSAR data is demonstrated and applied to different land cover types. Chapter 12 describes the use of spaceborne SAR and ground-based radar interferometry for mapping landslide displacements in the Swiss Alps. New methods for the detection of small-scale land surface feature changes in complex humanitarian crisis situations are demonstrated, transferring machine learning algorithms to environmental remote sensing.

With an increasing likelihood that mankind is unable or unwilling to respond effectively to the causes of climate change, there is a widening recognition that we will have to adapt to its impacts. In Part V on Earth Observation for climate adaptation, a study on remote sensing of wetland dynamics as indicator of water availability in semi-arid Africa is presented, using time series of optical and SAR satellite imagery. Satellite observations of drought events and crop stress in Europe conclude this chapter.

The book presents a collection of original research findings interspersed with selected review chapter and intends to serve as a compendium on the state-of-the-art in remote sensing in support of land and emergency monitoring going beyond the current operational monitoring services in Copernicus.

I am grateful to the GIONET team for the inspiring and productive work over the past four years, in particular to my colleagues at the University of Leicester, UK, Airbus Defence and Space (formerly Astrium GEO-Information Services), UK, Gamma Remote Sensing AG, Switzerland, the Institute of Geodesy and Cartography (IGIK) in Warsaw, Poland, Friedrich-Schiller-University in Jena, Germany, the Hungarian Academy of Sciences – Centre for Ecological Research and the German Aerospace Center (DLR), and the associated partners in the Joint Research Centre of the European Commission in Ispra, Italy, University of Stirling, UK, University of Padova, Italy, the National Observatory of Athens, Greece, Chalmers University in Sweden, and the companies Trimble, Germany, EXELIS Visual Information Solutions Ltd., UK, SpectoNatura, UK, BlackBridge, Germany, DANKO Plant Breeding in Poland, Envirosense, Hungary, and Earth Observation Services in Germany.

It has been a privilege and a pleasure to coordinate the international team of 14 early-stage researchers who were working towards their doctoral degrees in this unique international research environment.

Part I

Forest Monitoring

2

Methodology for Regional to Global Mapping of Aboveground Forest Biomass: Integrating Forest Allometry, Ground Plots, and Satellite Observations

P. Rodriguez-Veiga[1,2], S. Saatchi[3], J. Wheeler[1], K. Tansey[1] and Heiko Balzter[1,2]

[1] *University of Leicester, Centre for Landscape and Climate Research, Department of Geography, Leicester, UK*
[2] *National Centre for Earth Observation, University of Leicester, Leicester, UK*
[3] *NASA Jet Propulsion Laboratory, Pasadena, California, USA*

2.1 Forests and Carbon

Earth is undergoing significant global environmental change. The processes linked to global change are affecting the whole climate system and impacting human civilization. Understanding the effects and causes of these processes will assist human societies in devising adaptation and mitigation strategies. The United Nations Framework Convention on Climate Change (UNFCCC) and the Kyoto Protocol address the importance of reducing and monitoring greenhouse gas emissions (GHG), with CO_2 being the most significant trace gas. Changes in the amount of atmospheric CO_2 due to anthropogenic activities are altering the biogeochemical cycles that allow the recycling and reuse of carbon on Earth (global carbon cycle), and produce changes in weather patterns [1].

   How the global carbon cycle stores and exchanges carbon within the system is crucial to understand interactions and feedbacks with the climate system. The locations where the carbon is stored within the global carbon cycle are called carbon pools, and the rates of carbon exchanged between pools are known as fluxes, and are classified in sources (emission to the atmosphere) and sinks (uptake from the atmosphere). Knowledge of both carbon pools and fluxes is essential to understand the global carbon cycle. Terrestrial ecosystems play a vital role in the global carbon cycle. The terrestrial carbon pool is about three times bigger than the atmospheric pool [1], and removes 30% of anthropogenic emissions from fossil fuel combustion from the atmosphere [2]. The primary source of terrestrial carbon emissions is from anthropogenic land use change; especially deforestation in the tropics, while afforestation, reforestation and growth of existing forest is the major contribution to the terrestrial sink term. Terrestrial ecosystems appear to act as a net sink [3], but there are significant uncertainties on the carbon fluxes between land and atmosphere in comparison with the other fluxes, still making terrestrial carbon pools and fluxes one of the major remaining uncertainties in climate science [4–8].

Global Forests play an important role in the global carbon cycle as they cover approximately 30% of the land surface and store 45% of terrestrial carbon in the form of biomass via photosynthesis, which sequesters large amounts of carbon per year [8]. Forest accumulates carbon primarily in the form of living aboveground biomass of trees (*AGB*). Forest *AGB* is living organic plant material composed of 50% carbon [9] as well as hydrogen and oxygen, and it is usually defined for a given area. When forests are degraded, cleared or burned, large amounts of this carbon are released into the atmosphere as carbon dioxide and other compounds. Deforestation is the second largest anthropogenic source of carbon dioxide to the atmosphere, after fossil fuel combustion and the largest source of greenhouse gas emissions in most tropical countries [10]. Thus, monitoring *AGB* stored in the world's forests is essential for efforts to understand the processes related to the global carbon cycle and reducing carbon emissions originating from deforestation and forest degradation.

Biomass is an Essential Climate Variable (ECV) required by the Global Climate Observing System (GCOS) to support the work of the UNFCCC and the Intergovernmental Panel on Climate Change (IPCC) in monitoring climate change. Accurately monitoring and reporting the biomass or carbon content of forests (carbon stocks) is a requirement of different international mechanisms based on economic incentives that have been launched by the international community aiming to mitigate climate change, such as "Reducing Emissions from Deforestation and forest Degradation" (REDD+). Global estimates of *AGB* carbon stocks have been produced in the past to support the monitoring of CO_2 emissions from deforestation and land use change. However, the size and spatial distribution of forest *AGB* is still uncertain in most parts of the planet due to the difficulties measuring *AGB* at the ground level [11]. Very few global *AGB* carbon estimates are spatially explicit. Approaches that make full use of remote sensing techniques to estimate *AGB* are therefore needed.

This chapter will discuss current efforts to monitor forest and *AGB* at a global scale using traditional methods such as forest inventory ground measurements and more advanced methods based on Earth Observation data. Earth Observation is a very powerful tool to measure forest resources worldwide in an objective, efficient, and affordable manner. Earth Observation satellites use remote sensors that have different advantages and limitations to measure forest biomass. A synergistic use of different datasets and sensors is presented in this chapter as the key to extract the full potential from earth observation methods.

2.2 Using Earth Observation Imagery to Measure Aboveground Biomass

Three broad types of remote sensors are used by Earth Observation platforms: Optical, Synthetic Aperture Radar (SAR), and LiDAR. Each type of sensor has different characteristics, which make them suitable for monitoring forest vegetation. Detailed explanation and examples on the use of different earth observation sensors and techniques to estimate *AGB* can be found in Chapter 3 of this book or in other literature [e.g. 12]. The following is a brief summary of techniques and sensors available to measure biophysical parameters of vegetation.

Through optical remote sensing, it is possible to estimate a series of different vegetation indices such as Leaf Area Index (LAI) and Normalized Difference Vegetation Index (NDVI), which are mainly related to the photosynthetic components of the vegetation and therefore indirectly to *AGB*. This relies on an empirical relationship between green foliage and total *AGB*, however. In reality, forest *AGB* is primarily composed of the non-photosynthetic parts of trees like trunks and branches. Forest *AGB* can nevertheless be indirectly estimated from optical sensors based on the sensitivity of the reflectance to variations in canopy structure. Most optical approaches are based on this relationship in which signal retrieval is calibrated with ground measurements to model the spatial distribution of *AGB* across the landscape. Several studies have mapped *AGB* at different scales (medium to coarse resolution) relating ground measurements to the signal retrieved from optical sensors such as Landsat or MODIS [e.g. 13–15].

Optical sensors have great advantages for global vegetation monitoring. Vegetation can be easily differentiated from other surfaces due to its strong reflectance in near infrared and visible green, as well as absorption in the red and blue sections of the visible spectrum [16]. Optical sensors have been operating for a long time and have a rich archive that can be used to study vegetation changes. For example, the Landsat mission has global coverage of observations over the last 40 years. Another advantage of optical sensors is that coarse and medium resolution imagery they produce can usually be obtained for free or at a low cost. The main shortcoming of optical imagery is cloud cover, as the sensors cannot "see" through clouds. This is not crucial in boreal or temperate latitudes, but can be a problem in tropical areas where there are few days a year without cloud cover. Moreover, as passive sensors, they can only operate during daylight, which reduces the number of potential revisit times in comparison with active sensors like SAR or LiDAR. Thus, the chances to obtain a cloud-free image are also diminished. The way to overcome this problem is through the use of radiometrically consistent multi-temporal datasets, but this is costly, technically demanding, and time-consuming [13,17]. Estimation of *AGB* by optical sensors also faces the saturation of the signal retrieval at low *AGB* stocks [10] as the signal retrieved from vegetation depends on the absorption of light from the photosynthetic parts of the plants. Optical imagery is suitable for forest area mensuration, vegetation health monitoring, and forest classification, but presents limited correlation with *AGB* after canopy closure.

Radars are active sensors, which generate their own electromagnetic signal. They are independent of solar illumination of the target area, being able to obtain day and night observations, as well as to penetrate through haze, clouds and smoke. SAR is an airborne or spaceborne side-looking radar system that uses its relative motion, between the antenna and its target region, to provide distinctive long-term coherent signal variations used to generate high-resolution remote sensing imagery (Figure 2.1).

Each SAR satellite works within a specific radar frequency bandwidth (with corresponding wavelength), which is used to classify them in increasing wavelength size as X-, C-, S-, L- or P-band sensors. Several SAR satellites are currently operating (in orbit), including the new L-band ALOS-2 PALSAR, which was launched in May 2014 (Table 2.1).

The radar backscatter (the amount of scattered microwave radiation received by the sensor) is related to *AGB* as the electromagnetic waves interact with tree scattering elements like leaves, branches and stems, but their sensitivity to *AGB* depends on the radar wavelength [18]. Shorter wavelengths are sensitive to smaller canopy elements (X- and C-band), while longer wavelengths (L- and P-band) are sensitive to branches

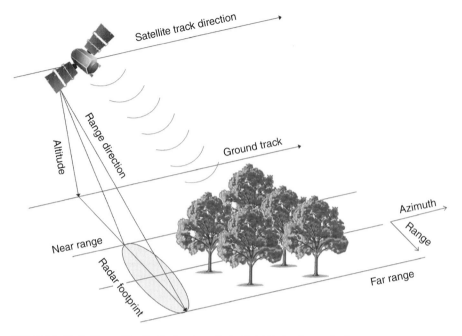

Figure 2.1 Illustration of Synthetic Aperture Radar Satellite basic terminology.

Table 2.1 Operating or planned satellites used for forest monitoring.

Sensor	Wavelength	*AGB saturation	Operating satellites	Planned satellites
LIDAR	Visible/near-infrared (532 & 1064 nm)	No limit		ICESat 2, GEDI (sensor attached to ISS)
SAR	P Band (30–100 cm)	100–200 t ha^{-1}		BIOMASS
	L band (15–30 cm)	40–150 t ha^{-1}	ALOS-2 PALSAR	SAOCOM 1A, 1B NISAR
	S band (7.5–15 cm)	Not reported	Huanjing 1C	NovaSAR-S NISAR
	C band (3.8–7.5 cm)	20–50 t ha^{-1}	Radarsat 1 Radarsat 2 Sentinel 1	RADARSAT Constellation
	X band (2.4–3.8 cm)	<20 t ha^{-1}	TerraSAR X Cosmo/SkyMed Tandem X	Paz
OPTICAL	Visible/near-infrared (380 nm–1 mm)	15–70 t ha^{-1}	High Resolution satellites, Terra/Aqua MODIS, Terra ASTER, SPOT 6 & 7, Landsat 7 & 8, EO-I, DMC constellation, Sentinel 2 & 3, PROBA V, and others	High Resolution satellites, Landsat 9, Ingenio, Amazonia, CBERS 4 & 4B, and others

* Range of *AGB* saturation thresholds found in the literature [21,22,25–28].

and stems [19]. Longer wavelengths are theoretically more suitable for estimation of *AGB* as tree branches and stems comprise the highest percentage of *AGB* in forests. SAR backscatter sensitivity using L-band usually saturates at around $100–150\,\mathrm{t\,ha^{-1}}$ [20,21]. However, other authors have found higher saturation values of more than $250\,\mathrm{t\,ha^{-1}}$ for L-band [22], and even more than $300\,\mathrm{t\,ha^{-1}}$ when combined with other SAR datasets such as X-band [23]. Nevertheless, there is no current satellite sensor in orbit (neither optical nor radar) that can offer a reasonable relationship between the observations and the high values of *AGB* often found in tropical areas ($>400\,\mathrm{t\,ha^{-1}}$). Even though a P-band sensor is very promising [24], at the moment there is only one planned satellite, the ESA Earth Explorer 8 BIOMASS mission [11], which will not be launched before 2021. The future P-Band BIOMASS mission by ESA has the following accuracy requirements at pixel level ($200\,\mathrm{m} \times 200\,\mathrm{m}$): an RMSE of $\pm10\,\mathrm{t\,ha^{-1}}$ for *AGB* below $50\,\mathrm{t\,ha^{-1}}$, and a relative error of $\pm20\%$ for *AGB* above $50\,\mathrm{t\,ha^{-1}}$.

LiDAR technology consists of optical active sensors transmitting laser pulses to measure the distance to the target. LiDAR remote sensing systems can be classified according to:

- platforms: spaceborne, airborne, or ground-based
- returned signals: discrete return or wave form
- scanning pattern: profiling or scanning
- footprint[1] size: small footprint: (<1 m diameter), medium footprint (10–30 m diameter), and large footprint (>50 m diameter).

Airborne imaging LiDAR provides direct and very accurate measurements of canopy height. LiDAR sensors do not suffer from signal saturation, as optical and radar sensors do, because the signal can penetrate the canopy. Nevertheless, the vertical extent of each waveform increases as a function of terrain slope and footprint size, making this information insufficient over sloped terrain to estimate canopy height [29]. However, the use of airborne and ground platforms would be too costly and impractical at national, continental or global level [10].

The only spaceborne profiling LiDAR sensor was the Geoscience Laser Altimeter System (GLAS) that was aboard the NASA Ice, Cloud, and land Elevation (ICESat). This satellite operated between 2003 and 2010. The GLAS LiDAR sensor on board of ICESat scanned the globe following a profiling pattern, and produced a global coverage of large full waveform signal footprints. ICESat sampled millions of approximately 65 m diameter footprints every 172 m along track in between 2003 and 2009. However, the vertical extent of each GLAS waveform increases as a function of terrain slope and footprint size, making this information insufficient over sloped terrain to estimate canopy height [29]. There is no current LiDAR satellite in orbit at the moment. ICESat-2 will be launched in 2017, and the Global Ecosystem Dynamics Investigation LiDAR (GEDI) mission, which will attach a LiDAR profiling sensor to the International Space Station (ISS), will not be operative until 2020. These profiling sensors cannot be used alone to produce wide area *AGB* mapping, but they are very useful in combination with other Earth Observation datasets [e.g. 30,31].

---

1 Area illuminated by the laser and from which the waveform-return signal gives information.

2.3   Global Forest Monitoring

The first challenge to monitoring forests at a global scale is the definition of forest itself, and consequently the definitions of deforestation and forest degradation. Forests are ecosystems dominated by trees and other woody vegetation, but there are approximately 1500 definitions of forest worldwide based on administrative, cover, use or ecological characteristics [32]. These different definitions are based on different concerns and interests of people and states. Legal definitions greatly differ from ecological or traditional definitions, though the characteristics and thresholds are more clearly defined. These definitions are mostly focused on setting the minimum physical thresholds for a vegetated ecosystem to be considered as a forest. Unfortunately, there is no universally agreed definition of forest (Figure 2.2). This situation makes any study at global scale using data generated at national level very complicated.

The remote sensing approaches allow the study of forest vegetation from a physical perspective. Therefore, the same vegetation thresholds defining forest, deforestation and forest degradation can be applied globally. The downside of this physical approach is that other types of woody vegetation such as oil palm plantations, which are responsible for large-scale deforestation in tropical areas, are sometimes included in the forest class, especially when using coarse or medium-resolution optical sensors to monitor the forest [e.g. 34]. To overcome this challenge, ancillary data with the location of these plantations or accurate remote sensing methods to differentiate

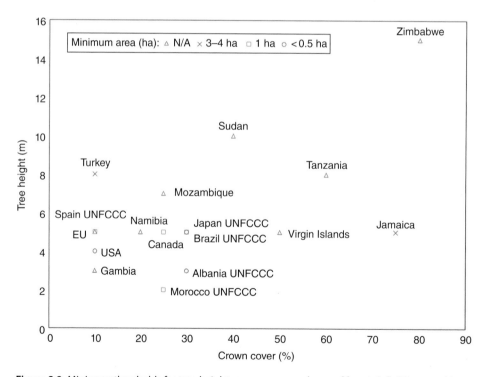

Figure 2.2 Minimum thresholds for tree height, crown cover and area of forest definitions used in different countries. Modified from [33]. Data from [32].

these plantations from natural forest have to be implemented. Long-wavelength Synthetic Aperture Radar (SAR) sensors could play an important role as the radar signal is sensitive to forest structure such as the regular spacing patterns observed in oil palm plantations [35].

Several products are globally or continentally produced to monitor forest changes (Table 2.2). These projects are generally based on in situ data and satellite optical sensors with medium to coarse spatial resolutions. The use of airborne and high-resolution sensors is restricted to sub-national level or project level, as it could be impractical and the cost prohibitive at country, continental or global scale. Nevertheless, the use of these sensors on demand can be extremely valuable for monitoring specific hot spots where deforestation is a major issue, as well as to assist in the validation of medium-resolution products. The products developed by these programmes are mostly created using optical imagery, which requires a complex and extensive data processing chain in order to produce consistent global products. The main parameters measured by these data projects are forest cover and forest type. Most of these products lack the capabilities to produce spatial *AGB* estimates.

The Forest Resource Assessments (FRAs) are based on the analysis of forest inventory information supplied by each country and supported by expert judgements, remote sensing and statistical modelling. A National Forest inventory is the most widely used method for in situ forest monitoring due to its historic roots in national forestry administrations, its accuracy and low technical requirements. A forest inventory is a systematic collection of forest data for assessment or analysis. The approach consists of sample-based statistical methods, sometimes in combination with remote sensing and aerial imagery. In developing countries where the labour cost is low, the use of forest inventories could be a relatively cost-effective approach. The FRAs analyse information on forest cover, forest state, forest services and non-wood forest products. However, it was not until 2000 that a single technical definition for forest was used (10% crown cover). Changes in baseline information, inconsistent methods and definitions through the different FRAs make their comparison difficult [39]. Several authors have questioned the country-level estimates of forest carbon stocks reported by the FRAs due to inadequate sampling for the national scale, inconsistent methods, and in most tropical countries figures that were based on 'best guesses' instead of actual measurements [7,10,47].

The Global Remote Sensing Survey (RSS) implemented in 2009 was a systematic sampling based on units located at longitude and latitude intersections worldwide. Each sample unit consist of Landsat imagery covering an area of 10 km × 10 km, which was automatically classified into forest/non-forest areas. The survey reported estimates of forest area, deforestation and afforestation at global, continental and ecological zone level for 1990, 2000 and to 2005.

The Moderate Resolution Imaging Spectroradiometer (MODIS) sensor on board the Terra and Aqua satellites provides biophysical parameter datasets, which allow monitoring of biosphere dynamics. MODIS Vegetation Continuous Fields (VCF) is a sub-pixel-level representation of surface vegetation cover estimates globally [48]. The percent canopy cover per MODIS pixel refers to the amount of sky obstructed by tree canopies equal to or greater than 5 m in height [48], which agrees with the UN Food and Agriculture Organization (FAO) definition of forest. The current version (collection 5) has been published with 250 m resolution globally. Initial results show that this version of the product

Table 2.2 Global Forest Monitoring Programmes.

Programme or study	Agency	Data source	Spatial resolution	Temporal coverage	Key issues and reference
Forest Resource Assessment (FRA)	FAO	National Forest Inventories	N/A	Every 5 years	Data sources are not globally available [36–39]
Global Remote Sensing Survey (RSS)	FAO	Landsat	N/A	1990, 2000, 2005	Systematic sample (10 km × 10 km) of Landsat imagery worldwide [40]
ALOS Kyoto and Carbon Initiative	JAXA	ALOS PALSAR	25 m/50 m/100 m	Annual 2007–10	L-band SAR imagery & Forest/Non-Forest mosaics [41–44]
Tree Cover Loss & Gain	University of Maryland	Landsat	30 m	Annual 2000–12	Identifies areas of tree cover loss (annual) and gain (12 years cumulative) [34]
Global Forest Watch	World Resources Institute	Landsat, MODIS, and others	30 m – 5 km [1]	Monthly quarterly, and annual from year 2000	Forest change, cover and use, alerts, crowdsourcing, etc.
Vegetation Continuous Fields	University of Maryland, NASA	MODIS	250 m, 500 m, 1 km	Annual 2000–10	Tree cover percentage at sub-pixel-level [45]
Tree Cover Continuous Fields	University of Maryland, NASA	Landsat	30 m	2000	Tree cover percentage at sub-pixel-level [46]
GlobCover	ESA	ENVISAT	300 m	2005/06 & 2009	Labelled according to the UN Land Cover Classification System
MODIS Land Cover Type	NASA	MODIS	500 m	Annual 2001–12	5 Global Land Cover Classification Systems
COPERNICUS Global Land Service	GEOLAND-2, ESA	SPOT, and others	1 km	Several intervals from 1999	Vegetation Biophysical parameters
Biomass Geo-Wiki	Several	Several [2]	30 m–0.01 grad	2000–10	Comparison AGB maps

[1] Landsat based products present 30 m spatial resolution while deforestation alerts, based on MODIS, go from 250 m up to 5 km resolution.
[2] Each product present different data sources, spatial coverage, and methods.

is substantially more accurate (50% improvement in RMSE) than the previous 500 m version [45]. The pixel size of 250 m (ca. 6.25 ha) is still far from a pixel size of 71 m, which would be the minimum resolution that could detect a minimum unit area of forest (0.5 ha) according to the main forest definitions [49]. Nevertheless, VCF collection 5 currently has the best temporal coverage (from 2000) among the coarse resolution global forest monitoring products that are free of charge. Following the success of MODIS VCF, a recent 30 m resolution Tree Cover Continuous Fields (VCF) dataset has been developed, re-scaling the 250 m MODIS VCF with Landsat imagery [46] (Figure 2.3).

A data mining approach of the Landsat archive by means of the Google Earth Engine was also used to globally quantify annual forest loss (2000–12) as well as 12 years of cumulative forest gain at 30 m spatial resolution [34]. This dataset together with others such as FORMA alerts [50], which provide tree cover lost alerts every 16-days interval, can be freely downloaded and visualised on the website of the Global Forest Watch (www.globalforestwatch.org/). This site is a web-platform that aims to provide reliable

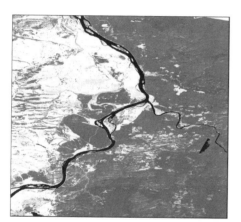
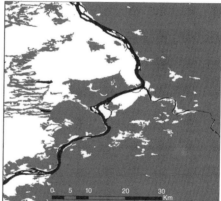
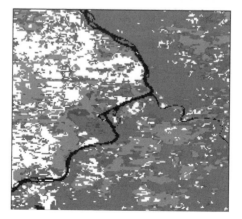

Figure 2.3 Area in Central Siberia. Left: Landsat Tree Cover Continuous Fields 30 m resolution (2000) (Raw data: [46]), Centre: Forest/Non Forest K&C Initiative Product 50 m resolution (2010) (Source data provided by JAXA as the ALOS high level product © JAXA, METI), Right: GlobCover 300 m resolution (2009) (Source Data: © ESA / ESA GlobCover Project, led by MEDIAS-France/POSTEL). Green colours (grey colours in printed version) denote forest, black colour water bodies, and white colour non-forest area.

information about forest to interested stakeholders such as governments, NGOs and companies by combining satellite technology, open data and crowdsourcing. The site also includes a forest carbon map for the year 2000 covering the tropical areas [31].

ESA's Copernicus Global Land Service provides vegetation biophysical parameters at global level such as Fraction of green Vegetation Cover (FCOVER), Leaf Area Index (LAI), Normalized Difference Vegetation Index (NDVI), Fraction of Absorbed Photosynthetically Active Radiation (FAPAR), and others. The products have 1 km spatial resolution.

Biomass Geo-Wiki is a partnership project between the International Institute for Applied Systems Analysis (IIASA), University of Applied Sciences Wiener Neustadt, and the University of Freiburg. The project uses a crowdsourcing approach to compare and validate forest *AGB* products generated from different providers (e.g. NASA, IIASA, Friedrich-Schiller University of Jena, etc.) at different spatial resolutions, for different areas and temporal coverage.

Land cover mapping provides a static representation of land cover. It does not show change in forest area, but serves as a baseline for assessment of forest cover change. Two main projects are the most representative and widely used at the moment: GlobCover and MODIS land products. GlobCover is a project from the European Space Agency (ESA) whose goal is to develop an Earth's global land cover product [51] (Figure 2.3). Data from the Medium Resolution Imaging Spectrometer and Advanced Synthetic Aperture Radar (MERIS) on board Environmental Satellite (ENVISAT) is used to develop a Land Cover product labelled according to the UN Food and Agriculture Organisation's Land Cover Classification System. Two GlobCover products based on ENVISAT MERIS data at full resolution (300 m) were released by ESA for the years 2005–06 and for 2009.

The MODIS Land Cover Type Product (MCD12Q1) provides data characterizing five global land cover classification systems and is offered free of charge. The land cover product is an annual 500 m spatial resolution product derived through a supervised decision-tree classification method.

The ALOS Kyoto & Carbon (K&C) Initiative is an international project led by Japan Aerospace Exploration Agency (JAXA). Coordinated by JAXA Earth Observation Research Centre (EORC), the programme focuses on producing data products primarily from the Phased Array L-band Synthetic Aperture Radar (PALSAR) sensor on-board the Advanced Land Observing Satellites (ALOS and ALOS-2). The main products from the K&C programme are the 25 m, 50 m and 100 m spatial resolution forest/non-forest (FNF) area mosaics from resampled 10 m data every year (2007–2010 and 2015 onwards) (Figure 2.3). The method for developing this FNF product is based on a decision-tree classification that applies different backscatter intensity thresholds [41].

2.4 Remote Sensing and Biomass Allometry

AGB is accurately and directly measured through in situ destructive sampling methods. Through these methods entire trees are felled and the different tree components are separated and weighted in situ, resulting in a significantly laborious, expensive and impractical approach at a large scale [52]. Non-destructive in situ methods such as forest inventories make use of allometric models to predict *AGB*. In situ non-destructive measurements are broadly used for *AGB* monitoring as their accuracy lies between

20% and 2% [53]. Biophysical parameters like tree height or diameter are commonly measured in forest inventories and other studies, and used to estimate *AGB* through allometric equations.

Derivation of allometric relationships is based on the allometry of living organisms. Allometry is the condition of geometric similitude, which results when geometry and shape are conserved among organisms differing in size [54]. It works as a "rule of proportions" between organism components and their whole. Allometric biomass regressions are developed by measuring the biomass of entire trees or their components and regressing the data against some more easily measured variables [55]. The use of allometric equations has been shown to be a cost efficient technique due to the use of existing and easily measured variables. Common examples of these variables are tree height, basal area, wood specific gravity, or diameter at breast height.

The most commonly used mathematical model for *AGB* estimation uses the form of a non-linear function (Eq. 2.1), where *Y* is the total aboveground tree dry biomass or any other tree component, b_0 and b_1 are parameters, and *X* is the biophysical parameter used for prediction [56]:

$$Y = b_0 \cdot X^{b_1} \tag{2.1}$$

Allometric models have been traditionally developed to be used in national forest inventories or specific studies. The samples used to create these models are usually delimited to the area under study. Such models are generally developed for specific species and sites [57–60]. In temperate and boreal forested areas, there is a large availability of allometric equations [61]. Unfortunately, these equations are not easily available for developing countries in tropical regions with large areas of natural forests due to the geographical remoteness, lack of research studies, data paucity, high tree diversity or armed conflict situations. The Congo Basin is a clear example of the scarcity of ground samples. Even though the Congo basin is one of the largest forested areas in the world, only a small number of allometric equations have been developed for the forests of this region [62]. Several studies found that allometric models could be generalised by the incorporation of additional variables that explain the regional variability, such as wood density, and developed models for specific regions or forest biomes based on a large number samples [52,63,64]. Generalized equations are frequently used in tropical areas, but are just recommended in cases where no local models are available [65].

Few allometric models relating remote sensing-derived biophysical parameters (usually canopy height) to *AGB* are presently available [e.g. 31,66,67–69]. This kind of relationship at the plot or pixel scale is conceptually similar to relationships at tree level. The main difference is that the relationship is established between the biophysical parameter and the *AGB* of all trees inside the area of interest. At tree level, *AGB* can be accurately calculated from tree height, diameter and specific wood density [64,70,71] using generalized models. Remote sensing can measure tree height but cannot directly measure wood density. The use of allometric models calibrated with regional ground data can circumvent this problem and provide accurate estimates of *AGB* [72]. Moreover, the use of additional forest structure variables can also improve the estimates [73,74]. Forest biomes or ecological regions present different tree allometries depending on climatic conditions, vegetation structure, species, soil types, and other characteristics, which ultimately affect the correlation between *AGB* and biophysical parameters like mean canopy height at plot and pixel level. It seems therefore logical to develop regional models that capture the regional

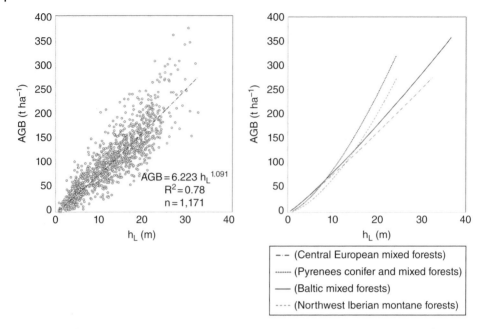

Figure 2.4 Left: Plot-level allometric model relating *AGB* to Lorey's mean canopy height (h_L) for the Central European Mixed Forest ecological region. Right: Allometric models for different ecological regions in Europe.

variability, as the slopes of these allometric functions will differ from region to region (Figure 2.4). Regional allometry has not been sufficiently explored to be used with remote sensing, but its use could improve *AGB* estimation worldwide.

New techniques applied to SAR and LiDAR sensors can be used to estimate biophysical parameters such as tree canopy height [31,75,76]. Biophysical parameters can be used with regional allometric models to estimate forest *AGB*. This approach does not suffer from the *AGB*/radar backscatter saturation problem. *AGB* can be mapped by SAR from interferometric height models in combination with allometry [75]. This approach requires a ground Digital Terrain Model (DTM), which is not always easy to obtain. Polarimetric Interferometry is another SAR technique, which in contrast to single-polarisation interferometry, does not rely on an external DTM, as it estimates terrain and canopy height from the different polarimetric scattering mechanisms [67,77,78]. It relies on the coherence of two SAR scenes taken over the same site, either within a short time window or simultaneously from two slightly different positions within a certain distance range or baseline. SAR Tomography goes beyond the polarimetric interferometry technique by using a multi-baseline of interferometric SAR images to generate a 3D vertical structure of the vegetation based on the variation of backscatter scattering as a function of height [24,79].

Several authors have studied LiDAR-derived biophysical canopy metrics such as maximum canopy height, Lorey's mean height (h_L) and the height of median energy (HOME) to characterize forest vertical structure [75,80–85]. In recent studies [80,86], spaceborne profiling LiDAR from the GLAS sensor was used to create global maps of forest canopy height. The maps estimated top canopy height [86] and Lorey's mean canopy height (h_L) [80] from the full waveform of the GLAS footprints (area illuminated by the laser and from which the waveform-return signal gives information). Lorey's mean canopy height is the basal area weighted height of all trees. At plot and LiDAR footprint level h_L shows a robust relationship with *AGB* [80]. The size of the GLAS footprints (<0.4 ha) is comparable to most forest plots sizes (0.02–1 ha). Therefore, there are plenty of data available for developing regional models, which could relate canopy height to *AGB* as seen in [31,66,69].

2.5 Synergistic Use of Regional Allometry, in situ Measurements, and Spaceborne Profiling LiDAR, with Optical and SAR Imagery for Biomass Mapping

There is no sensor that can currently be used for *AGB* estimation across larger regions, either because of limitations in signal saturation, cloud cover persistence, or complex signal retrieval due to topography. Several studies have aimed to map *AGB* at global, biome, or continental levels using a variety of methods. Products mapping forest *AGB* and carbon stocks globally [87], continentally [14,88], in the tropical [30,31], temperate and boreal regions [89], as well as growing stock volume continentally [90], and in boreal regions [91] have recently been published. Together with the limitations of remote sensing imagery to map forest *AGB*, all these products also face important challenges regarding ground data availability to calibrate their approaches. Most of these studies use methods for combination of multiple datasets in order to circumvent such limitations. Data synergy approaches make possible to exploit the specific strengths of each sensor. For example, LiDAR sensors can be used for estimation of *AGB* samples across the landscape, while SAR sensors in combination with optical sensors can be used for forest area estimation and extrapolation of the measurements.

There are a number of parametric and non-parametric approaches to extrapolate values of *AGB* to larger spatial scales using remote sensing imagery. Multiple regression analysis, k-nearest neighbour technique (k-NN), co-kriging, random forests, and neural networks are some examples. Parametric approaches make assumptions on the shape (i.e., normal distribution), and on the parameters or form of the sample distribution, while non-parametric approaches only make few or no assumptions. Parametric models present bigger challenges for extrapolating *AGB* data, as there are no current satellite observations that can be reasonably related to *AGB* across the whole landscape. Moreover, the assumptions in parametric models of independence and multivariate-normality are often violated [92]. As complex ecological systems like forests show non-linear relationships, autocorrelation, and variable interaction across temporal and spatial scales, the use of non-parametric algorithmic methods often outperform parametric methods [93].

Two recent papers mapped the spatial distribution of *AGB* in the tropics using synergistic approaches based on the use of GLAS footprints for the estimation of *AGB* [30,31]. The approach described by Baccini *et al.* [30] relates GLAS waveforms to *AGB* using a model calibrated by ground plots directly located under the GLAS footprints, while Saatchi *et al.* [31] uses three continental allometric models derived from ground data to relate GLAS-derived Lorey's mean canopy height to *AGB*. As discussed in the previous section, the use of a model for each continent might better explain the allometric regional variability than a single model, but might still introduce a great amount of uncertainty when applied to very different forest biomes such as temperate coniferous and tropical rainforest. The studies use non-parametric approaches such as Random Forest [94] and MaxEnt [95,96] for extrapolation of the *AGB* across wide areas, to produce 463 m and 1 km resolution maps respectively. One of the most innovative features of using MaxEnt is the possibility of mapping the uncertainty of the *AGB* estimation on a pixel-by-pixel basis. Both approaches use MODIS and SRTM products, and in the case of [31] also Quickscatterometer data (QSCAT). None of these products can solely explain the variability of *AGB* across the landscape, but the methods used by these studies aim to take advantage of the full potential of the information contained in each product.

2.5.1 Global Biomass Monitoring Approach

Based on the previous examples, it is possible to define a general concept for *AGB* mapping (Figure 2.5) using a combination of datasets from different remote sensors by means of a non-parametric approach such as MaxEnt to extrapolate the *AGB* calibration data. These *AGB* data could be directly obtained from forest inventory datasets, or by means of regional allometric models relating *AGB* to remote sensing-derived biophysical parameters such as mean canopy height calculated from Spaceborne LiDAR.

A baseline *AGB* map can be developed and updated at predefined time intervals (e.g. every 3–5 years). If annual products are needed for periods in between updates, those

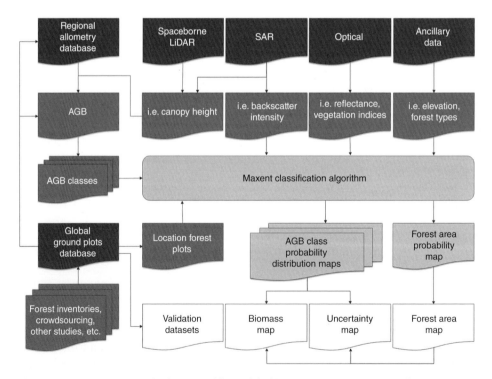

Figure 2.5 *AGB* mapping method proposed for a Global Biomass Monitoring Approach.

Table 2.3 Mapping levels and proposed imagery.

Mapping levels	Spatial resolution	Satellite sensors		
		Optical	SAR	LIDAR
Global	1 km–100 m	MODIS, SPOT VGT, PROBA-V, MERIS Archive	ALOS PALSAR Archive, ALOS-2 PALSAR, BIOMASS	ICESat archive, ICESat-2, GEDI
Sub-National	<100 m	Landsat archive, Landsat (TM, ETM+, OLI), Sentinel-2	ALOS PALSAR Archive, ALOS-2 PALSAR, BIOMASS	ICESat archive, ICESat-2, GEDI

could be easily generated on demand using forest area change maps developed from optical or SAR imagery [e.g. 34,41].

Remote sensing imagery suitable for this *AGB* monitoring approach can be classified by sensor according its spatial resolution, defining two mapping levels: Global-Regional, and Sub-National (Table 2.3). The output of the approach consists not only in *AGB* estimations in a pixel basis, but also a wall-to-wall map of its associated uncertainty and a forest area map.

2.5.2 Case Study Regional Level: Forest AGB, Uncertainty, and Forest Probability for Mexico

Forest inventory sampling units comprising four 0.04 ha sub-plots representing an area of about 1 ha from the Mexican Forest Inventory (INFyS) were used for calibration of the MaxEnt algorithm to map *AGB* in combination with three different remote sensing products: MODIS 250 m reflectance bands and vegetation indices, SRTM digital elevation model, and ALOS PALSAR L-band dual polarization imagery [97]. L-band SAR backscatter intensity signal is relatively sensitive to *AGB* up to $150\,t\,ha^{-1}$ and provides a different type of information than SRTM and MODIS. The Area Under the Receiver Operator Curve – AUC [95,96] was used to examine the performance of the final *AGB* maps using different combinations of these datasets. The average AUC increased from a minimum of 0.79 (single product) to a maximum of 0.93 using the three products combined [97]. Wall-to-wall 250 m resolution (6.25 ha) *AGB*, uncertainty, and forest probability maps (Figures 2.6) were developed using this method.

The variable importance for generation of the *AGB* map by the MaxEnt algorithm was analysed as groups of variables belonging to the same remote sensing product. The overall per cent contribution of ALOS PALSAR explains approximately 50.9%, while MODIS and SRTM products were 32.9%, and 16.2% respectively. The analysis of variable contributions by different biomass classes shows that ALOS PALSAR product was the most important variable to predict biomass for the most abundant biomass classes (up to 100–$120\,t\,ha^{-1}$) and was still very relevant up to $180\,t\,ha^{-1}$ (Figure 2.7). The decline in the contribution of ALOS PALSAR in the estimation for *AGB* above 100–$120\,t\,ha^{-1}$ is in agreement with the saturation effect that can be seen in the literature for L-band backscatter [20,21]. After this point the weight of the estimation fluctuates among products. It seems that none of the products can solely explain the amount of *AGB* after this point.

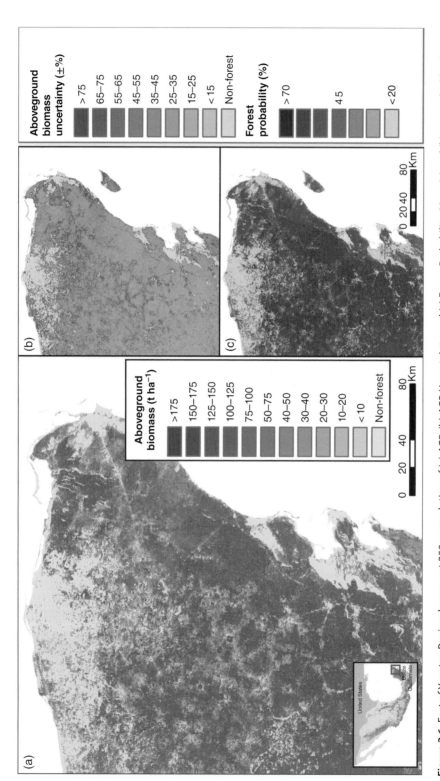

Aboveground biomass uncertainty (±%)

- >75
- 65–75
- 55–65
- 45–55
- 35–45
- 25–35
- 15–25
- <15
- Non-forest

Forest probability (%)

- >70
- 45
- <20

Aboveground biomass (t ha⁻¹)

- >175
- 150–175
- 125–150
- 100–125
- 75–100
- 50–75
- 40–50
- 30–40
- 20–30
- 10–20
- <10
- Non-forest

0 20 40 80 Km

United States
Belize
Guatemala

Figure 2.6 East of Yucatan Peninsula maps at 250 m resolution of (a) *AGB*, (b) *AGB*-Uncertainty, and (c) Forest Probability. Maps (a) and (b) are masked by the forest probability map (c) thresholded by 25%.

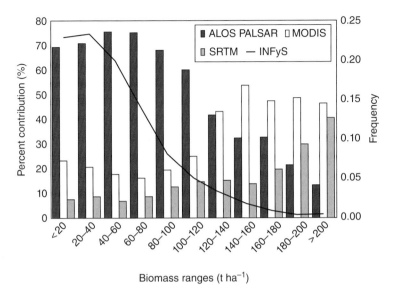

Figure 2.7 Columns: Percent contributions to the *AGB* map per biomass range. Line: frequency of observed biomass values from the INFyS forest inventory dataset.

It stands out the high contribution of SRTM for the highest biomass ranges (above $160\,t\,ha^{-1}$), which can be explained due to the specific distribution of forest *AGB* in Mexico across the topographic gradient. The highest values of *AGB* per ha in Mexico occur at the highest altitudes and slopes, while the lowest *AGB* usually occur at the lower areas with relatively flat terrain. The only exception to this gradient can be observed in the Yucatan peninsula, where high *AGB* values can be found at low elevation and flat terrain.

The forest probability map is developed by calibrating the MaxEnt algorithm with the location of forest inventory plots with more than 10% tree cover and remote sensing imagery. The map indicates the proportional positive conditional probability of each pixel to be forest. A probability threshold is then used to generate a binary forest area map.

MODIS VCF and ALOS PALSAR products are commonly used to create binary forest/non-forest area maps in several studies by means of a threshold approach [e.g. 31,41,98]. A visual comparison of these methods to the forest area defined by the Land Use and Vegetation map of Mexico [99] revealed evident problems for both products to delineate the appropriate forest area in Mexico (Figure 2.8). This is not surprising in the case of MODIS VCF, which appears insufficiently accurate in regions with sparse vegetation where it tends to overestimate tree cover [46] (Figure 2.8c). The ALOS PALSAR product clearly improves over MODIS VCF, but still presents some inaccuracies in mountainous areas non-vegetated or covered by shrub (Figure 2.8d). The forest probability map was validated in comparison to several threshold-based MODIS VCF and ALOS PALSAR forest/non-forest binary maps. The κ coefficient of agreement for the Forest Probability map was 0.83, while for ALOS PALSAR and VCF forest maps were 0.78 and 0.66 respectively. The forest probability product combines the advantages of optical and SAR imagery bypassing their individual problems to differentiate forest vegetation (Figure 2.8b).

(a)　　　　　　　　　　　　　　　　　　(c)

(b)　　　　　　　　　　　　　　　　　　(d)

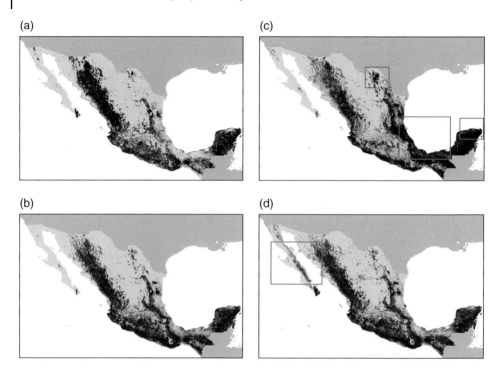

Figure 2.8 Forest area mask delineated by (a) Mexico Land Use and Vegetation map, (b) Forest Probability map (25% threshold), (c) MODIS Vegetation Continuous Fields (10% tree cover threshold), and (d) ALOS PALSAR HV backscatter (−14.5 dB threshold). Squares in (c) show areas of sparse vegetation and agricultural land misclassified as forest in the MODIS VCF product, and the square in (d) shows an area of shrub land in a mountainous region misclassified as forest in the ALOS PALSAR product.

The uncertainty of the *AGB* estimations varies by regions and by biomass ranges showing lower uncertainty in areas with high biomass levels, and higher uncertainty in areas with low biomass density. The root mean square error at pixel scale was estimated in $36.1 \, t \, ha^{-1}$ by validation against an independent dataset. The highest relative errors primarily occur in the lowest *AGB* ranges where the sub-pixel variability plays an important role. Pixels with *AGB* below $20 \, t \, ha^{-1}$ are areas with a small amount of woody vegetation encroachment, which is not enough to be considered a forest. However, there is no clear definition of forest cover based on forest biomass estimated by remote sensing. The use of the forest/non-forest binary mask developed by this study (Figure 2.6c) reduces the number of pixels with very low *AGB* values (non-forest), and at the same time decreases the overall uncertainty of the map. As mentioned before, the BIOMASS mission by ESA is expected to achieve better accuracy levels once is operative beyond 2021. Nevertheless, these maps achieve good accuracy levels in areas that can be easily identified by the use of the corresponding uncertainty map.

AGB, Uncertainty and Forest Probability maps can be globally generated using the presented approach. Even though ground data are not globally available, LiDAR footprints from the ICESat GLAS archive, future ICESat-2 and GEDI could be used to obtain millions of *AGB* estimates, which will be added to the ground data to train the algorithm [e.g. 30,31]. Crowdsourcing is also becoming an important source of data for the research community, and projects such as Geo-Wiki (www.geo-wiki.org/) shows how geographical research can benefit. Therefore any global approach aiming to map forest biomass should take advantage of these sources of data.

Global annual mosaics of L-Band ALOS PALSAR backscatter slope corrected, orthorectified, and radiometrically calibrated have been recently made available by the K&C Initiative with a spatial resolution of 25 m for the period 2007–10 [43]. Data from ALOS-2 PALSAR (since 2014) is available from 2015 onwards. The continuity of this mission ensures the availability of long wavelength SAR imagery.

MODIS products are globally available and can be used in combination with those datasets. Moreover, the use of Landsat imagery instead of MODIS, in combination with ALOS PALSAR, opens the possibility of generating higher spatial resolution maps [e.g. 100] aiming to satisfy sub-national level monitoring requirements.

Nevertheless, this approach presents certain requirements to be carried out by governments, spaces agencies, and the research community for its feasibility:

1) The development of a geodatabase of regional allometric models that could relate remote sensing-derived biophysical parameters to biomass at LiDAR-footprint or pixel level.
2) Access to an extensive global dataset of *AGB* ground data for calibration and validation purposes collected from forest inventories, research studies and other sources such as crowdsourcing.
3) Continuation of the ICESat GLAS measurements interrupted in 2010 by means of the ICESat-2 satellite and the GEDI platform (to be operational by 2017 and 2020 respectively).
4) Continuation of optical missions such as MODIS, PROBA-V, Sentinel-2 and Landsat, which allow the generation of radiometrically consistent global datasets.
5) Continuation of long wavelength SAR satellite missions such as the L-band ALOS-2 PALSAR (launched in May 2014) and the future P-band BIOMASS expected not before 2021.

2.6 Conclusions

This chapter has discussed current approaches to the mapping and monitoring of aboveground forest biomass from optical, radar and LiDAR sensors, and their limitations. A method using a maximum entropy approach for the estimation of *AGB* from a combination of selected geospatial input datasets is presented. Case studies showed the different spatial resolutions that can be achieved based on different inputs and the relative importance of the different types of sensors in the generation of the *AGB* maps. Additionally, it was shown that improved forest area maps can also be obtained within the same approach. The accuracy of the *AGB* estimates and the forest area was studied using error propagation methods, and validated by independent forest inventory datasets.

A conceptual approach for a Global Biomass Information System is presented. It is based on two hierarchical mapping scales and uses the synergies between data products from optical, radar and LiDAR data. It is applicable to the current and forthcoming series of new satellites and sensors from the ESA Copernicus Programme (i.e. Sentinels), JAXA (i.e. ALOS-2 PALSAR), and NASA (i.e. Landsat, ICESat-2, GEDI). Such approaches clearly require a strong collaboration and data sharing agreement between the Japanese, European and US space agencies. It is possible within limits to generate global *AGB* maps with accuracies that make them meaningful. They do not read the expected accuracies achievable from the BIOMASS mission by ESA, but appear as the best way to obtain quality *AGB* estimations for the period before this mission.

References

1 IPCC, Climate change 2007: The Physical Science Basis: contribution of Working Group I to the Fourth Assessment Report of the Intergovernmental Panel on Climate Change, Cambridge University Press, Cambridge (United Kingdom) and New York, NY (USA), (2007).

2 J.G. Canadell, C. Le Quéré, M.R. Raupach, C.B. Field, E.T. Buitenhuis, P. Ciais, T.J. Conway, N.P. Gillett, R.A. Houghton and G. Marland, Contributions to accelerating atmospheric CO_2 growth from economic activity, carbon intensity, and efficiency of natural sinks, *Proceedings of the National Academy of Sciences*, **104**, 18866–18870, (2007).

3 K. Andersson, National forest carbon inventories: policy needs and assessment capacity, *Climatic Change*, **93**, 69–101, (2009).

4 R.T. Watson, I.R. Noble, B. Bolin, N.H. Ravindranath, D.J. Verardo and D.J. Dokken, *Land use, land-use change, and forestry*, Cambridge University Press, UK, (2000).

5 R.A. Houghton, K.T. Lawrence, J.L. Hackler and S. Brown, The spatial distribution of forest biomass in the Brazilian Amazon: a comparison of estimates, *Global Change Biology*, **7**, 731–746, (2001).

6 J.I. House, I.C. Prentice, N. Ramankutty, R.A. Houghton and M. Heimann, Reconciling apparent inconsistencies in estimates of terrestrial CO_2 sources and sinks, *Tellus B*, **55**, 345–363, (2003).

7 R.A. Houghton, Above Ground Forest Biomass and the Global Carbon Balance, *Global Change Biology*, **11**, 945–958, (2005).

8 G.B. Bonan, Forests and Climate Change: Forcings, Feedbacks, and the Climate Benefits of Forests, *Science*, **320**, 1444–1449, (2008).

9 IPCC, *Good Practice Guidance for Land Use, Land Use Change and Forestry*, Prepared by the National Greenhouse Gas Inventories Programme, IGES, Japan, (2003).

10 H.K. Gibbs, S. Brown, J.O. Niles and J.A. Foley, Monitoring and estimating tropical forest carbon stocks: making REDD a reality, *Environmental Research Letters*, **2**, 045023, (2007).

11 S. Quegan, T. Le Toan, J. Chave, J. Dall, K. P. Papathanassiou, and F. Rocca, *BIOMASS Report for mission selection*, European Space Agency – ESA (2012).

12 T. Fatoyinbo, *Remote Sensing of Biomass – Principles and Applications*, INTECH, http://www.intechopen.com/books/remote-sensing-of-biomass-principles-and-applications, (2012).

13 V. Avitabile, A. Baccini, M.A. Friedl and C. Schmullius, Capabilities and limitations of Landsat and land cover data for aboveground woody biomass estimation of Uganda, *Remote Sensing of Environment*, **117**, 366–380, (2012).

14 A. Baccini, N. Laporte, S.J. Goetz, M. Sun and H. Dong, A first map of tropical Africa's aboveground biomass derived from satellite imagery, *Environmental Research Letters*, **3**, (2008).

15 J.A. Blackard, M.V. Finco, E.H. Helmer, G.R. Holden, M.L. Hoppus, D.M. Jacobs, A.J. Lister, G.G. Moisen, M.D. Nelson, R. Riemann, B. Ruefenacht, D. Salajanu, D.L. Weyermann, K.C. Winterberger, T.J. Brandeis, R.L. Czaplewski, R.E. Mcroberts, P.L. Patterson and R.P. Tymcio, Mapping U.S. forest biomass using nationwide forest inventory data and moderate resolution information, *Remote Sensing of Environment*, **112**, 1658–1677, (2008).

16 T.M. Lillesand, R.W. Kiefer and J.W. Chipman, *Remote sensing and image interpretation*, 6th edn. John Wiley & Sons Inc., Hoboken, NJ, (2007).

17 S.O. Los, N.H. Pollack, M.T. Parris, G.J. Collatz, C.J. Tucker, P.J. Sellers, C.M. Malmström, R.S. Defries, L. Bounoua and D.A. Dazlich, A Global Nine Year Biophysical Land Surface Dataset from NOAA AVHRR Data, *Journal of Hydrometeorology*, **1**, 183–199, (2000).

18 T. Le Toan, S. Quegan, I. Woodward, M. Lomas, N. Delbart and G. Picard, Relating radar remote sensing of biomass to modelling of forest carbon budgets, *Climatic Change*, **67**, 379–402, (2004).

19 S. Goetz, A. Baccini, N. Laporte, T. Johns, W. Walker, J. Kellndorfer, R. Houghton and M. Sun, Mapping and monitoring carbon stocks with satellite observations: a comparison of methods, *Carbon Balance and Management*, **4**, 2, (2009).

20 W. Wagner, A. Luckman, J. Vietmeier, K. Tansey, H. Balzter, C. Schmullius, M. Davidson, D. Gaveau, M. Gluck, T. Le Toan, S. Quegan, A. Shvidenko, A. Wiesmann and J.J. Yu, Large-scale mapping of boreal forest in SIBERIA using ERS tandem coherence and JERS backscatter data, *Remote Sensing of Environment*, **85**, 125–144, (2003).

21 E.T.A. Mitchard, S.S. Saatchi, I.H. Woodhouse, G. Nangendo, N.S. Ribeiro, M. Williams, C.M. Ryan, S.L. Lewis, T.R. Feldpausch and P. Meir, Using satellite radar backscatter to predict aboveground woody biomass: A consistent relationship across four different African landscapes, *Geophysical Research Letters*, **36**, L23401, (2009).

22 R. Lucas, J. Armston, R. Fairfax, R. Fensham, A. Accad, J. Carreiras, J. Kelley, P. Bunting, D. Clewley, S. Bray, D. Metcalfe, J. Dwyer, M. Bowen, T. Eyre, M. Laidlaw and M. Shimada, An Evaluation of the ALOS PALSAR L-Band Backscatter – Aboveground Biomass Relationship Queensland, Australia: Impacts of Surface Moisture Condition and Vegetation Structure, *Selected Topics in Applied Earth Observations and Remote Sensing, IEEE Journal of*, **3**, 576–593, (2010).

23 S. Englhart, V. Keuck and F. Siegert, Aboveground biomass retrieval in tropical forests – The potential of combined X- and L-band SAR data use, *Remote Sensing of Environment*, **115**, 1260–1271, (2011).

24 T. Le Toan, S. Quegan, M.W.J. Davidson, H. Balzter, P. Paillou, K. Papathanassiou, S. Plummer, F. Rocca, S. Saatchi, H. Shugart and L. Ulander, The BIOMASS mission: Mapping global forest biomass to better understand the terrestrial carbon cycle, *Remote Sensing of Environment*, **115**, 2850–2860, (2011).

25 M.L. Imhoff, Radar backscatter and biomass saturation: ramifications for global biomass inventory, *IEEE Transactions on Geoscience and Remote Sensing*, **33**, 511–518, (1995).

26 J.M.B. Carreiras, M.J. Vasconcelos and R.M. Lucas, Understanding the relationship between aboveground biomass and ALOS PALSAR data in the forests of Guinea-Bissau (West Africa), *Remote Sensing of Environment*, **121**, 426–442, (2012).

27 E. Naesset, Airborne laser scanning as a method in operational forest inventory: Status of accuracy assessments accomplished in Scandinavia, *Scandinavian Journal of Forest Research*, **22**, 433–442, (2007).

28 M.C. Dobson, F.T. Ulaby, T. Letoan, A. Beaudoin, E.S. Kasischke and N. Christensen, Dependence of radar backscatter on coniferous forest biomass, *IEEE Transactions on Geoscience and Remote Sensing*, **30**, 412–415, (1992).

29 M.A. Lefsky, M. Keller, Y. Pang, P.B. De Camargo and M.O. Hunter, Revised method for forest canopy height estimation from Geoscience Laser Altimeter System waveforms, *Journal of Applied Remote Sensing*, **1**, 013537, (2007).

30 A. Baccini, S.J. Goetz, W.S. Walker, N.T. Laporte, M. Sun, D. Sulla-Menashe, J. Hackler, P.S.A. Beck, R. Dubayah, M.A. Friedl, S. Samanta and R.A. Houghton, Estimated carbon dioxide emissions from tropical deforestation improved by carbon-density maps, *Nature Climate Change*, **2**, 182–186, (2012).

31 S.S. Saatchi, N.L. Harris, S. Brown, M. Lefsky, E.T.A. Mitchard, W. Salas, B.R. Zutta, W. Buermann, S.L. Lewis, S. Hagen, L. Petrova, L. White, M. Silman and A. Morel, Benchmark map of forest carbon stocks in tropical regions across three continents, *Proceedings of the National Academy of Sciences*, **108**, 9899–9904, (2011).

32 H.G. Lund, (Revised) *Definitions of Forest, Deforestation, Afforestation, and Reforestation.* Online, https://ecaths1.s3.amazonaws.com/proteccionforestalfcf/1351694455. Definitiones%20of%20forest,%20deforestation,%20afforestation,%20and%20 reforestation.%20Gyde%20Lund%20marzo%202014.pdf Gainesville, VA, (2014).

33 R. Wadsworth, H. Balzter, F. Gerard, C. George, A. Comber and P. Fisher, An environmental assessment of land cover and land use change in Central Siberia using quantified conceptual overlaps to reconcile inconsistent data sets, *Journal of Land Use Science*, **3**, 251–264, (2008).

34 M.C. Hansen, P.V. Potapov, R. Moore, M. Hancher, S.A. Turubanova, A. Tyukavina, D. Thau, S.V. Stehman, S.J. Goetz, T.R. Loveland, A. Kommareddy, A. Egorov, L. Chini, C.O. Justice and J.R.G. Townshend, High-Resolution Global Maps of 21st century Forest Cover Change, *Science*, **342**, 850–853, (2013).

35 A.C. Morel, S.S. Saatchi, Y. Malhi, N.J. Berry, L. Banin, D. Burslem, R. Nilus and R.C. Ong, Estimating aboveground biomass in forest and oil palm plantation in Sabah, Malaysian Borneo using ALOS PALSAR data, *Forest Ecology and Management*, **262**, 1786–1798, (2011).

36 FAO, *FRA 2000 – On definitions of forest and forest cover change.* FRA Working paper 33, FAO-FRA programme, Rome, Italy, 2000.

37 FAO, *Global forest resources assessment 2005*, UN Food and Agriculture Organization of the United Nations, 2005.

38 FAO, *Global forest resources assessment 2010*, UN Food and Agriculture Organization of the United Nations, 2010.

39 FAO, *Global forest resources assessment 2015: How are the world's forests changing?* UN Food and Agriculture Organization of the United Nations, 2015.

40 FAO, JRC, SDSU and UCL, *The 2010 Global Forest Resources Assessment Remote Sensing Survey: an outline of the objectives, data, methods and approach.* FRA Working paper 155, Rome, 2009.

41 M. Shimada, O. Isoguchi, T. Motooka, T. Shiraishi, A. Mukaida, H. Okumura, T. Otaki and T. Itoh, Generation of 10 m resolution PALSAR and JERS-SAR mosaic and forest/non-forest maps for forest carbon tracking, *in Geoscience and Remote Sensing Symposium (IGARSS), 2011 IEEE International*, 3510–3513, 2011.

42 M. Shimada, O. Isoguchi, R. Preesan, N. Longépé, A. Mukaida, H. Okumura, T. Otaki and T. Itoh, Generation of decadal global high resolution L-band SAR dataset for forest

carbon tracking, *International Archives of the Photogrammetry, Remote Sensing and Spatial Information Science XXXVIII*, **8**, (2010).

43 M. Shimada and T. Ohtaki, Generating Large-Scale High-Quality SAR Mosaic Datasets: Application to PALSAR Data for Global Monitoring, *Journal of Selected Topics in Applied Earth Observations and Remote Sensing*, **3**, 637–656, (2010).

44 M. Shimada and T. Otaki, Generating Continent-scale High-quality SAR Mosaic Datasets: Application to PALSAR Data for Global Monitoring, *Journal of Selected Topics in Applied Earth Observations and Remote Sensing, Special Issue on Kyoto and Carbon Initiative*, **3**, 637–656, (2010).

45 J.R.G. Townshend, M. Hansen, M. Carroll, C. Dimiceli, R. Sohlberg and C. Huang, *User Guide for the MODIS Vegetation Continuous Fields Product Collection Five version One*, University of Maryland, College Park, Maryland, 2011.

46 J.O. Sexton, X-P. Song, M. Feng, P. Noojipady, A. Anand, C. Huang, D-H. Kim, K.M. Collins, S. Channan and C. Dimiceli, Global, 30-m resolution continuous fields of tree cover: Landsat-based rescaling of MODIS vegetation continuous fields with lidar-based estimates of error, *International Journal of Digital Earth*, **6**, 427–448, (2013).

47 P.E. Waggoner, Forest Inventories: Discrepancies and Uncertainties, Resources For the Future, Washington DC, 2009.

48 M.C. Hansen, R.S. Defries, J.R.G. Townshend, M. Carroll, C. Dimiceli and R.A. Sohlberg, Global Percent Tree Cover at a Spatial Resolution of 500 Meters: First Results of the MODIS Vegetation Continuous Fields Algorithm, *Earth Interactions*, **7**, 1–15, (2003).

49 D. Schoene, W. Killmann, H. Von Lüpke and M.L. Wilkie, Definitional Issues Related to Reducing Emissions from Deforestation in Developing Countries, Forests and Climate Change Food and Agriculture Organization of the United Nations, 2007.

50 D. Hammer, R. Kraft and D. Wheeler, FORMA Alerts, World Resources Institute and Center for Global Development, 2013.

51 O. Arino, H. Trebossen, F. Achard, M. Leroy, C. Brockman, P. Defourny, R. Witt, J. Latham, C. Schmullius, S. Plummer, H. Laur, P. Goryl and N. Houghton, The Globcover Initiative, *in Proceedings of the MERIS-AATSR workshop*, European Space Agency (ESA), 36, Frascati, Italy, 2005.

52 Q.M. Ketterings, R. Coe, M. Van Noordwijk, Y. Ambagau and C.A. Palm, Reducing uncertainty in the use of allometric biomass equations for predicting aboveground tree biomass in mixed secondary forests, *Forest Ecology and Management*, **146**, 199–209, (2001).

53 A. Bombelli, V. Avitabile, H. Balzter, L.B. Marchesini *et al.*, Biomass-assessment of the status of the development of the standards for the terrestrial essential climate variables, *Biomass*, 2009.

54 K.J. Niklas, *Plant allometry: the scaling of form and process*, University of Chicago Press, 1994.

55 J. Pastor, J.D. Aber and J.M. Melillo, Biomass prediction using generalized allometric regressions for some northeast tree species, *Forest Ecology and Management*, **7**, 265–274, (1984).

56 N. António, M. Tomé, J. Tomé, P. Soares and L. Fontes, Effect of tree, stand, and site variables on the allometry of Eucalyptus globulus tree biomass, *Canadian Journal of Forest Research*, **37**, 895–906, (2007).

57 P.W. West, K.F. Wells, D.M. Cameron, S.J. Rance, C.R.A. Turnbull and C.L. Beadle, Predicting tree diameter and height from aboveground biomass for four eucalypt species, *Trees-Structure and Function*, **5**, 30–35, (1991).

58 L. Saint-André, A. Mbou, A. Mabiala, W. Mouvondy, C. Jourdan, O. Roupsard, P. Deleporte, O. Hamel and Y. Nouvellon, Age-related equations for above- and below-ground biomass of a hybrid in Congo, *Forest Ecology and Management*, **205**, 199–214, (2005).

59 P. Muukkonen, Generalized allometric volume and biomass equations for some tree species in Europe, *European Journal of Forest Research*, **126**, 157–166, (2007).

60 J. Návar, Biomass component equations for Latin American species and groups of species, *Annals of Forest Science*, **66**, 208–208, (2009).

61 D. Zianis, P. Muukkonen, R. Mäkipää and M. Mencuccini, Biomass and stem volume equations for tree species in Europe, *Silva Fennica Monographs*, **4**, 1–63, (2005).

62 M. Henry, N. Picard, C. Trotta, R.J. Manlay, R. Valentini, M. Bernoux and L. Saint-André, Estimating tree biomass of sub-Saharan African forests: a review of available allometric equations, *Silva Fennica*, **45**, 477–569, (2011).

63 S. Brown, *Estimating biomass and biomass change of tropical forests: a primer*, Food and Agriculture Organization (FAO), 1997.

64 J. Chave, C. Andalo, S. Brown, M. Cairns, J. Chambers, D. Eamus, H. Fölster, F. Fromard, N. Higuchi, T. Kira, J.P. Lescure, B. Nelson, H. Ogawa, H. Puig, B. Riéra and T. Yamakura, Tree allometry and improved estimation of carbon stocks and balance in tropical forests, *Oecologia*, **145**, 87–99, (2005).

65 J. Chave, R. Condit, S. Aguilar, A. Hernandez, S. Lao and R. Perez, Error propagation and scaling for tropical forest biomass estimates, *Philosophical Transactions of the Royal Society of London. Series B: Biological Sciences*, **359**, 409–420, (2004).

66 G. Asner, J. Mascaro, H. Muller-Landau, G. Vieilledent, R. Vaudry, M. Rasamoelina, J. Hall and M. Van Breugel, A universal airborne LiDAR approach for tropical forest carbon mapping, *Oecologia*, **168**, 1147–1160, (2012).

67 S.R. Cloude, I.H. Woodhouse and J.C. Suarez, *Polarimetry and Polarimetric Interferometry for Forestry Applications: Final Report*, (2011).

68 T. Mette, K. Papathanassiou, I. Hajnsek, H. Pretzsch and P. Biber, Applying a common allometric equation to convert forest height from Pol-InSAR data to forest biomass, *IGARSS 2004: IEEE International Geoscience and Remote Sensing Symposium Proceedings*, Vols 1–7: Science for Society: Exploring and Managing a Changing Planet, 2004.

69 E.T.A. Mitchard, S.S. Saatchi, L.J.T. White, K.A. Abernethy, K.J. Jeffery, S.L. Lewis, M. Collins, M.A. Lefsky, M.E. Leal, I.H. Woodhouse and P. Meir, Mapping tropical forest biomass with radar and spaceborne LiDAR in Lopé National Park, Gabon: overcoming problems of high biomass and persistent cloud, *Biogeosciences*, **9**, 179–191, (2012).

70 T.R. Feldpausch, J. Lloyd, S.L. Lewis, R.J.W. Brienen *et al.*, Tree height integrated into pan-tropical forest biomass estimates, *Biogeosciences*, **9**, 3381–3403, (2012).

71 G. Vieilledent, R. Vaudry, S.F.D. Andriamanohisoa, O.S. Rakotonarivo, H.Z. Randrianasolo, H.N. Razafindrabe, C.B. Rakotoarivony, J. Ebeling and M. Rasamoelina, A universal approach to estimate biomass and carbon stock in tropical forests using generic allometric models, *Ecological Applications*, **22**, 572–583, (2011).

72 M. Fagan and R. Defries, *Measurement and Monitoring of the World's Forests: A Review and Summary of Remote Sensing Technical Capability, 2009–15*, Resources for the Future, 2009.

73 S.C. Popescu, R.H. Wynne and R.F. Nelson, Measuring individual tree crown diameter with LiDAR and assessing its influence on estimating forest volume and biomass, *Canadian Journal of Remote Sensing*, **29**, 564–577, (2003).

74 M. Palace, M. Keller, G.P. Asner, S. Hagen and B. Braswell, Amazon Forest Structure from IKONOS Satellite Data and the Automated Characterization of Forest Canopy Properties, *Biotropica*, **40**, 141–150, (2008).

75 H. Balzter, C.S. Rowland and P. Saich, Forest canopy height and carbon estimation at Monks Wood National Nature Reserve, UK, using dual-wavelength SAR interferometry, *Remote Sensing of Environment*, **108**, 224–239, (2007).

76 M.A. Lefsky, D.J. Harding, M. Keller, W.B. Cohen, C.C. Carabajal, F.D.B. Espirito-Santo, M.O. Hunter, R. De Oliveira and P. De Camargo, Estimates of forest canopy height and aboveground biomass using ICESat, *Geophysical Research Letters*, **32**, L22S02, (2005).

77 K.P. Papathanassiou, F. Kugler, S. Lee, L. Marotti and I. Hajnsek, Recent advances in Polarimetric SAR Interferometry for forest parameter estimation, *RADAR Conference, 2008. IEEE*, **1–6**, (2008).

78 S.R. Cloude and K.P. Papathanassiou, Polarimetric SAR interferometry, *IEEE Transactions on Geoscience and Remote Sensing*, **36**, 1551–1565, (1998).

79 S.R. Cloude, Polarization coherence tomography, *Radio Science*, **41**, RS4017, (2006).

80 M.A. Lefsky, A global forest canopy height map from the Moderate Resolution Imaging Spectroradiometer and the Geoscience Laser Altimeter System, *Geophysical Research Letters*, **37**, L15401, (2010).

81 G. Sun, K.J. Ranson, D.S. Kimes, J.B. Blair and K. Kovacs, Forest vertical structure from GLAS: An evaluation using LVIS and SRTM data, *Remote Sensing of Environment*, **112**, 107–117, (2008).

82 H. Balzter, A. Luckman, L. Skinner, C. Rowland and T. Dawson, Observations of forest stand top height and mean height from interferometric SAR and LiDAR over a conifer plantation at Thetford Forest, UK, *International Journal of Remote Sensing*, **28**, 1173–1197, (2007).

83 S.A. Hinsley, R.A. Hill, P.E. Bellamy and H. Balzter, The application of LiDAR in woodland bird ecology: climate, canopy structure, and habitat quality, *Photogrammetric Engineering and Remote Sensing*, **72**, 1399–1406, (2006).

84 R.B. Bradbury, R.A. Hill, D.C. Mason, S.A. Hinsley, J.D. Wilson, H. Balzter, G.Q.A. Anderson, M.J. Whittingham, I.J. Davenport and P.E. Bellamy, Modelling relationships between birds and vegetation structure using airborne LiDAR data: a review with case studies from agricultural and woodland environments, *Ibis*, **147**, 443–452, (2005).

85 J.B. Drake, R.O. Dubayah, R.G. Knox, D.B. Clark and J.B. Blair, Sensitivity of large-footprint LiDAR to canopy structure and biomass in a neotropical rainforest, *Remote Sensing of Environment*, **81**, 378–392, (2002).

86 M. Simard, N. Pinto, J.B. Fisher and A. Baccini, Mapping forest canopy height globally with spaceborne LiDAR, *Journal of Geophysical Research: Biogeosciences*, **116**, G04021, (2011).

87 G.E. Kindermann, I. Mccallum, S. Fritz and M. Obersteiner, A global forest growing stock, biomass and carbon map based on FAO statistics, *Silva Fennica*, **42**, 387–396, (2008).

88 D. Schepaschenko, I. Mccallum, A. Shvidenko, S. Fritz, F. Kraxner and M. Obersteiner, A new hybrid land cover dataset for Russia: a methodology for integrating statistics, remote sensing and in situ information, *Journal of Land Use Science*, **6**, 245–259, (2011).

89 M. Thurner, C. Beer, M. Santoro, N. Carvalhais, T. Wutzler, D. Schepaschenko, A. Shvidenko, E. Kompter, B. Ahrens, S.R. Levick and C. Schmullius, Carbon stock and density of northern boreal and temperate forests, *Global Ecology and Biogeography*, **23**, 297–310, (2014).

90 H. Gallaun, G. Zanchi, G.-J. Nabuurs, G. Hengeveld, M. Schardt and P.J. Verkerk, EU-wide maps of growing stock and above-ground biomass in forests based on remote sensing and field measurements, *Forest Ecology and Management*, **260**, 252–261, (2010).

91 M. Santoro, C. Beer, O. Cartus, C. Schmullius, A. Shvidenko, I. Mccallum, U. Wegmüller and A. Wiesmann, Retrieval of growing stock volume in boreal forest using hyper-temporal series of Envisat ASAR ScanSAR backscatter measurements, *Remote Sensing of Environment*, **115**, 490–507, (2011).

92 L. Breiman, Statistical modeling: The two cultures (with comments and a rejoinder by the author), *Statistical Science*, **16**, 199–231, (2001).

93 J.S. Evans and S.A. Cushman, Gradient modeling of conifer species using random forests, *Landscape Ecology*, **24**, 673–683, (2009).

94 L. Breiman, Random forests, *Machine Learning*, **45**, 5–32, (2001).

95 S.J. Phillips, R.P. Anderson and R.E. Schapire, Maximum entropy modeling of species geographic distributions, *Ecological Modelling*, **190**, 231–259, (2006).

96 S.J. Phillips, M. Dud, and R.E. Schapire, *A maximum entropy approach to species distribution modeling, Proceedings of the Twenty-First International Conference on Machine Learning*, Banff, Alberta, Canada, 2004.

97 P. Rodriguez-Veiga, S. Saatchi, K. Tansey and H. Balzter, Magnitude, spatial distribution and uncertainty of forest biomass stocks in Mexico, *Remote Sensing of Environment*, **183**, 265–281. ISSN 0034-4257, http://dx.doi.org/10.1016/j.rse.2016.06.004. (http://www.sciencedirect.com/science/article/pii/S0034425716302425).

98 C.J. Thiel, C. Thiel and C.C. Schmullius, Operational Large-Area Forest Monitoring in Siberia Using ALOS PALSAR Summer Intensities and Winter Coherence, *IEEE Transactions on Geoscience and Remote Sensing*, **47**, 3993–4000, (2009).

99 INEGI, *Cartografia uso de suelo y vegetación*, Instituto Nacional de Estadistica y Geografia, Mexico, 2009.

100 P. Rodriguez-Veiga, M. Stelmaszczuk-Górska, C. Hüttich, C. Schmullius, K. Tansey and H. Balzter, Aboveground Biomass Mapping in Krasnoyarsk Kray (Central Siberia) using Allometry, Landsat, and ALOS PALSAR, *Proceedings RSPSoc Annual Conference*, 2nd–5th September 2014, Aberystwyth, Wales, 2014.

3

Remote Sensing for Aboveground Biomass Estimation in Boreal Forests

M.A. Stelmaszczuk-Górska, C.J. Thiel and C.C. Schmullius

Friedrich-Schiller-University Jena, Department of Earth Observation, Jena, Germany

3.1 Introduction

In order to understand the state and dynamics of ecosystems, their roles within global cycles and their responses to climate change, it is important to quantify and monitor different biophysical parameters. Of these, it is particularly essential to measure forest aboveground biomass (*AGB*) [1]. *AGB* is the mass of all living matter above the soil including stem, stump, branches, bark, seeds and foliage in a particular area [2]. Another important variable is growing stock volume (GSV), which is directly related to *AGB*. GSV is defined as the stem volume of all living species per unit area, including bark and excluding branches and stumps. For the sake of simplicity, the term "biomass" will be used throughout the text to refer collectively to both variables.

When properly related to biophysical tree parameters, remote sensing data can provide information on forest biomass from local to global scales. This information is of special interest in global carbon budget studies since forest biomass estimates still introduce significant source of uncertainty [3–5]. Moreover, accurate estimates of biomass are important in terms of commitments to the reduction of greenhouse gas emissions within international conventions on climate change. An understanding of biomass changes in the boreal forest ecosystem is important for studies of climate change since the boreal forest system covers the largest land area and constitutes a substantial carbon sink. Successful monitoring and modelling of forest biomass depends on the availability of frequent, local to global scale measurements. These can only be provided by means of remotely acquired data e.g., from satellite or airborne sensors. There are different remote sensing sensors: optical, light detection and ranging (LiDAR) and radio detection and ranging (radar). In this chapter the emphasis has been put on radar remote sensing in particular due to the all-weather, global coverage capacity and new, upcoming radar missions. At least seven new radar missions are planned before the end of the 2020s. In April 2016, the Sentinel-1B has been successfully put into orbit, already fourth of the Sentinel satellites. The satellites are designed to increase European Earth Observation capability as a part of the European Copernicus program. The Sentinel data are available without charge [6].

Earth Observation for Land and Emergency Monitoring, First Edition. Edited by Heiko Balzter.
© 2017 John Wiley & Sons Ltd. Published 2017 by John Wiley & Sons Ltd.

Despite more than 30 years of research and algorithm development using radar data, there is still no operational method for global biomass estimation. The main objective of this book chapter is to summarize and synthesize existing estimates of boreal forest biomass obtained using radar sensors. In order to analyse the potential of radar remote sensing, an extensive body of literature has been gathered and reviewed by authors. We systematically compared the methods, taking into account different radar sensor parameters as well as the accuracy of the estimates achieved.

This book chapter is organized in three parts. First, the characteristics of the boreal forest and future challenges with regards to climate change are described. Second, a general overview is given of the remote sensing techniques for biomass estimation from optical, LiDAR and radar sensors. Third, quantified data of the performance of radar biomass estimation in boreal forests are presented and summarized.

3.2   The Boreal Forest Ecosystem

The boreal forest, also known as "taiga", is the world's largest terrestrial biome, covering more than 1.5 billion hectares [7] – just under 11% of the total land surface and 37.5% of the world's forested area [8]. It spreads around the circumpolar region, occupying an area in the northern hemisphere down to a latitude of approximately 45°N in Russia and 50–60°N in northern Europe, Canada and Alaska.

The boreal forest is of great importance in economic, biodiversity and climate change terms alike. It is a major source of forest products such as wood, paper and pulp, which are crucial for manufacturing, energy and the construction industry. It provided circa 23% of forest and other wooded land growing stock in 2010 [8]. Moreover, it creates employment directly in the forestry sector for approximately 1.5 million people [9]. In terms of biodiversity, boreal forests have a low level of species variability. They are characterized by the presence of four main coniferous tree species: spruce (*Picea*), pine (*Pinus*), fir (*Abies*) and larch (*Larix*). Deciduous species are also present: mainly birch (*Betula*), aspen and poplar (*Populus*), which are typical for early succession stages of the forest. Boreal forest is also a habitat for many other plants and fungi, as well as many species of mammals, birds, insects and other types of animals. In terms of climate change, it is important due to (i) global greenhouse gas modelling and remaining uncertainties in the spatial distribution of the terrestrial carbon sink [10,11]; (ii) the sensitivity of the boreal ecosystem to climate warming and (iii) monitoring and reporting obligations within international legislation. Based on the recent results [10], the boreal carbon stock is estimated to be 272 ± 23 Pg C (32%). That makes the boreal biome the second most important terrestrial carbon pool after the tropical forest.

The majority of the carbon (60%) in boreal forests is accumulated in the soil, characterizing this region as having one of the most carbon-rich soils in the world. 20% of the carbon is stored in biomass, which plays a crucial role in regulating carbon as well as other greenhouse gas concentrations in the atmosphere [10]. Forest fires are especially important in nutrient cycling and the release of carbon to the atmosphere [12]. Moreover, fires are the most significant cause of boreal forest loss [13]. Current model predictions indicate that the number of fire events in the boreal zone may double by the end of 2100 [14]. Fires are a major natural disturbance in the Russian boreal forest in particular, where coniferous stands dominate and most of the forest territory is unmanaged and unprotected [15]. Due to the vastness of the Russian territory, comprising different ecosystem types, Russia is expected to suffer the most significant transformation due to climate change [16]. The average annual temperature has already increased substantially, by up to 2 °C in the period from 1971 to 2010, whereas the global annual average increase is less than 1 °C over this same period [12].

The forest in Russia is of special interest because of the more frequent, catastrophic fires expected as a result of climate warming and heat waves such as that which occurred in 2010 [15]. It is also the most forest-rich country in the world (809 million ha [8]), providing more than 90% of the carbon sink of the world's boreal forests in the 2000 to 2007 period [10]. At the same time, it provides the highest uncertainty in global carbon stock calculations. Asian part of Russia provided uncertainty of ± 66 Tg C yr^{-1} and European part of Russia provided ± 50 Tg C yr^{-1} of annual change in carbon stock calculated for 2000–2007 [10]. For this time period the uncertainty for the boreal biome was calculated to be ± 83 Tg C yr^{-1} [10]. Additionally, illegal logging poses a huge problem for the monitoring of Russian forests. According to different sources (WWF Russia, World Bank), illegal clear-cutting in Russia is estimated to comprise approximately 25% of all logging activity. Between 2000 and 2013 Russia has lost more than 37 million hectares of forest [13], more forest than any other country in the world. Moreover, due to the lack of financial support, some forested regions in Siberia have not inventoried for more than 20 years [17]. All of the aforementioned aspects combined make the Russian boreal forest of special monitoring interest.

3.3 Remote Sensing for Biomass Estimation

The ground-based methods for biomass estimation are expensive and labour extensive. Hence, remotely sensed data, especially in combination with *in situ* data when possible, can offer a solution for forest mapping over large areas. Scientists have already demonstrated that remote sensing can contribute to measuring and monitoring current carbon sources and sinks through its ability to provide systematic, globally consistent observations of land cover (LC), land cover change (LCC), forest disturbance and biomass. There are extensive publications available on the topics of optical, LiDAR and radar earth observation (EO) data for LCC and biomass estimation. Several publications provide a comprehensive review of use of remote sensing techniques for biomass estimation including sourcebooks of recommended methods and data sources [18–28]. However, since now no detailed review on accuracy of radar-based biomass estimates for boreal forests has been published.

Optical images record the interaction of solar radiance with Earth surface. The optical satellite sensors provide systematic observations of LC at scales, from global to local. However, direct biomass measurements are unfeasible [29]. This is due to the fact that the sunlight hardly penetrates vertically into the forest cover. Nevertheless, it must be underlined that very high and high resolution data in combination with geostatistical or modelling-based approaches have already been implemented for forest inventory purposes (e.g., for forest stand density, and volume estimation) [30–37]. Furthermore, spectral information from optical EO data are of particular use in the mapping of different vegetation types and vegetation conditions, as well as in the detection of forest fires or other disturbances [38,39]. Indirect estimates of biomass from optical data are feasible at low biomass levels using vegetation indices, bidirectional reflectance distribution function (BRDF) and texture [40–47]. The latest results of biomass estimation based on Landsat data revealed that accuracy of ±36% can be measured in boreal zone [48].

The most accurate, but most financially demanding estimates of biomass are based on airborne LiDAR. Because of the system properties, LiDAR can directly measure the three-dimensional vegetation structure, unlike conventional radar and optical sensors. Airborne small footprint (approximately 20 cm in diameter) [21,49–54] and satellite large footprint LiDAR systems (approximately 70 m in diameter) [55–58] have been already successfully used for biomass estimation. Hyyppä *et al.* [59] demonstrated that Airborne Laser Scanning (ALS) can provide accurate information on the stem volume in boreal zones, comparable with field survey measurements. However, in the case of the ALS system, measurements are too expensive for larger areas. For global applications, archive datasets from only one satellite sensor are available, namely the Ice, Cloud, and land Elevation (ICESat) Geoscience Laser Altimeter System (GLAS). In contrast to the airborne systems, the GLAS LiDAR had some limitations related to the large footprint, sparse coverage and sensitivity to terrain variability. Due to the large footprint, the interference of topographic impacts can hardly be corrected [57,60–63]. The assessment of global vegetation biomass was a secondary mission priority. To reduce these limitations, the ICESat-2 satellite, which is planned for launch in 2017, will use a micro-pulse, multi-beam photon counting approach to provide footprint of 10 m in diameter [64]. Another planned LiDAR mission is the Global Ecosystem Dynamics Investigation (GEDI). The GEDI LiDAR will be mounted on the International Space Station (ISS) with its main objective to create detailed 3D maps and measure the biomass of forests. The mission is scheduled for deployment on the ISS in 2019 [65].

Since no LiDAR mission is in operation, radar sensors are expected to provide the most accurate biomass estimates [20]. Similar to the LiDAR sensors, radar systems are sensitive to the geometrical properties of the observed objects. In addition, the signals are also affected by the dielectric properties of the targets. Moreover, radar systems are weather independent and relatively cost-efficient, which makes the Synthetic Aperture Radar (SAR) sensors suitable for forest biomass estimation; they can provide data on regular basis. Furthermore, new radar satellite missions are planned to be launched (see Figure 3.1). The basic SAR terminology is given in the previous chapter.

It has been already demonstrated that radar data can be physically related to vegetation structure until the saturation point, allowing the retrieval of forest parameters, and that longer wavelengths are superior to shorter ones for this purpose [66–73]. Many SAR properties can be exploited to quantify biomass: backscattering intensity (e.g., [74,75]), interferometric phase (e.g., [76,77]), interferometric correlation (e.g., [78,79]), polarimetric signature (e.g., [80,81]), SAR tomography and radargrammetry (e.g., [82,83]).

Description of global forest monitoring can be found in the chapter titled: Methodology for regional to global mapping of aboveground forest biomass: Integrating forest allometry, ground plots, and satellite observations, whereas the aspects of tropical forest monitoring, case study of the central African forests, are given in Chapter 4: Forest Mapping of the Congo Basin using Synthetic Aperture Radar (SAR).

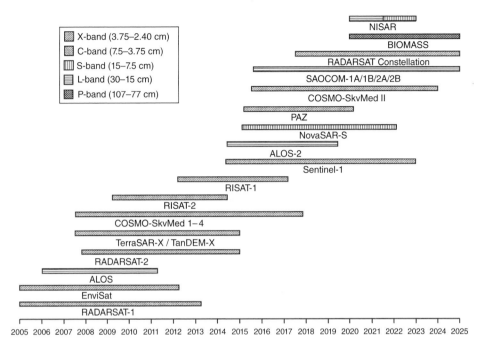

Figure 3.1 SAR missions up to year 2025.

3.4 Radar Remote Sensing for Biomass Estimation in Boreal Forests

3.4.1 Estimation Methods

A variety of approaches have already been developed to estimate biomass in boreal forests using radar remote sensing. Most of the methods that have been implemented are based on empirical and semi-empirical models. The semi-empirical models employ radar data as a function of biomass. It is a combination of physical modelling and empirical observations. The scatterers can be described as a layer, water cloud model [84] or layer with gaps (water cloud model with gaps), which are especially important in studies in the boreal forests [85], or randomly distributed scatterers, Random Volume over Ground model (RVoG) [80]. The empirical models are based on regression analysis. Using linear, multiple-linear or non-linear functions, the biomass is related to the SAR remote sensing data. There are advantages and disadvantages of both approaches. The main drawback of the semi-empirical method is that the models can be either over-simplified, leading to great estimation errors, or too complex, resulting in inversion problems. The main disadvantage in the case of empirical approaches is that they strongly depend on the data set used for model selection and training, and therefore result in poor model transferability. However, both methods have already demonstrated very good biomass estimation accuracy (e.g., [77,86]). It was found that the biomass retrievals based on the empirical model provide equally good estimations compared to the semi-empirical approach [75]. Currently, machine learning algorithms have been also successfully implemented for biomass estimation in boreal forests [87–90].

In order to synthetize the biomass estimation errors reported in the literature using the semi-empirical and empirical approaches, selected statistics from almost 50 peer-reviewed journal articles have been analysed. The papers up to mid-2014 were reviewed. In Table 3.1, a summary of recently reported biomass retrieval results is presented. The accuracy is represented by the coefficient of determination (R^2) or coefficient of correlation (r), root mean square error (RMSE), and relative root mean square error (rRMSE). rRMSE is normalized to the mean of the observed values. In some cases the biomass estimation error RMSE was corrected by the systematic, sampling errors (e.g., for Sweden) [91]. In cases where the rRMSE was not given, it was calculated based on the provided biomass value.

In general, it is difficult to compare all implemented methods due to the differences between the test sites and stand characteristics, stand size, weather conditions during the acquisitions, validation procedures (training, testing), and the model used. However, authors have made efforts to describe the results according to the radar data used, the method implemented and the test site characteristics so that general conclusions can be drawn in order to summarize the retrieval procedures.

3.4.2 Results Using Backscatter, Polarimetric Parameters and Radargrammetry

One of the first models implemented in the boreal zone employing backscatter information for stem volume estimation was a semi-empirical model developed based on the large number of measurements recorded with a high-resolution ranging scatterometer (HUTSCAT). The scatterometer operated at C-band and X-band frequencies. This model presumes that the signal scattered back from the forest canopy is a function of stem volume, soil moisture, and vegetation water content. The results from the HUTSCAT airborne measurements at C-band VV polarization showed high correlation (r = 0.75) between the backscattering coefficient and stem volume for a forest test site in southern Finland [74]. The same region was later studied using the satellite multi-temporal C-band ERS–1 data [97]. Employing the satellite data, a Pearson's coefficient of 0.64 between estimated biomass and *in situ* measurements was calculated. The RMSE was reported to be $90\,\mathrm{m^3\,ha^{-1}}$ (rRMSE = 58%). Both experiments showed that scattering from forested areas is a mixture of parameters related to surface topography and roughness as well as soil moisture and weather conditions. It was concluded that the retrieval error depends on the forest stand size. For the stands greater than 10 ha, the error become negligible. The model was also used for L-band JERS-1 data [98]. The L-band data showed more sensitivity to stem volume and less to environmental conditions, and resulted in a correlation coefficient of 0.65 between the estimated stem volume and the ground truth and a relative error approximately 35% for data acquired in the month of March. It was also emphasized that in the case of the multi-temporal approach, the accuracy can be improved (rRMSE = 25% and r = 0.73 for forest blocks >10 ha). In [99] it was demonstrated that for stands larger than 20 ha, a correlation coefficient between the estimates and the reference values of up to 0.85 and 25% relative error can be achieved using L-band and 0.65 with 27% relative error using C-band. The estimates were improved when the C- and L-band data were used together (rRMSE <25%).

A similar semi-empirical model was introduced by Askne *et al.* [85]. The model is based on the water cloud model including gaps in the layer and expresses the total forest

Table 3.1 Summary of recently published biomass estimation results in boreal forests using SAR data.

Reference	Sensor	Wavelength Polarization	Location/site/No. of stands (No. of plots)	No. of images	Estimated variable	Range of estimated variable (mean)/ range of stand size (mean) [ha]	Radar data used	Method to estimate variable	R² (r)	RMSE [unit as estimated variable] (rRMSE [%])**
[81]	E-SAR*	L-band P-band	Sweden/Krycklan/27	Multiple acquisitions	AGB [t ha⁻¹]	23–183 (94.3)/2.4–26.3	PolInSAR	Multiple linear regression	(0.86) (0.84)	21.3–36.2 (17–27) 22.7–40.7 (5–43)
[92]	E-SAR*	P-band HH, HV, VV	Sweden/ Remningstorp/50(10) Sweden/ Krycklan/29(97)	9 7	AGB [t ha⁻¹]	52–267 (181)/0.5–9.4 8–257 (95)/1.5–22	Backscattering coefficient	Multiple linear regression	0.22–0.64 0.31–0.64	40–59 (22–33) 27–40 (28–42)
[93]	ALOS PLASAR	L-band HH, HV	Finland/Kuortane/123 Finland/Heinävesi/103	3 3	Stem volume [m³ ha⁻¹]	0–314 (95)/(3) 0–425 (110)/(4.8)	Backscattering coefficient	Semi-empirical model	0.65 0.71	41.2 (43.4) 47.0 (42.8)
[94]	ENVISAT ASAR	C-band HH, VV	Sweden/country/11425 Russian Fed./Central Siberia/46487 Canada/Québec/731	Hyper-temporal dataset	Stem volume [m³ ha⁻¹]	(125) (180)/(26) (63)	Backscattering coefficient	Semi-empirical model (BIOMASAR algorithm)	(0.73) (0.86) (0.93)	(22) (15) (27)
[77]	TanDEM-X	X-band HH	Sweden/ Remningstorp/201	18 pairs	AGB [t ha⁻¹]	6–267 (150)/>0.25	InSAR height	Semi-empirical models: Interferometric Water Cloud Model (IWCM) Random Volume over Ground (RVoG) Penetration depth (PD)	– – –	(16.7–33) (16.7–39.7) (17.9–33.1)
[95]	TerraSAR-X Stereo	X-band	Finland/Espoo-Kirkkonummi/94 (207)	2	AGB [t ha⁻¹]	3.9–347.1 (116.5)/ 0.5–12.2(4.1)	Backscattering coefficient (radargrammetry)	Multiple linear regression	–	15.6–16.7 (16.1–17.3)

* Airborne sensor;
** Relative error of less than 25% is considered to be satisfactory in terms of biomass estimation [96].

backscatter as the sum of direct scattering from the ground through canopy gaps, ground scattering attenuated by the tree canopy and direct scattering from the vegetation. Based on this semi-empirical model, Santoro *et al.* [100] reported a very good relative estimation error of 25% ($36\,m^3\,ha^{-1}$) using the JERS-1 L-band data over a Swedish test site located northwest of Stockholm. Additionally, no saturation was observed up to $350\,m^3\,ha^{-1}$. The relationship was almost linear. The data were acquired under unfrozen conditions. In the case of frozen conditions and L-band HH ALOS PALSAR single image an estimation error of 30% was reported (south-eastern Sweden) [101]. The model used in aforementioned studies was implemented by Santoro *et al.* [102] as part of the BIOMASAR algorithm. The algorithm employs hyper-temporal SAR data acquired by the Envisat ASAR C-band sensor and the Vegetation Continuous Field (VCF) product for model training. The retrieved biomass values showed results reaching 34.2% estimation error for Central Siberia. The authors stated that "the key to such results was the optimal weighting of the GSV from individual observations based on the forest/non-forest backscatter sensitivity". The weight is calculated as the difference between the two model parameters, namely the backscatter from vegetation and from the ground [98]. The methodology proved to be robust and consistently resulted in an estimation error in the range of 20 to 30% at a 0.5° resolution. Using five ALOS PALSAR L-band images, Peregon and Yamagata [103] reported estimation error in the range of 25 to 32% for a test site in Western Siberia. Using the same model inversion but multi-temporal filtering of ALOS PALSAR L-band, Antropov *et al.* [93] reported 43% estimation error for a Finnish test site.

Research by Fransson and Israelsson [75] showed that backscatter can be modelled by a semi-empirical water cloud model as well as through linear regression, providing equally good results. Correlation coefficient of 0.78 was calculated between SAR L-band estimated stem volume and ground measurements. Rauste [104] confirmed that multiple linear regression performs satisfactorily using the L-band data showing no saturation up to $150\,m^3\,ha^{-1}$.

Based on a first regression analysis for biomass estimation, second order polynomial regression was implemented by Rignot *et al.* [105]. The model was developed using data acquired by the NASA/JPL radar system (airborne AIRSAR) that operated in three frequencies and polarizations. An estimation error of approximately 20% using P-band HH, HV, VV and L-band HH, HV, VV was achieved. Moreover, the authors demonstrated that the radar acquisition geometry influences biomass retrieval. Similar regression results were found for a temperate forest (Landes, a forest plantation in France), obtaining 11% relative error using P-band, and 13% using L-band HV polarization. These results already showed the potential of implementing the same empirical approach across different ecosystems and large area biomass mapping at an early stage of investigations using backscattering coefficient data. In the case of [106], an empirical model based on multiple linear regression analysis was used. The model resulted in very good agreement between estimated biomass from an airborne P-band acquisition and *in situ* measurements. 99% of the variability in radar estimated biomass was explained by *in situ* biomass measurements. This result was obtained for a summer acquisition. Ranson *et al.* [107] also reported very good model performance ($R^2 = 0.91$, 90 m spatial resolution) for hemi-boreal forest. The estimation was made using multiple linear regression analysis for L- and C-band data acquired by shuttle mission SIR-C/X-SAR. The L- and P-band data from the airborne Experimental-SAR (E-SAR) system were studied in [108]. Results obtained for southern Sweden indicated that P-band in HH polarization can be successfully used for biomass estimation at the same level of accuracy as the *in situ* measurements (rRMSE of less than 20%). Even better results were found by [109] from the same sensor implementing polari-metric parameters, forest height and ground-to-volume ratio. The estimation error was found to be between 25 to 30% using L-band and 18 to 39% using P-band for northern Sweden. Ulander *et al.* [110] evaluated six different regression models for southern and northern Sweden employing E-SAR airborne data acquired at the P-band. The best bio-mass estimation, with 20 to 30% RMSE of the mean biomass was achieved by implement-ing a multiple linear regression including HV backscatter, the VV/HH backscatter ratio, and the ground slope. Further results using E-SAR P-band data and a similar model were obtained by Soja *et al.* [92]. The estimation error was reported to be in the range of 22 to 33% for a test site in south-western Sweden, which is similar to the upper error limit for *in situ* and LiDAR biomass estimates in Sweden (up to 21%).

The applications above describe the use of X- to P-band SAR data. However, even better biomass estimation is expected from a SAR system operating in longer wave-lengths (e.g., Very High Frequency (VHF, 30–300 MHz, metric waves)). This is due to the fact that the long waves penetrate deeper to the foliage and thus the radar signal scatters back mainly from the stem, where the majority of the biomass is accumulated [111]. No saturation for stem volumes up to $900 \, \mathrm{m^3 \, ha^{-1}}$ has been reported for a temper-ate forest in France [112], acquired by the airborne system CARABAS-II, owned by the Swedish Defence Research Agency. However, the long wavelength also faces some limi-tations related to the change in wave properties when going through the ionosphere. In [113], where a Swedish forest is studied, the best coefficient of determination between estimated biomass and reference measurements was reported to be 0.93, with an esti-mation error of $48 \, \mathrm{m^3 \, ha^{-1}}$ (rRMSE of approximately 25%).

Promising results of biomass estimation based on height information were also obtained by employing radargrammetry. This method is based on stereoscopic height measurements using SAR images. For the plot-level estimation based on the TerraSAR-X data, Karjalainen *et al.* [114] reported a 34% relative estimation error for a Finnish test site using the regression trees method. Additionally, no saturation up to $400 \, \mathrm{m^3 \, ha^{-1}}$ was observed. Even better results employing the same source of data for a Swedish test site were reported by Persson *et al.* [83]. Based on the multiple linear regression, the bio-mass was calculated with a 22.9% estimation error, which is comparable to that of much more expensive LiDAR measurements. The latest results over a Finnish test site reported an estimation error as low as 16.1% with −9.5% bias [95].

3.4.3 Results Using SAR Interferometry (InSAR) and Polarimetric Interferometry SAR (PolInSAR)

Complementary information on forested areas can be obtained through the analysis of interferometric SAR data. One of the first studies using repeat-pass InSAR on ERS-1 C-band data from over boreal forest [115] demonstrated the high potential of the inter-ferometric phase information and coherence for forestry applications. The authors obtained very good results based on winter acquisitions over northern Sweden. The tree height for dense forest was estimated with a 10% estimation error using the phase information. In later studies, [116] and [117], authors used interferometric correlation information, which showed very good potential for land-use and forest type mapping, due to high coherence values for non-forest and low values for forested areas. The authors concluded that coherence is seasonally dependant and very sensitive to differences in the acquisition times, showing for example a great potential for clear-cut detection. In [118], the authors indicated that low and stable temperatures without precipitation events are more favourable in InSAR studies. Koskinen *et al.* [78] demonstrated that the highest coherence can be obtained under winter conditions in the presence of snow cover. The authors calculated a correlation coefficient of −0.76 between stem volume and observed coherence for a test site in southern Finland, using a tandem ERS-1/2 image pair acquired under frozen conditions. Moreover, the authors developed an empirical model that expresses coherence information using backscatter information described through a semi-empirical model.

One of the first results presenting the accuracy of stem volume estimation was published by Smith *et al.* [119]. The study over Swedish test sites utilizing C-band tandem ERS-1/2 coherence data estimated the stem volume based on linear regression analysis. The results show high correlation between the coherence and stem volume. In the best case, the coefficient of determination was 0.86 (data from March) with the accuracy of stem volume retrieval of $31 \, m^3 \, ha^{-1}$ (relative error = 23%, assuming $135 \, m^3 \, ha^{-1}$ mean stem volume from [79]). Additionally, no saturation was observed up to $400 \, m^3 \, ha^{-1}$. For the same test site, Santoro *et al.* [79] and [120] showed that the retrieval accuracy can be improved through the implementation of a multi-temporal combination of interferometric data. Using the coherence and backscatter values from ERS-1/2 C-band data and the semi-empirical interferometric water cloud model (IWCM), $21 \, m^3 \, ha^{-1}$ (15.5% error, assuming $135 \, m^3 \, ha^{-1}$ mean stem volume from [79]) and $10.0 \, m^3 \, ha^{-1}$ (7.4% error, $R^2 = 0.94$) accuracy was achieved based on four coherence pairs (March, April). The model was introduced by Askne *et al.* [85]. The model describes coherence as a function of stem volume, perpendicular baseline and tree height, which can be expressed as a function of stem volume. In addition to the study of C-band data, Askne *et al.* [121] investigated four JERS-1 L-band coherence images over the same Swedish test site using the same semi-empirical model. The authors achieved an accuracy in the range of 30 to $35 \, m^3 \, ha^{-1}$ (21.4 to 25%). The achieved accuracy was in agreement with the previous studies by Santoro *et al.* [120] over the same test site, but using a different number of stands for model training. In the study by Pulliainen *et al.* [122], tandem ERS-1/2 data were used over a Finnish test site. 14 image pairs were employed to estimate stem volume based on a simplified formulation of the semi-empirical water cloud model. The authors confirmed that images acquired under frozen (winter) conditions are superior to those obtained under unfrozen conditions. The authors measured an rRMSE of 48%. Over the same Finnish test site semi-empirical, IWCM was also implemented using tandem ERS-1/2 data [123]. The best achieved accuracy of 22% was obtained when only large, homogenous stands were considered. It was shown at Finnish and Swedish test sites in the next study [124] that using large homogenous forest stands and data acquired under frozen conditions, a relative RMSE on the order of 20% can be achieved. Moreover, no saturation of C-band ERS-1/2 was observed for stem volume. The authors demonstrated that small stands cause larger uncertainties than very steep slopes. The accuracy was improved to 17% when only large, homogenous stands were included in the retrieval accuracy analysis [125]. When all stands were included, the retrieval accuracy was 63%. The authors concluded that for selected forest stands and based on multi-temporal filtering, an accuracy in the range of 20 to 30% can be measured.

Based on coherence information, a classification of the Central Siberia was performed [69,126]. Four stem volume classes (0–20, 20–50, 50–80, $>80 \, m^3 \, ha^{-1}$) were delineated by employing C-band ERS-1/2 and backscatter information from L-band JERS-1with high resulting accuracy (>81% for user's and producers's accuracy).

Santoro *et al.* [127] used 10 tandem ERS-1/2 pairs to retrieve stem volume in Central Siberia using a semi-empirical water cloud model. The best estimation accuracy achieved was in the range of 20 to 25%. However, such RMSE was obtained only for stands larger than 3 ha and a relative stocking of at least 50%.

In the study of Solberg *et al.* [128] the usability of interferometric X-band SAR for the inventory of boreal forest over southern Norway was investigated. The authors employed Digital Surface Model (DSM) data from the 11 day repeat-pass Shuttle Radar Topography Mission (SRTM) in conjunction with detailed Digital Terrain Models (DTM). The relationship between biomass and InSAR height was reported to be linear with coefficient of determination in the range of 0.45 to 0.81 showing no saturation. The estimation error was between 18 to 36%, indicating promising results of a new, at that time, TanDEM-X mission. The best result was obtained for homogeneous stands (>90% species composition) larger than 2 ha. In the next study, Solberg *et al.* [86] investigated two interferometric pairs acquired by TanDEM-X for a site in southeast Norway. At the stand level, the accuracy for stem volume estimation was reported to be 19–20%. At the plot level (circular plots of 250 m^2), the error doubled, resulting in 43–44% relative RMSE. In the most recent study [77], 18 bistatic image pairs acquired by the TanDEM-X mission were utilized to estimate biomass over a southern Sweden test site. The results are in agreement with the previous studies, namely error in the range of 17 to 33% was reported. When all 18 pairs were used and weighed inversely proportional to the square of the height of ambiguity (HOA), RMSE was 16% and the coefficient of determination was 0.93 using IWCM retrieval (stands larger than 1 ha). The authors also used other models: RVoG and a simple model based on penetration depth (PD). However, the results were comparable to those obtained using IWCM or resulted in a slightly higher estimation error.

In the study by Neumann *et al.* [81], the performance of L-band and P-band PolInSAR data in estimating boreal forest biomass was assessed. The data were acquired by an airborne E-SAR system over a test site located in southern Sweden. From multiple linear regression, an accuracy in the range of 17 to 25% at L-band and 5 to 27% at P-band was achieved when combining polarimetric information and estimated structure information (forest height and ground–volume ratio). In addition, non-parametric methods (support vector machines and random forests algorithms) were implemented and showed no improvement in estimation accuracy, although a more realistic distribution of biomass values was reported.

3.4.4 Biomass Estimation Without *in situ* Data

One of the difficulties in estimating biomass over vast areas e.g., Central Siberia is the lack of any and/or reliable *in situ* data. Therefore, approaches that do not make use of *in situ* data have been developed. It is important to note that such approaches require some basic knowledge of the biomass distribution in the investigated area. The methods exploit statistics of backscatter and coherence values for non-forest and dense forest based on other remote sensing products or based on the temporal consistency of radar data. The BIOMASAR algorithm introduced by Santoro *et al.* [102] made use of synergy between the optically sensed tree cover product (Moderate Resolution Imaging Spectroradiometer Vegetation Continuous Field) and Envisat ASAR coherence.

The other approach introduced by Santoro [129] and Askne [130] makes use of a consistency plot between two coherence observations. In the latter study, tandem ERS-1/2 data acquired in March and April over a Swedish test site were used. Implementing four tandem pairs in IWCM model, a relative error of 18% and a coefficient of determination of 0.9 was measured. However, it must be underlined that this method requires some manual adjustment to the ridges of the density plots.

3.5 Summary

Biomass estimation using radar remote sensing is a complex task. It includes radar data acquisition, processing and analysis. Furthermore, each step requires an in depth understanding of radar data properties and interpretation. In this review chapter, mainly the last step of the SAR data analysis has been discussed. By means of the retrieved accuracies of biomass: RMSE and R^2, the research findings have been reported and compared.

Historically, the first studies focused on airborne data, backscattering coefficient correlated with biomass over pine plantations and monoculture forests (e.g., [67,131]). Thereafter, more complex, heterogeneous forests were chosen as test sites (e.g., [105,132]). It has already been demonstrated that the radar backscatter increases with increasing biomass until it saturation (e.g., [133]). The relationship has already been successfully modelled by means of empirical and semi-empirical models (e.g., [97–100]). It was also found that the radar signal saturates at a certain biomass level depending on the wavelength used, polarisation, forest structure and environmental conditions. The L- and P-bands are considered to be the most suitable for the estimation of forest parameters due to the deeper penetration into canopy (e.g., [66,67]). It was also shown that the integration of many SAR images can increase the point at which saturation of the radar signal occurs to higher biomass levels up to no saturation (e.g., [102]). The saturation point also increases with the usage of different polarisations and radar frequencies (e.g., [132]). The HV and HH polarizations were proven to be the most sensitive to biomass (e.g., [108]).

InSAR provides new information, complementary to the backscatter intensity. Both interferometric height information and coherence were used for biomass retrieval. The first studies found that the SAR interferometry can be used to derive height information and for classification purposes (e.g., [116]). It has already been demonstrated that the interferometric coherence is inversely correlated to the biomass due to the volume decorrelation. The relationship between biomass and coherence has been described using empirical and semi-empirical models (e.g., [69,121]). It has been shown that the interferometric information provides better biomass estimation results than does backscattering coefficient (e.g., [86,130]). It was stated that environmental conditions need to be stable between data acquisitions to keep the temporal decorrelation at a low level. Nevertheless, the magnitude of temporal decorrelation is related to biomass-increasing biomass typically results in increased temporal decorrelation. It could be shown by means of ALOS PALSAR data that this relationship is most significant during frozen conditions. Therefore, frozen conditions are preferable for biomass estimation based on PALSAR InSAR coherence [79, 134, 135]. Thawing periods or the usage of images from dissimilar seasons showed complete decorrelation (e.g., [135]). Like backscatter intensity, the interferometric coherence also saturates at a certain level of biomass and the multi-temporal approach increases the biomass level at which saturation occurs and the accuracy of the retrieval (e.g., [79]). In many research results, it has been shown that under optimal environmental conditions, the saturation level of interferometric correlation is higher than it would be using backscatter (e.g., [78]). It was also found that in addition to the environmental conditions, SAR data are affected by radar system acquisition geometry, topography and stand characteristics (e.g., [127]).

Most of the studies (53%) employed data acquired by C-band satellite missions and were focused on European sites, 35% using L-band, and 12% using X-band data (Figure 3.2). Only 12% used data acquired over Siberian sites (Figure 3.3).

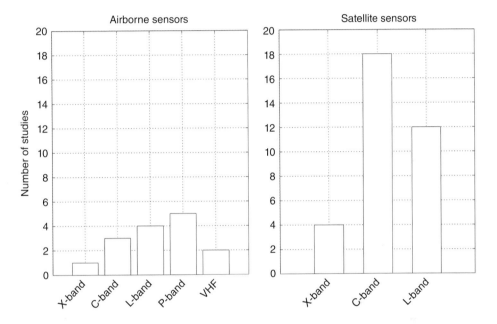

Figure 3.2 Type of radar remote sensing sensors for biomass estimation in boreal forests.

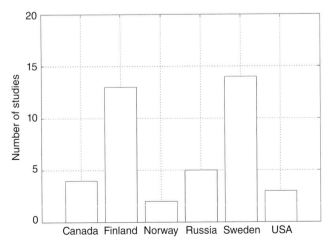

Figure 3.3 Summary of studies per test site in boreal forests.

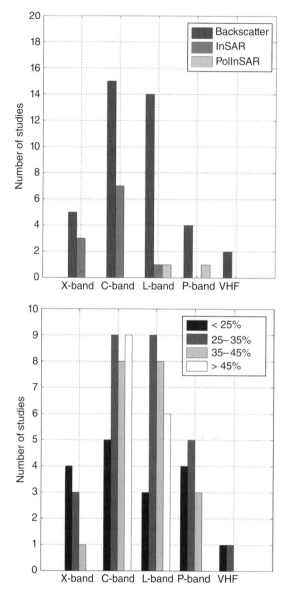

Figure 3.4 Type of data used for biomass estimation (top plot); biomass estimation errors reported depending on the SAR frequency used (bottom plot).

Biomass estimation error varied from more than 100% to 7.4%, when multi-temporal combination was used [79]. Most of the reported models used semi-empirical models (54%). However, Fransson and Israelsson [75] showed that backscatter can be modelled by a semi-empirical model as well as through linear regression, providing equally good results. Most of the reported models employed ground reference data for biomass estimation. Thus, these methods cannot be used when no reliable *in situ* data are available, which is an issue especially for vast territories such as Russia or Canada. Hence, researchers have implemented other strategies for model training. Two studies proposed the use of complementary optical remote sensing tree cover products [102,136]. Another study implemented a consistency plots approach [130]. In this study an estimation error of 18% was calculated.

Although greater wavelengths are preferable for biomass estimation in general, the most useful results have been achieved using C- and X-band data. Figure 3.4 presents a summary of reported estimation errors using aforementioned SAR products: backscatter intensity, interferometric coherence (InSAR), and polarimetric interferometric coherence (PolInSAR).

The lowest estimation errors were reported employing semi-empirical models based on multi-temporal C-band coherence and backscatter (7.4% [79]) and X-band data using InSAR height (16% [77]) and radargrammetry method (16% [95]) for European forest sites. For Siberian forests an estimation error of 25% was achieved [127]. Nevertheless, expedient results can be expected from the forthcoming missions, e.g. P-band BIOMASS mission, which are highly anticipated within the forest biomass remote sensing community.

Acknowledgments

We would like to thank Stephanie Palmer for English proofreading of the first version of the manuscript.

References

1 R.A. Houghton, F. Hall, and S.J. Goetz, Importance of biomass in the global carbon cycle, *Journal of Geophysical Research: Biogeosciences*, **114**(G2), 1–13 (2009).

2 FAO, *Global Forest Resources Assessment 2005: Progress towards sustainable forest management*, Rome, 2006.

3 P.A. Harrell, L.L. Bourgeau-Chavez, E.S. Kasischke, N.H.F. French, and N.L. Christensen Jr., Sensitivity of ERS-1 and JERS-1 radar data to biomass and stand structure in Alaskan boreal forest, *Remote Sensing of Environment*, **54**, 247–260 (1995).

4 FAO, Terrestrial Essential Climate Variables. For Climate Change Assessment, Mitigation and Adaptation – BIOMASS, FAO, Rome, 2009.

5 S. Nilsson, A. Shvidenko, M. Jonas, I. McCallum, A. Thomson, and H. Balzter, Uncertainties of a regional terrestrial biota full carbon account: A systems analysis, *Water, Air Soil Pollution, Focus*, **7**, 425–441 (2007).

6 EU, *Regulation (EU) No 377/2014 of the European Parliament and of the Council of 3 April 2014 establishing the Copernicus Programme and repealing Regulation (EU) No 911/2010 (Text with EEA relevance)*, 44–66, 2014.

7 WHRC, *The boreal forest ecosystem, 2014*. Online. Available: www.whrc.org/ecosystem/ highlatitude/.

8 FAO, *Global Forest Resources. Assessment 2010*. Main report, FAO, Rome, 2010.

9 FAO, *State of the World's Forests: Enhancing the socioeconomic benefits from forests*, FAO, Rome, 2014.

10 Y. Pan, R.A. Birdsey, J. Fang, R. Houghton *et al.*, A large and persistent carbon sink in the world's forests., *Science*, **333**, 988–93 (2011).

11 R.A. Houghton, Aboveground forest biomass and the global carbon balance, *Global Change Biology*, **11**, 945–958 (2005).

12 IPCC, Climate Change 2013: *The physical science basis.* Working Group I contribution to the fifth assessment report of the Intergovermental Panel on Climate Change, 2013.

13 M.C. Hansen, P.V Potapov, R. Moore, M. Hancher, S.A. Turubanova, A. Tyukavina, D. Thau, S.V Stehman, S.J. Goetz, T.R. Loveland, A. Kommareddy, A. Egorov, L. Chini, C.O. Justice, and J.R.G. Townshend, High-resolution global maps of 21st-century forest cover change, *Science*, **342**, 850–3 (2013).

14 A.Z. Shvidenko, D.G. Shchepashchenko, E.A. Vaganov, A.I. Sukhinin, S.S. Maksyutov, I. McCallum, and I.P. Lakyda, Impact of wildfire in Russia between 1998–2010 on ecosystems and the global carbon budget, *Doklady Earth Sciences*, **441**, 1678–1682 (2011).

15 A. Shvidenko, D. Schepaschenko, A. Sukhinin, I. McCallum, and S. Maksyutov, Carbon emissions from forest fires in boreal Eurasia between 1998–2010, *in Proceedings of the 5th International Wildland Fire Conference*, 2011.

16 FAO, *The Russian Federation forest sector outlook study to 2030*, FAO, Rome, 2012.

17 A. Shvidenko, D. Schepaschenko, I. McCallum, and S. Nilsson, Can the uncertainty of full carbon accounting of forest ecosystems be made acceptable to policymakers?, *in Greenhouse Gas Inventories*, Springer., Dordrecht, 137–157, 2011.

18 E.S. Kasischke, J. M. Melack, and M.C. Dobson, The use of imaging radars for ecological applications - A review, *Remote Sensing of Environment*, **59**, 141–156 (1997).

19 Å. Rosenqvist, A. Milne, R. Lucas, M. Imhoff, and C. Dobson, A review of remote sensing technology in support of the Kyoto Protocol, *Environmental Science and Policy*, **6**, 441–455 (2003).

20 G. Patenaude, R. Milne, and T.P. Dawson, Synthesis of remote sensing approaches for forest carbon estimation: reporting to the Kyoto Protocol, *Environmental Science and Policy*, **8**, 161–178 (2005).

21 I. McCallum, W. Wagner, C. Schmullius, A. Shvidenko, M. Obersteiner, S. Fritz, and S. Nilsson, Satellite-based terrestrial production efficiency modeling, *Carbon Balance and Management*, **4**, 8 (2009).

22 S.J. Goetz, A. Baccini, N.T. Laporte, T. Johns, W. Walker, J. Kellndorfer, R.A. Houghton, and M. Sun, Mapping and monitoring carbon stocks with satellite observations: a comparison of methods., *Carbon Balance and Management*, **4**, 2 (2009).

23 S. Frolking, M.W. Palace, D.B. Clark, J.Q. Chambers, H.H. Shugart, and G.C. Hurtt, Forest disturbance and recovery: A general review in the context of spaceborne remote sensing of impacts on aboveground biomass and canopy structure, *Journal of Geophysical Research: Biogeosciences*, **114**, G00E02.

24 C. Thiel, M. Santoro, O. Cartus, C. Thiel, T. Riedel, and C. Schmullius, Perspectives of SAR based forest cover, forest cover change and biomass mapping, *in The Kyoto Protocol: Economic Assessments, Implementation Mechanisms, and Policy implications*, C.P. Vasser (Ed.), New York: Nova Science Publishers, Inc., 13–56, 2009.

25 S. Goetz and R. Dubayah, Advances in remote sensing technology and implications for measuring and monitoring forest carbon stocks and change, *Carbon Management*, **2**, 231–244 (2011).

26 S.G. Zolkos, S.J. Goetz, and R. Dubayah, A meta-analysis of terrestrial aboveground biomass estimation using lidar remote sensing, *Remote Sensing of Environment*, **128**, 289–298 (2013).

27 GOFC-GOLD, *A sourcebook of methods and procedures for monitoring and reporting anthropogenic greenhouse gas emissions and removals caused by deforestation, gains and losses of carbon stocks in forests remaining forests, and forestation*, 2009.

28 T. Fatoyinbo (Ed), *Remote sensing of biomass – Principles and applications*. InTech, 2012.
29 R.A. Houghton, D. Butman, A.G. Bunn, O.N. Krankina, P. Schlesinger, and T.A. Stone, Mapping Russian forest biomass with data from satellites and forest inventories, *Environmental Research Letters*, **2** (2007).
30 H. Franco-Lopez, A.R. Ek, and M.E. Bauer, Estimation and mapping of forest stand density, volume, and cover type using the k-nearest neighbors method, *Remote Sensing of Environment*, **77**, 251–274 (2001).
31 E. Muinonen, M. Maltamo, and H. Hyppa, Forest stand characteristics estimation using a most similar neighbor approach and image spatial structure information, *Remote Sensing of Environment*, **78**, 223–228 (2001).
32 E. Tomppo, M. Nilsson, M. Rosengren, P. Aalto, and P. Kennedy, Simultaneous use of Landsat-TM and IRS–1C WiFS data in estimating large area tree stem volume and aboveground biomass, *Remote Sensing of Environment*, **82**, 156–171 (2002).
33 A. Pekkarinen, Image segment-based spectral features in the estimation of timber volume, *Remote Sensing of Environment*, **82**, 349–359 (2002).
34 P. Muukkonen and J. Heiskanen, Estimating biomass for boreal forests using ASTER satellite data combined with standwise forest inventory data, *Remote Sensing of Environment*, **99**, 434–447 (2005).
35 P. Muukkonen and J. Heiskanen, Biomass estimation over a large area based on standwise forest inventory data and ASTER and MODIS satellite data: A possibility to verify carbon inventories, *Remote Sensing of Environment*, **107**, 617–624 (2007).
36 J.E. Luther, R.A. Fournier, D.E. Piercey, L. Guindon, and R.J. Hall, Biomass mapping using forest type and structure derived from Landsat TM imagery, *International Journal of Applied Earth Observation and Geoinformation*, **8**, 173–187 (2006).
37 H. Fuchs, P. Magdon, C. Kleinn, and H. Flessa, Estimating aboveground carbon in a catchment of the Siberian forest tundra: Combining satellite imagery and field inventory, *Remote Sensing of Environment*, **113**, 518–531 (2009).
38 R. Fraser and Z. Li, Estimating fire-related parameters in boreal forest using SPOT VEGETATION, *Remote Sensing of Environment*, **82**, 95–110 (2002).
39 C. Souza, L. Firestone, L.M. Silva, and D. Roberts, Mapping forest degradation in the Eastern Amazon from SPOT 4 through spectral mixture models, *Remote Sensing of Environment*, **87**, 494–506 (2003).
40 S.A. Sader, R.B. Waide, W.T. Lawrence, and A.T. Joyce, Tropical Forest Biomass and Successional Age Class Relationships to a Vegetation Index Derived from Landsat TM Data, *Remote Sensing of Environment*, **28**, 143–156 (1989).
41 D.S. Boyd, G.M. Foody, and P.J. Curran, The relationship between the biomass of Cameroonian tropical forests and radiation reflected in middle infrared wavelengths (3.0–5.0 µm), *International Journal of Remote Sensing*, **20**, 1017–1023 (1999).
42 C. De Wasseige and P. Defourny, Retrieval of tropical forest structure characteristics from bi-directional reflectance of SPOT images, *Remote Sensing of Environment*, **83**, 362–375 (2002).
43 J. Dong, R.K. Kaufmann, R.B. Myneni, C.J. Tucker, P.E. Kauppi, J. Liski, W. Buermann, V. Alexeyev, and M.K. Hughes, Remote sensing estimates of boreal and temperate forest woody biomass: carbon pools, sources, and sinks, *Remote Sensing of Environment*, **84**, 393–410 (2003).
44 G. Zheng, J. M. Chen, Q.J. Tian, W.M. Ju, and X.Q. Xia, Combining remote sensing imagery and forest age inventory for biomass mapping, *Journal of Environmental Management*, **85**, 616–23 (2007).

45 C. Joshi, J. De Leeuw, A.K. Skidmore, I.C. Van Duren, and H. van Oosten, Remotely sensed estimation of forest canopy density: A comparison of the performance of four methods, *International Journal of Applied Earth Observation and Geoinformation*, **8**, 84–95 (2006).

46 V. Avitabile, M. Herold, M. Henry, and C. Schmullius, Mapping biomass with remote sensing: a comparison of methods for the case study of Uganda, *Carbon Balance and Management*, **6**, 7 (2011).

47 V. Avitabile, A. Baccini, M. A. Friedl, and C. Schmullius, Capabilities and limitations of Landsat and land cover data for aboveground woody biomass estimation of Uganda, *Remote Sensing of Environment*, **117**, 366–380 (2012).

48 L. Ji, B.K. Wylie, D.R. Nossov, B. Peterson, M.P. Waldrop, J.W. McFarland, J. Rover, and T.N. Hollingsworth, Estimating aboveground biomass in interior Alaska with Landsat data and field measurements, *International Journal of Applied Earth Observation and Geoinformation*, **18**, 451–461 (2012).

49 R. Nelson, W. Krabill, and J. Tonelli, Estimating forest biomass and volume using airborne laser data, *Remote Sensing of Environment*, **24**, 247–267 (1988).

50 M. Nilsson, Estimation of tree heights and stand volume using an airborne Lidar system, *Remote Sensing of Environment*, **56**, 1–7 (1996).

51 E. Næsset, Estimating timber volume of forest stands using airborne laser scanner data, *Remote Sensing of Environment*, **61**, 246–253 (1997).

52 E. Næsset and K.-O. Bjerknes, Estimating tree heights and number of stems in young forest stands using airborne laser scanner data, *Remote Sensing of Environment*, **78**, 328–340 (2001).

53 G. Patenaude, R. Hill, R. Milne, D.L.A. Gaveau, B.B.J. Briggs, and T.P. Dawson, Quantifying forest above ground carbon content using LiDAR remote sensing, *Remote Sensing of Environment*, **93**, 368–380 (2004).

54 Z.J. Bortolot and R.H. Wynne, Estimating forest biomass using small footprint LiDAR data: An individual tree-based approach that incorporates training data, *ISPRS Journal of Photogrammetry and Remote Sensing*, **59**, 342–360 (2005).

55 M.A. Lefsky, W.B. Cohen, S.A. Acker, G.G. Parker, T.A. Spies, and D. Harding, Lidar remote sensing of the canopy structure and biophysical properties of Douglas-Fir Western Hemlock forests, *Remote Sensing of Environment*, **70**, 339–361 (1999).

56 M.A. Lefsky, W.B. Cohen, S.W.J. Way, D.J. Harding, G.G. Parker, S.A. Acker, F. Avenue, and S.T. Gower, LiDAR remote sensing of aboveground biomass in three biomes, *International Archives of Photogrammetry and Remote Sensing*, **XXXIV–3/W4**, 22–24 (2001).

57 M.A. Lefsky, D.J. Harding, M. Keller, W.B. Cohen, C.C. Carabajal, F. Del Bom Espirito-Santo, M.O. Hunter, and R. de Oliveira, Estimates of forest canopy height and aboveground biomass using ICESat, *Geophysical Research Letters*, **32**, 22–25 (2005).

58 J.B. Drake, R.O. Dubayah, R.G. Knox, D.B. Clark, J.B. Blair, and C. Rica, Sensitivity of large-footprint lidar to canopy structure and biomass in a neotropical rainforest, *Remote Sensing of Environment*, **81**, 378–392 (2002).

59 J. Hyyppä, H. Hyyppä, D. Leckie, F. Gougeon, X. Yu, and M. Maltamo, Review of methods of small footprint airborne laser scanning for extracting forest inventory data in boreal forests, *International Journal of Remote Sensing*, **29**, 1339–1366 (2008).

60 R. Nelson, K.J. Ranson, G. Sun, D.S. Kimes, V. Kharuk, and P. Montesano, Estimating Siberian timber volume using MODIS and ICESat/GLAS, *Remote Sensing of Environment*, **113**, 691–701 (2009).

61 L.I. Duncanson, K.O. Niemann, and M.A. Wulder, Estimating forest canopy height and terrain relief from GLAS waveform metrics, *Remote Sensing of Environment*, **114**, 138–154 (2010).

62 M. Simard, N. Pinto, J.B. Fisher, and A. Baccini, Mapping forest canopy height globally with spaceborne lidar, *Journal of Geophysical Research: Biogeosciences*, **116**, G04021, (2011).

63 E. Khalefa, I.P.J. Smit, A. Nickless, S. Archibald, A. Comber, and H. Balzter, Retrieval of savanna vegetation canopy height from ICESat-GLAS spaceborne LiDAR with terrain correction, *IEEE Geoscience and Remote Sensing Letters*, **10**, 1439–1443 (2013).

64 D. Harding, P. Dabney, J. Abshire, T. Huss, G. Jodor, R. Machan, J. Marzouk, K. Rush, A. Seas, C. Shuman, X. Sun, S. Valett, A. Vasilyev, A. Yu, and Y. Zheng, The slope imaging multi-polarisation photon-counting LiDAR: an advanced technology airborne laser altimeter, *in Proceedings of the NASA Earth Science Technology Forum*, 2010.

65 GEDI Mission, 2014. Online. Available: http://geog.umd.edu/feature/gedi-lidar-piralph-dubayah-selected-earth-ventures-instrument-program.

66 M.C. Dobson, F.T. Ulaby, A. Beaudoin, E.S. Kasischke, and N. Christensen, Dependence of radar backscatter on coniferous forest biomass, *IEEE Transactions on Geoscience and Remote Sensing*, **30**, 412–415 (1992).

67 T. Le Toan, A. Beaudoin, J. Riom, and D. Guyon, Relating forest biomass to SAR data, *IEEE Transactions on Geoscience and Remote Sensing*, **30**, 403–411 (1992).

68 A. Beaudoin, T. Le Toan, S. Goze, E. Nezry, A. Lopes, E. Mougin, C.C. Hsu, H.C. Han, J.A. Kong, and R. T. Shin, Retrieval of forest biomass from SAR data, *International Journal of Remote Sensing*, **15**, 2777–2796 (1994).

69 W. Wagner, A. Luckman, J. Vietmeier, K. Tansey, H. Balzter, C. Schmullius, M. Davidson, D. Gaveau, M. Gluck, T. Le, S. Quegan, A. Shvidenko, A. Wiesmann, and J. Jiong, Large-scale mapping of boreal forest in SIBERIA using ERS tandem coherence and JERS backscatter data, *Remote Sensing*, **85**, 125–144 (2003).

70 K.J. Tansey, A.J. Luckman, L. Skinner, H. Balzter, T. Strozzi, and W. Wagner, Classification of forest volume resources using ERS tandem coherence and JERS backscatter data, *Forestry* (2004).

71 H. Balzter, Forest mapping and monitoring with interferometric synthetic aperture radar (InSAR), *Progress in Physical Geography*, **25**, 159–177 (2001).

72 H. Balzter, C.S. Rowland, and P. Saich, Forest canopy height and carbon estimation at Monks Wood National Nature Reserve, UK, using dual-wavelength SAR interferometry, *Remote Sensing of Environment*, **108**, 224–239 (2007).

73 A. Luckmann, J. Baker, and U. Wegmüller, Repeat-Pass Interferometric Coherence Measurements of Disturbed Tropical Forest from JERS and ERS Satellites, *Remote Sensing of Environment*, **73**, 350–360 (2000).

74 J.T. Pulliainen, K. Heiska, J. Hyyppa, and M.T. Hallikainen, Backscattering properties of boreal forests at the C- and X-bands, *IEEE Transactions on Geoscience and Remote Sensing*, **32**, 1041–1050 (1994).

75 J.E.S. Fransson and H. Israelsson, Estimation of stem volume in boreal forests using ERS–1 C- and JERS-1 L-band SAR data, *International Journal of Remote Sensing*, **20**, 123–137 (1999).

76 S. Solberg, R. Astrup, T. Gobakken, E. Næsset, and D.J. Weydahl, Remote sensing of environment estimating spruce and pine biomass with interferometric X-band SAR, *Remote Sensing of Environment*, **114**, 2353–2360 (2010).

77 J. Askne, J. Fransson, M. Santoro, M. Soja, and L. Ulander, Model-based biomass estimation of a hemi-boreal forest from multitemporal TanDEM-X acquisitions, *Remote Sensing*, **5**, 5574–5597 (2013).

78 J.T. Koskinen, J.T. Pulliainen, J. M. Hyyppä, M.E. Engdahl, and M.T. Hallikainen, The seasonal behavior of interferometric coherence in boreal forest, *IEEE Transactions on Geoscience and Remote Sensing*, **39**, 820–829 (2001).

79 M. Santoro, J. Askne, G. Smith, and J.E.S. Fransson, Stem volume retrieval in boreal forests from ERS-1/2 interferometry, *Remote Sensing of Environment*, **81**, 19–35 (2002).

80 K.P. Papathanassiou and S.R. Cloude, Single-baseline polarimetric SAR interferometry, *IEEE Transactions on Geoscience and Remote Sensing*, **39**, 2352–2363 (2001).

81 M. Neumann, S.S. Saatchi, L.M.H. Ulander, and J.E.S. Fransson, Assessing performance of L- and P-Band polarimetric interferometric SAR data in estimating boreal forest above-ground biomass, *IEEE Transactions on Geoscience and Remote Sensing*, **50**, 714–726 (2012).

82 S. Tebaldini and F. Rocca, Multibaseline polarimetric SAR tomography of a boreal forest at P- and L-bands, *IEEE Transactions on Geoscience and Remote Sensing*, **50**, 232–246 (2012).

83 H. Persson and J. Fransson, Forest variable estimation using radargrammetric processing of TerraSAR-X images in boreal forests, *Remote Sensing*, **6**, 2084–2107 (2014).

84 F.T. Ulaby, R.K. Moore, and A.K. Fung, *Microwave Remote Sensing Active and Passive. Volume II. Radar Remote Sensing and Surface Scattering and Emission Theory.* Norwood, MA, USA: Artech House, Inc., 1982.

85 J.I.H. Askne, P.B.G. Dammert, L.M.H. Ulander, and G. Smith, C-band repeat-pass interferometric SAR observations of the forest, *IEEE Transactions on Geoscience and Remote Sensing*, **35**, 25–35 (1997).

86 S. Solberg, R. Astrup, J. Breidenbach, B. Nilsen, and D. Weydahl, Monitoring spruce volume and biomass with InSAR data from TanDEM-X, *Remote Sensing of Environment*, **139**, 60–67 (2013).

87 P. Rodriguez-Veiga, M. Stelmaszczuk-Górska, C. Hüttich, C. Schmullius, K. Tansey, and H. Balzter, Aboveground Biomass Mapping in Krasnoyarsk Kray (Central Siberia) using Allometry, Landsat, and ALOS PALSAR, *in Proceedings of the RSPSoc Annual Conference*, 2014.

88 S. Wilhelm, C. Hüttich, M. Korets, and C. Schmullius, Large area mapping of boreal Growing Stock Volume on an annual and multi-temporal level using PALSAR L-band backscatter mosaics, *Forests*, **5**, 1999–2015 (2014).

89 C. Hüttich, M. Korets, S. Bartalev, V. Zharko, D. Schepaschenko, A. Shvidenko, and C. Schmullius, Exploiting Growing Stock Volume Maps for Large Scale Forest Resource Assessment: Cross-Comparisons of ASAR- and PALSAR-Based GSV Estimates with Forest Inventory in Central Siberia, *Forests*, **5**, 1753–1776 (2014).

90 M.A. Stelmaszczuk-Górska, P. Rodriguez-Veiga, N. Ackermann, C. Thiel, H. Balzter, and C. Schmullius, Non-Parametric Retrieval of Aboveground Biomass in Siberian Boreal Forests with ALOS PALSAR Interferometric Coherence and Backscatter Intensity, *Journal of Imaging*, **2**, 1 (2015).

91 J.E.S. Fransson, F. WaLter, and L.M.H. Ulander, Estimation of forest parameters using CARABAS-II VHF SAR data, *IEEE Transactions on Geoscience and Remote Sensing*, **38**, 720–727 (2000).

92 M.J. Soja, G. Sandberg, L.M.H. Ulander, and S. Member, Regression-based retrieval of boreal forest biomass in sloping terrain using P-band SAR backscatter intensity data, *IEEE Transactions on Geoscience and Remote Sensing*, **51**, 2646–2665 (2013).

93 O. Antropov, Y. Rauste, H. Ahola, and T. Häme, Stand-level stem volume of boreal forests from spaceborne SAR imagery at L-band, *IEEE Transactions on Geoscience and Remote Sensing*, **6**, 4776–4779 (2013).

94 M. Santoro, O. Cartus, J. Fransson, A. Shvidenko, I. McCallum, R. Hall, A. Beaudoin, C. Beer, and C. Schmullius, Estimates of Forest Growing Stock Volume for Sweden, Central Siberia, and Québec Using Envisat Advanced Synthetic Aperture Radar Backscatter Data, *Remote Sensing*, **5**, 4503–4532 (2013).

95 M. Vastaranta, M. Niemi, M. Karjalainen, J. Peuhkurinen, V. Kankare, J. Hyyppä, and M. Holopainen, Prediction of forest stand attributes using TerraSAR-X stereo imagery, *Remote Sensing*, **6**, 3227–3246 (2014).

96 H. Balzter, E. Talmon, W. Wagner, D. Gaveau, S. Plummer, J.J. Yu, S. Quegan, M. Davidson, T. Le Toan, M. Gluck, A. Shvidenko, S. Nilsson, K. Tansey, A. Luckman, and C. Schmullius, Accuracy assessment of a large-scale forest cover map of central Siberia from synthetic aperture radar, *Canadian Journal of Remote Sensing*, **28**, 719–737 (2002).

97 J.T. Pulliainen, P.J. Mikkela, M.T. Hallikainen, and J. Ikonen, Seasonal dynamics of C-band backscatter of boreal forests with applications to biomass and soil moisture estimation, *IEEE Transactions on Geoscience and Remote Sensing*, **34**, 758–770 (1996).

98 L. Kurvonen, J. Pulliainen, and M. Hallikainen, Retrieval of biomass in boreal forests from multitemporal ERS-1 and JERS-1 SAR images, *IEEE Transactions on Geoscience and Remote Sensing*, **37**, 198–205 (1999).

99 J.T. Pulliainen, L. Kurvonen, and M.T. Hallikainen, Multitemporal behavior of L- and C-band SAR observations of boreal forests, *IEEE Transactions on Geoscience and Remote Sensing*, **37**, 927–937 (1999).

100 M. Santoro, L. Eriksson, J. Askne, and C. Schmullius, Assessment of stand-wise stem volume retrieval in boreal forest from JERS-1 L-band SAR backscatter, *International Journal of Remote Sensing*, **27**, 3425–3454 (2006).

101 L.E.B. Eriksson, M. Magnusson, J.E.S. Fransson, G. Sandberg, and L.M.H. Ulander, Stem volume estimation for boreal forest using ALOS PALSAR, *in Proceedings of the the International Symposium on Retrieval of Bio- and Geophysical Parameters from SAR Data for Land Applications, Bari, Italy, September 25–28, 2007, ESA*, 2007.

102 M. Santoro, C. Beer, O. Cartus, C. Schmullius, A. Shvidenko, I. Mccallum, U. Wegmüller, and A. Wiesmann, Retrieval of growing stock volume in boreal forest using hyper-temporal series of Envisat ASAR ScanSAR backscatter measurements, *Remote Sensing of Environment*, **115**, 490–507 (2011).

103 A. Peregon and Y. Yamagata, The use of ALOS/PALSAR backscatter to estimate above-ground forest biomass: A case study in Western Siberia, *Remote Sensing of Environment*, **137**, 139–146 (2013).

104 Y. Rauste, Multi-temporal JERS SAR data in boreal forest biomass mapping, *Remote Sensing of Environment*, **97**, 263–275 (2005).

105 E. Rignot, J. Way, C. Williams, and L. Viereck, Radar estimates of aboveground biomass in boreal forests of interior Alaska, *IEEE Transactions on Geoscience and Remote Sensing*, **32**, 1117–1124 (1994).

106   E.J. Rignot, R. Zimmermann, and J.J. Van Zyl, Spaceborne applications of P-band imaging radars for measuring forest biomass, *IEEE Transactions on Geoscience and Remote Sensing*, **33**, 1162–1169 (1995).

107   K.J. Ranson, G. Sun, R.H. Lang, N.S. Chauhan, R.J. Cacciola, and O. Kilic, Mapping of boreal forest biomass from spaceborne synthetic aperture radar, *Remote Sensing of Environment*, **102**, 29599–29610 (1997).

108   G. Sandberg, L.M.H. Ulander, J.E.S. Fransson, J. Holmgren, and T. Le Toan, L- and P-band backscatter intensity for biomass retrieval in hemiboreal forest, *Remote Sensing of Environment*, **115**, 2874–2886 (2011).

109   M. Neumann, S.S. Saatchi, L.M.H. Ulander, and J.E.S. Fransson, Boreal forest biomass estimation Using PolInSAR Vertical and morphological structure indicators, *in Proceedings of the PolInSAR 5th Int. Workshop on Science and Applications of SAR Polarimetry and Polarimetric Interferometry*, 2011.

110   L.M.H. Ulander, G. Sandberg, and M.J. Soja, Biomass retrieval algorithm based on P-band biosar experiments of boreal forest, *in Proceedings of the 2011 IEEE International Geoscience and Remote Sensing Symposium*, 4245–4248, 2011.

111   A.Z. Shvidenko, E. Gustafson, A.D. Mcguire, V.I. Kharuk *et al.*, Terrestrial Ecosystems and Their Change, *in Regional Environmental Changes in Siberia and Their Global Consequences*, P. Y. Groisman and G. Gutman (Eds), Dordrecht: Springer Netherlands, 171–249, 2013.

112   P. Melon, J.M. Martinez, T. Le Toan, L.M.H. Ulander, and A. Beaudoin, On the retrieving of forest stem volume from VHF SAR data: observation and modeling, *IEEE Transactions on Geoscience and Remote Sensing*, **39**, 2364–2372 (2001).

113   K. Folkesson, G. Smith-jonforsen, and L.M.H. Ulander, Model-based compensation of topographic effects for improved stem-volume retrieval from CARABAS-II VHF-band SAR images, *IEEE Transactions on Geoscience and Remote Sensing*, **47**, 1045–1055 (2009).

114   M. Karjalainen, V. Kankare, M. Vastaranta, M. Holopainen, and J. Hyyppä, Prediction of plot-level forest variables using TerraSAR-X stereo SAR data, *Remote Sensing of Environment*, **117**, 338–347 (2012).

115   J. Hagberg, S. Member, L.M.H. Ulander, and J. Askne, Repeat-pass SAR interferometry over forested terrain, *IEEE Transactions on Geoscience and Remote Sensing*, **33**, 331–340 (1995).

116   U. Wegmuller and C.L. Werner, SAR interferometric signatures of forest, *IEEE Transactions on Geoscience and Remote Sensing*, **33**, 1153–1161 (1995).

117   J. Askne and G. Smith, Forest InSAR decorrelation and classification properties, *in Proceedings of the FRINGE 96 Workshop on ERS SAR Interferometry*, 1996.

118   B.G. Dammert and J. Askne, Interferometric tree height observations in boreal forests with SAR interferometry, *in Proceedings of the IEEE International Geoscience and Remote Sensing Symposium 1998*, 1363–1366, 1998.

119   G. Smith, P.B.G. Dammert, M. Santoro, J.E.S. Fransson, U. Wegmüller, J. Askne, and N. Federico, Biomass retrieval in boreal forest using ERS and JERS SAR, *in Proceedings of the Retrieval of Bio- and Geophysical Parameters from SAR Data for Land Applications*, 293–300, 1998.

120   M. Santoro, J. Askne, P.B.G. Dammert, J.E.S. Fransson, and G. Smith, Retrieval of biomass in boreal forest from multi-temporal ERS-1/2 interferometry, *in Proceedings of the FRINGE '99 Advancing ERS SAR Interferometry from Applications towards Operations*, 1999.

121 J. Askne, M. Santoro, G. Smith, and J.E.S. Fransson, Multitemporal repeat-rass SAR interferometry of boreal forests, *IEEE Transactions on Geoscience and Remote Sensing*, **41**, 1540–1550 (2003).

122 J. Pulliainen, M. Engdahl, and M. Hallikainen, Feasibility of multi-temporal interferometric SAR data for stand-level estimation of boreal forest stem volume, *Remote Sensing of Environment*, **85**, 397–409 (2003).

123 J. Askne and M. Santoro, Multitemporal repeat pass SAR interferometry of boreal forests, *IEEE Transactions on Geoscience and Remote Sensing*, **43**, 1219–1228 (2005).

124 J. Askne and M. Santoro, Boreal forest stem volume estimation from multitemporal C-band InSAR observations, *in Proceedings of the Envisat Symposium 2007*, 2007.

125 J. Askne and M. Santoro, Selection of forest stands for stem volume retrieval from stable ERS tandem InSAR observations, *IEEE Geoscience and Remote Sensing Letters*, **4**, 46–50 (2007).

126 W. Wagner, J. Vietmeier, C. Schmullius, M. Davidson, T. Le Toan, S. Quegan, J.J. Yu, A. Luckman, K. Tansey, H. Balzter, and D. Gaveau, The use of coherence information from ERS Tandem Pairs for determining forest stock volume in Siberia, *in Proceedings of the IGARSS 2000 Geoscience and Remote Sensing Symposium*, 1396–1398, 2000.

127 M. Santoro, A. Shvidenko, I. Mccallum, J. Askne, and C. Schmullius, Properties of ERS-1/2 coherence in the Siberian boreal forest and implications for stem volume retrieval, *Remote Sensing of Environment*, **106**, 154–172 (2007).

128 S. Solberg, R. Astrup, T. Gobakken, E. Næsset, and D.J. Weydahl, Estimating spruce and pine biomass with interferometric X-band SAR, *Remote Sensing of Environment*, **114**, 2353–2360 (2010).

129 M. Santoro, J. Askne, C. Beer, O. Cartus, C. Schmullius, U. Wegmüller, and A. Wiesmann, Automatic model inversion of multi-temporal C-band coherence and backscatter measurements for forest stem volume retrieval, *in Proceedings of the Geoscience and Remote Sensing Symposium, 2008. IGARSS 2008*, 124–127, 2008.

130 J.I.H. Askne and M. Santoro, Automatic model-based estimation of boreal forest stem volume from repeat pass C-band InSAR coherence, *IEEE Transactions on Geoscience and Remote Sensing*, **47**, 513–516 (2009).

131 M.C. Dobson, S. Member, F.T. Ulaby, A. Beaudoin, E.S. Kasischke, and N. Christensen, Dependence of radar backscatter on coniferous forest biomass, *IEEE Transactions on Geoscience and Remote Sensing*, **30**, 412–415 (1992).

132 M.L. Imhoff, Radar backscatter and biomass saturation: ramifications for global biomass inventory, *IEEE Transactions on Geoscience and Remote Sensing*, **33**, 511–518 (1995).

133 K.J. Ranson and G. Sun, Mapping biomass of a northern forest using multifrequency SAR data, *IEEE Transactions on Geoscience and Remote Sensing*, **32**, 388–396 (1994).

134 C. Thiel and C. Schmullius, Investigating ALOS PALSAR interferometric coherence in central Siberia at unfrozen and frozen conditions: implications for forest growing stock volume estimation, *Canadian Journal of Remote Sensing*, **39**, 232–250 (2013).

135 L.E.B. Eriksson, M. Santoro, A. Wiesmann, and C.C. Schmullius, Multitemporal JERS repeat-pass coherence for growing-stock volume estimation of Siberian forest, *IEEE Transactions on Geoscience and Remote Sensing*, **41**, 1561–1570 (2003).

136 O. Cartus, M. Santoro, C. Schmullius, P. Yong, C. Erxue, and L. Zengyuan, Creation of large area forest biomass maps for Northeast China using ERS-1/2 tandem coherence, *in Proceedings of the Dragon 1 Programme Final Results 2004–2007*, 2008.

4

Forest Mapping of the Congo Basin using Synthetic Aperture Radar (SAR)

J. Wheeler[1], P. Rodriguez-Veiga[1,2], Heiko Balzter[1,2], K. Tansey[1] and N.J. Tate[1]

[1] University of Leicester, Centre for Landscape and Climate Research, Department of Geography, Leicester, UK
[2] National Centre for Earth Observation, University of Leicester, Leicester, UK

4.1 Introduction

Ground-based studies of the Congo Basin's tropical forest are few and mostly representative of its periphery rather than its centre [1]. Deforestation and forest degradation are a major source of anthropogenic greenhouse gas emissions [2]. Stored carbon is released into the atmosphere when forests are burned to create land available for agriculture [3]. Exposed and especially drained soil also releases carbon, which for certain soil types such as peat has been stored in greater volumes (as soil organic carbon, SOC) than in the living woody material (as Aboveground Biomass, AGB; see Chapter 2 for a more detailed description of AGB and its estimation) [4]. Conversion of forest to other land cover and land use types reduces the area available to store carbon and therefore reduces the ability to offset the increases of atmospheric carbon from fossil fuel emissions. Deforestation and forest degradation result in loss of biodiversity, and are specifically mentioned in the Convention for Biological Diversity's (CBD) "Aichi Biodiversity Targets" as part of their 2011–2020 plan [5].

A key intergovernmental programme that has been created in recognition of the important role of tropical forests in climate change, and the potential benefits of their conservation, is the UN REDD programme (Reducing Emissions from Deforestation and forest Degradation in developing countries), which aims to incentivize reducing carbon emissions from forest loss in developing countries through financial means [6]. REDD now supports REDD+, which adds "conservation of forest stocks, sustainable management of forests, and enhancement of forest carbon stocks" to the climate change mitigating actions of REDD [7]. In order to achieve the goals of REDD, a system of monitoring, reporting and verification (MRV) is necessary to assess forest area and stocks of carbon contained in AGB in participating countries.

The Congo Basin contains the second largest continuous area of tropical forest in the world, yet is one of the least studied areas [8,9]. For reasons such as political instability, including numerous civil wars, and resulting lack of infrastructure, forest loss is not as severe in Africa compared with the two other major tropical forested continents, South

America and Asia [10,11]. For the same reasons, as well as sheer scale, a ground-based survey of forest cover for the entire Congo Basin would be hugely impractical, especially for the Democratic Republic of Congo (DRC), which recent estimates suggest contains over 60% of the Central African forest area [12].

This chapter examines the current and potential role of remote sensing, in particular spaceborne Synthetic Aperture Radar (SAR) in providing forest information at a regional scale covering the entire Congo Basin, as well as in more detail in a case study of a smaller area of the Congo Basin. The advantages of SAR over optical sensors, as well as the relative limitations of SAR to date are also discussed.

4.2 Remote Sensing of Central African Forests

Ground-based forest inventories provide the most direct and accurate method of assessing tree cover, as well as other forest parameters such as species distribution and *AGB*, and are necessary for interpreting remote sensing data and validating results, but alone they are not practical for regular wide area surveys, particularly in dense tropical forest. Remote sensing from airborne, and especially spaceborne sensors has the potential for consistent "wall-to-wall", regular surveys of forest cover. Forest cover is measurable from the two most widely used earth observation systems: passive optical (visible, near infrared and short wave infrared frequencies) and active Synthetic Aperture Radar (SAR) (microwave frequencies). SAR also has the potential to inform on aspects of forest structure, depending on the modes of acquisition used. A reliable wall-to-wall forest/non-forest map can also be used to improve methods of *AGB* modelling by facilitating interpolation between known or modelled values of *AGB* where there is spacing between samples, an example of which is *AGB* derived from tree-height measurements from the Ice, Cloud and land Elevation Satellite's Geoscience Laser Altimeter System (IceSAT-GLAS) instrument [13].

4.2.1 Optical Remote Sensing of Tropical Forests

Pan-tropical optical remote sensing approaches to forest mapping began with coarse spatial resolution (≥ 1 km) data, combined samples of medium (what was then considered high) resolution (30 m) imagery to estimate rates of deforestation, or used both of these techniques. The Tropical Ecosystem Environment Monitoring by Satellites (TREES) programme [25], the UN Food and Agriculture Organization (FAO) Forest Resources Assessment (FRA) 2000 [26], and the Advanced Very High Resolution Radiometer (AVHRR) Pathfinder programme [14] all incorporated one or both of these approaches during the 1990s. Improvements in processing power, new sensors, and availability of data, have resulted in a steady refinement of spatial resolution in optical remote sensing studies since then, with the resolution of wall-to-wall forest cover maps in Central Africa dropping to 500 m [18], 300 m [22], 250 m [19] and more recently 30 m [20,21]. Persistent cloud cover over tropical forests remains a barrier to frequent mapping of forest cover at finer spatial resolutions, resulting in wide-area optical cloud free mosaics generated from data spanning several years. This can reduce the accuracy of classifications, and also hinders rapid detection of fine-scale forest change over large areas.

Table 4.1 Major remote sensing projects with complete coverage of Central African forest cover.

Project name	Sensor; sensor type	Spatial resolution	Year(s)	Project organization; reference
AVHRR Pathfinder	AVHRR; optical	8 km	Mid-1980s– mid-1990s	University of Maryland (UMD), Woods Hole Research Centre; [14]
CAMP	ERS-1; C-band SAR	100 m	1994	Joint Research Centre (JRC); [15]
GRFM	JERS-1; L-band SAR	100 m	1996 (×2)	NASDA (now JAXA); [16]
GLC2000/ TREES II	SPOT-VEGETATION; optical	1 km	2000	JRC; [17]
Vegetation Continuous Fields (VCF)	Terra-MODIS; optical	250 m, 500 m, 1 km	2000–2010 (Annual)	UMD, NASA; [18,19]
Tree Cover Continuous Fields	Landsat; optical	30 m	2000	UMD; [20]
Global Forest Change	Landsat; optical	30 m	2000–2012	UMD, NASA, Google; [21]
GlobCover	ENVISAT-MERIS; optical	300 m	2005/6, 2009	ESA; [22,23]
ALOS Kyoto & Carbon Initiative	ALOS-PALSAR; L-band SAR	50 m, 500 m	2007–2010 (Annual)	JAXA, JRC; [24]

4.2.2   SAR Remote Sensing of Tropical Forests

SAR is a side-looking active remote sensing system which transmits microwaves and receives measurements from the signal backscattered from a surface, to produce an image after appropriate processing. The longer wavelengths of microwaves allow imaging of the ground through cloud cover, and as it is an active system, images can also be gathered at night.

There have been three major projects to map Central African forests using space-borne SAR sensors. The Central African Mapping Project (CAMP) used data collected in 1994 from the European Space Agency's (ESA) ERS-1 (European Remote Sensing Satellite) [15]. The Global Rainforest Mapping project (GRFM) used two 1996 mosaics (wet and dry season acquisitions) from the Japanese Space Exploration Agency's (JAXA) Japanese Earth Resources Satellite (JERS-1) [16]. Most recently, the Kyoto and Carbon Initiative (K&C) has produced four annual forest/non-forest (FNF) products from 2007–10 using data from JAXA's Advanced Land Observing Satellite – Phased Array L-band SAR (ALOS-PALSAR) [24].

The utility of a SAR system for wide-area wall-to-wall forest mapping depends on several factors, the most important being the band wavelength used, the polarisation of the sent and received signal, and consistency of atmospheric and ground conditions (particularly surface wetness) at time of data acquisition.

The wavelengths of SAR range from around 2.4 cm up to 100 cm for bands X to P, as seen in Table 4.2 [27]. Longer wavelength SARs (i.e. L- and P- band) have a greater ability to

Table 4.2 Historical, current, and future spaceborne SAR satellites/sensors, with their wavelength and frequency ranges, adapted from [27].

Wavelength (cm)	2.4	3.75	7.5	15	30	100
Radar Band		X	C	S	L	P
Completed, *currently* active, *planned future* and *proposed future** SAR spaceborne sensors	1990–2000		ERS-1/2	Almaz	JERS-1	
	2000–2010	**TerraSAR-X, Cosmo- Skymed**	ENVISAT-ASAR, SRTM, **Radarsat -1/2**		ALOS-PALSAR	
	2010–2020	**TanDEM-X,**	**Sentinel 1** A/B, *Radarsat Constellation*	*NovaSAR-S*	**ALOS-2,** *SAOCOM-1A/1B,* *TanDEM-L**	*BIOMASS*
Frequency (GHz)	12.5	8	4	2	1	0.3

penetrate the surface and canopy cover. The signal interacts with objects at the same scale or larger than its wavelength, with smaller objects not affecting the backscatter. As a result, longer wavelength SAR sensors pass through leaves and small branches in the upper canopy and offer more information about differences in larger woody material such as stems and large branches, making them more suitable for forestry applications.

Emitted SAR signals are polarized in either horizontal (H) or vertical (V) planes, and the returned signal is similarly received in either horizontal or vertical planes. Co-polarised SAR data (VV – vertical send, vertical receive, and HH – horizontal send, horizontal receive) is generally less useful than cross-polarised (HV and VH) SAR data for forest measurements; a cross-polarised sensor configuration is sensitive to the changes in polarisation produced by volume scattering elements within tree canopy [28]. Differences in ground and atmospheric conditions at the time of acquisition (a mosaicked image is comprised of many smaller near-square scenes, or in some cases a smaller number of long data strips, all acquired at different times by an orbiting sensor or constellation of sensors) cause problems for automated classification, since SAR is sensitive to the dielectric constant of a target, meaning the more moisture, the stronger a reflection will be. Wet conditions in one scene and dry conditions in an adjacent scene will result in an overall backscatter difference that will need to be accounted for during classification, which is especially difficult if the wet conditions are uneven throughout the scene. The closer in acquisition time (and season) the scenes or data strips are that comprise a mosaic, the better the chances are that atmospheric and ground conditions will be more stable.

The first two SAR forest mapping projects, CAMP and GRFM, both produced maps at a down-sampled 100 m spatial resolution (due to limited processing power at the time [15,29]) using a single co-polarized SAR band (VV and HH, respectively), while the K&C Initiative has released down-sampled 50 m maps (and has produced but not yet released 25 m maps) generated from dual co- and cross-polarised data (HH and HV) [24]. CAMP used shorter wavelength C-band data, whereas the GRFM project and the K&C Initiative used longer wavelength L-band SAR data. The acquisition strategy of the JERS-1 sensor was such that adjacent data strips were acquired within one day of each other. This allowed the GRFM project to use one wet season and one dry season mosaic from the same year [29]. CAMP used a similarly short time window to acquire all images (44 days) [15]. The K&C Initiative acquired ALOS-PALSAR data strips over a longer period (June-October) for each year, with missing data occasionally replaced by data from the previous or following years [24].

Despite the longer acquisition time for each mosaic, the dual-polarization mode, long wavelength, higher spatial resolution, and four-year time series of the K&C Initiative mosaics make them the most promising and complete dataset to work with for forest mapping in Central Africa to date. In addition, JAXA's ALOS-2 satellite is continuing production of the K&C Initiative annual mosaics after a four-year hiatus.

4.3 Case Study

This case study aimed to generate annual forest cover and forest change (loss and gain) maps for a large section of the Congo Basin for 2007 to 2010 from a supervised classification of the K&C Initiative SAR mosaics, and compared the results with the K&C Initiative's own forest/non-forest maps produced from the same dataset.

4.3.1 Study Area

The Congo rainforest often refers to all of the forested areas of Central Africa, rather than specifically those contained within the drainage area of the Congo river [10,30] and it either wholly or partially covers the following six countries: Cameroon, the Central African Republic, Equatorial Guinea, Gabon, the Democratic Republic of the Congo (DRC) and the Republic of the Congo. Of these, the DRC and the Republic of Congo are countries with active UN-REDD national programmes, and Cameroon, the Central African Republic, and Gabon are REDD partner countries.

Analysis was performed on a five degree by five degree square with an upper left coordinate of Lat. 5°N, Long. 15°E (Figure 4.1). K&C Initiative JAXA ALOS PALSAR data is organized into grids of five degree lat/long mosaicked squares. The selected area covers part of four countries: Cameroon, the Central African Republic, the Republic of Congo, and the Democratic Republic of Congo, and contains a wide range of land cover types, including: rain fed croplands, mosaic cropland/vegetation, broadleaved

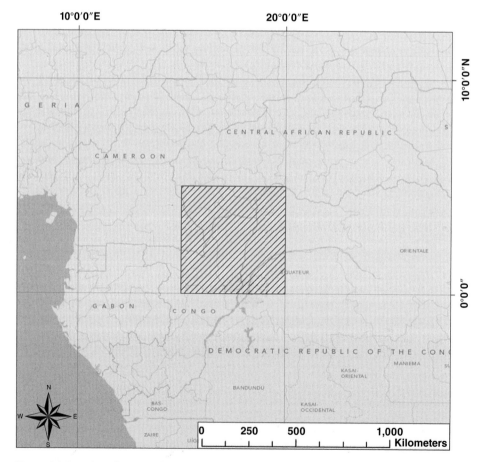

Figure 4.1 Location of case study area within Central Africa (hatched square). Background map © Esri, HERE, DeLorme, Mapmylndia, OpenStreetMap contributors, and the GIS user community.

deciduous forest, shrub land, herbaceous vegetation (savannah), permanently and regularly flooded forest, urban areas and water bodies[1].

4.3.2   Data

Dual polarized (HH and HV) K&C Initiative ALOS-PALSAR data was used, at a spatial resolution of 50 m. The data was pre-processed and released in a mosaic format, and geocoded consistently and accurately, allowing for easier comparison with other forest cover maps. The HH and HV channels were provided as normalized radar cross-section, gamma-nought, meaning the backscatter intensity has been adjusted for topography using the cosine of the local incidence angle as described in [31]. Four mosaics from 2007–2010 were acquired. Metadata at a pixel level is provided in the form of local incidence angle (the angle from nadir at which the sensor images a pixel, including the contribution of topography) and number of days after launch of image acquisition, allowing calculation of acquisition date (Figure 4.2). A global forest/non-forest product based on the ALOS-PALSAR data is provided, at the same scale.

4.3.3   Supervised Classification of K&C Initiative ALOS-PALSAR Mosaics

A training and validation dataset for the study was created from a random sample of 1000 square (200 m × 200 m) plots covering the area of interest from Figure 4.1. Medium and high resolution Google Earth imagery (acquired between 2012 and 2014) was used to determine the class in each sample square, and there was a loss of 102 samples due to mixing of classes, cloud cover, some poorly georeferenced high resolution imagery, and lack of clarity in the available training data. The resulting two sample sets (forest and non-forest) were each divided 50:50 into training and validation vector data.

HH and HV K&C Initiative SAR images were compared with the Landsat-derived datasets from [21] for projection and georeferencing consistency, and the error was found to be below one pixel. A support vector machine (SVM) classification, using default ENVI parameters based on [32], was applied to the HH and HV data for each individual data year. SVM is an advanced supervised classification algorithm that takes training data, typically from two classes, and classifies the image based on a decision surface referred to as an optimal hyperplane, where the separability between classes is greatest. A filtering algorithm using a majority analysis kernel was applied to the resulting class image to remove and reclassify lone pixels to the dominant surrounding class, effectively creating a minimum mapping unit of 3–4 pixels (~1 ha).

Following classification, an accuracy assessment was carried out using the validation samples from both classes. This accuracy assessment was also applied to the K&C Initiative FNF product. Forest area was calculated for each year with both methods, and post-classification change analysis was performed to assess year on year loss/gain of forest, and the change over the entire 2007–2010 period.

4.3.4   Case Study Results

Figure 4.3 shows the forest area by year calculated from the two classification methods examined. A rise in forest area is observed using both methods, though a smoother

---

1 Landcover classes taken from ESA's Globcover 2009 product [23].

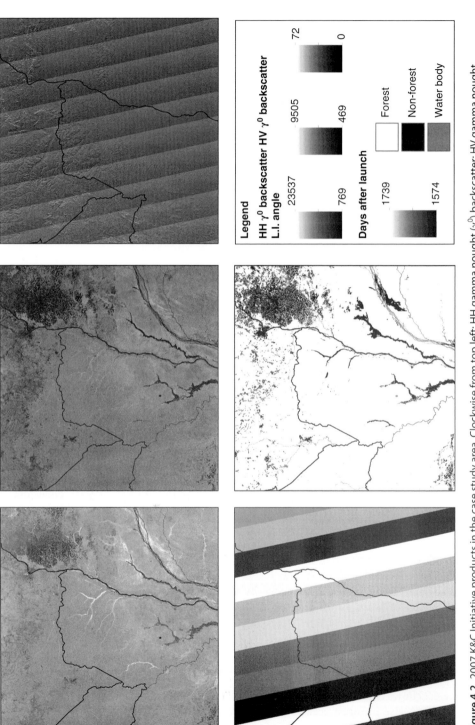

Figure 4.2 2007 K&C Initiative products in the case study area. Clockwise from top left: HH gamma nought (γ^0) backscatter; HV gamma nought backscatter; local incidence angle; forest/non-forest map; date of data strip acquisition (days after launch).

Legend

HH γ^0 backscatter HV γ^0 backscatter
L.I. angle

23537   9505   72

769   469   0

Forest

Non-forest

Water body

Days after launch

1739

1574

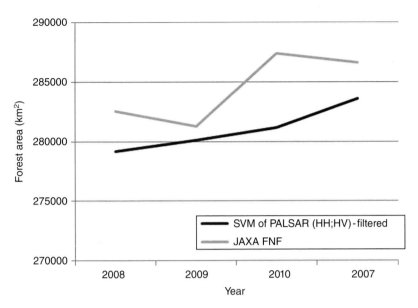

Figure 4.3 Graph of forest area by year showing differences between this study's supervised classification (SVM) and the approach used for JAXA's K&C Initiative FNF map.

change is seen using the SVM classifier (Figure 4.3). Annual fluctuations described by the K&C Initiative FNF method appear to be more indicative of seasonal changes in moisture or flood conditions from different acquisition dates from year to year. This is illustrated by the comparison between the two approaches in Figure 4.4. Analysis of the 2007–10 forest loss/gain product generated (Figure 4.5) gave a net increase in forest area of approximately 4412 km^2, the equivalent of 1.4% of the study area. While these figures are difficult to verify without ground data, and may therefore not be a reliable indicator of actual forest area change, the class changes (both gain and loss) from year to year seen in Figure 4.4, and in closer detail for the entire period in Figure 4.5, often occur in a linear pattern consistent with clearing along logging and other access trails. The scale of disturbance seen in this data does not appear in coarser resolution datasets, and is largely absent from JAXA's K&C Initiative FNF maps, possibly due to the segmentation approach used.

Figure 4.6 shows the results of the accuracy assessments using the SVM classification method described in section 4.3.3. Although the validation and training samples were generated by manual interpretation of recent (2012–2014) medium and high resolution imagery from Google Earth, a downward trend in accuracy was observed with time, from 2007 to 2010. The homogeneity selection criteria for the validation samples may account for this, as this tended to preclude forest/non-forest border zones, where difficulties in classification are more likely to occur. From the confusion matrices in Table 4.3 the high overall accuracies are due to the dominance of the forest class. Looking at the producer's accuracies for the non-forest class, there is a tendency within both approaches to misclassify non-forest as forest.

Forest Change Year by Year

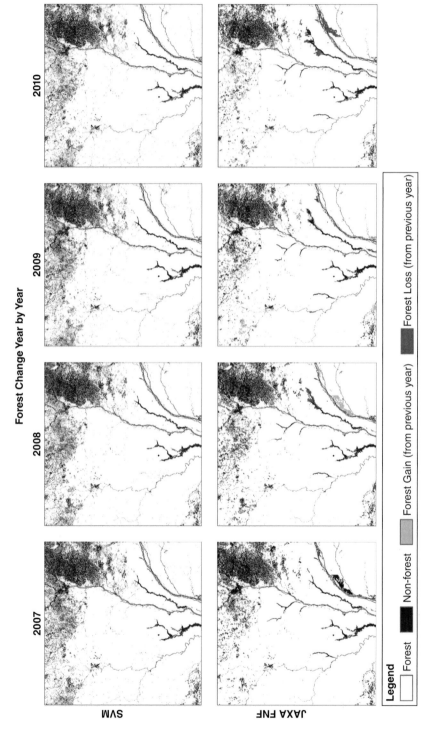

Figure 4.4 Comparison of Forest/Non-Forest classifications from SVM and JAXA's K&C Initiative FNF map. Top image: 2007–2010 stable forest/non-forest and change (loss and gain) from previous year over the entire study area.

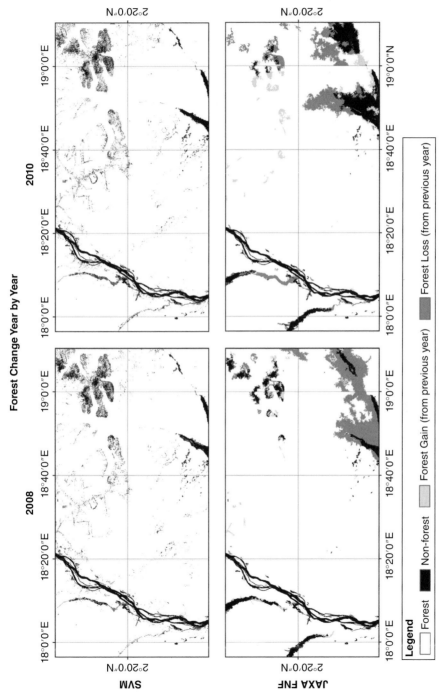

Figure 4.4 (Continued) Bottom image: 2008 and 2010 stable forest/non-forest and change (loss and gain) from previous year in finer detail over a central subset of the study area.

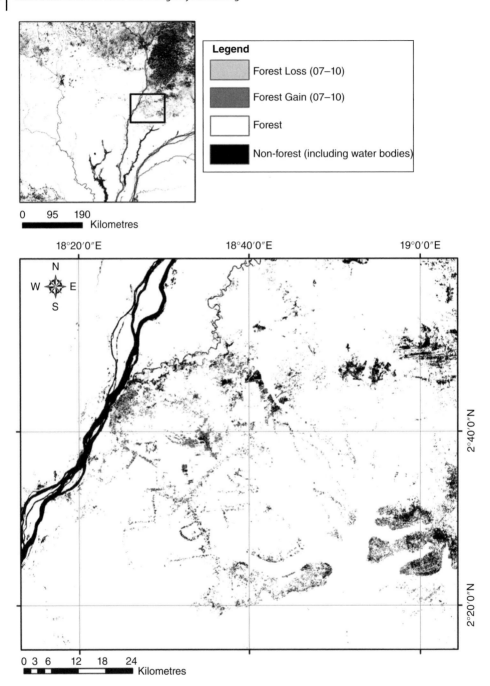

Figure 4.5 2007–10 map of forest gain/loss derived from SVM classification of HH and HV data. Large subset shows pattern of change consistent with clearing along logging and other access trails.

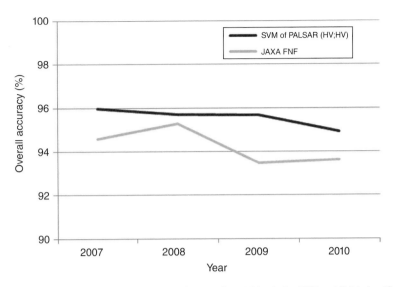

Figure 4.6 Graph of overall accuracy by year for K&C Initiative FNF and SVM classifier.

4.4 Discussion

There is an improvement in accuracy and consistency of the classification method used in this case study over JAXA's K&C Initiative FNF product in the same area. This may be attributed to their application of a global algorithm, compared with the supervised regional approach described in this study, particularly in forested areas that are prone to flooding. While their forest class uses a regionally changing threshold from the HV channel, their decision tree methodology uses a global HH backscatter threshold for a non-forest "settlement" class that is within the seasonal range of backscatter from Congolese swamp forest (from preliminary results in [33]). As a result, large areas of forest are classified as non-forest in wet conditions and forest in dry conditions (see Figure 4.4). In addition, the use of segmentation in their approach limits the detection of finer scale forest change that the 50 m pixel resolution would otherwise reveal. While the exact figures generated by the case study for forest area, and by extension annual forest change, are unreliable without verification from ground data, the method described could be a useful indicator of finer scale change. The results from the case study provide justification for applying this study's methodology across the whole Congo Basin, with refinement of the supervised classification parameters, and incorporating available field data from forestplots.net [34] in training and validation where possible. This will allow a broader analysis of forest change across the entire region, and it will provide further insight into the results of this study, depending on the change seen across the entire Congo Basin.

4.5 Conclusion

4.5.1 Future Work

Analysis of the case study results show that more work is necessary to improve the training dataset of the supervised classification, particularly for non-forest areas. The

Table 4.3 Confusion Matrices in pixel numbers for each class – Forest (F) and Non-forest (NF) – from this study's SVM classifier and JAXA's K&C Initiative FNF data against the training dataset. Percentage overall accuracy, and producer's and user's accuracies (for both classes) are also presented.

Classification	Data	2007 Training Data			2008 Training Data			2009 Training Data			2010 Training Data		
		F	NF	Total	F	NF	Total	F	NF	Total	F	NF	Total
SVM	F	6198	237	6435	6210	276	6486	6201	270	6471	6207	332	6539
	NF	26	723	749	14	684	698	23	690	713	17	628	645
	Total	6224	960	7184	6224	960	7184	6224	960	7184	6224	960	7184
	Prod. Acc	99.58	75.31		99.78	71.25		99.63	71.88		99.73	65.42	
	User. Acc	96.32	96.53		95.74	97.99		95.83	96.77		94.92	97.36	
	Overall Accuracy			96.34			95.96			95.92			95.14
JAXA FNF	F	6125	290	6415	6168	284	6452	6173	417	6590	6187	422	6609
	NF	99	670	769	56	676	732	51	543	594	37	538	575
	Total	6224	960	7184	6224	960	7184	6224	960	7184	6224	960	7184
	Prod. Acc	98.41	69.79		99.10	70.42		99.18	56.56		99.41	56.04	
	User. Acc	95.48	87.13		95.60	92.35		93.67	91.41		93.61	93.57	
	Overall Accuracy			94.59			95.27			93.49			93.61

application of water and urban masks, as well as the inclusion of a separate class for flooded forest could remove a source of error. Daily rainfall estimate data could be useful for establishing areas of flooded forest in the mosaics, as well as selecting scenes from the mosaics with more uniform acquisition conditions, using a method similar to that of [35].

4.5.2   Future Direction of SAR Remote Sensing in the Congo Basin

Table 4.2 shows that the number of spaceborne SAR sensors is set to increase, particularly in the domain of longer wavelength L- and P-band sensors. Forest remote sensing has not been the primary objective of any spaceborne SAR sensors to date, so the modes of acquisition and acquisition strategy have not always been optimal for this application. However, this will change with the BIOMASS mission, which will carry the first spaceborne P-band sensor and is dedicated to the monitoring of forest *AGB* [36]. The Copernicus Sentinel 1 constellation, of which one satellite was launched in 2014, will provide a potential six-day repeat acquisition cycle. While the C-band SAR sensor is not as suited to monitoring of small changes in *AGB* or precise measuring of carbon stocks, it will be a useful tool for rapid monitoring of forest disturbance [37]. The launch of ALOS-2 with its contribution towards the K&C Initiative [38], and the planned launch of SAOCOM-1A [39] will continue the role of spaceborne L-band SAR for forest remote sensing.

It is generally the case that newer SAR sensors (since 2000) have the capability for acquisition in any of the four polarisation configurations, although acquisition of simultaneous fully polarimetric scenes remains limited to experimental applications. More advanced techniques of SAR that exploit the interferometric analysis of two or more similarly acquired images of the same scene can offer further potential for SAR, and this has been demonstrated in large-scale studies of boreal forest [40–42]. This will be enhanced by shorter repeat pass acquisitions and longer wavelengths of SAR to overcome the difficulties this technique faces in wetter, higher *AGB*-containing tropical forests [36].

References

1 E. Kearsley, T. de Haulleville, K. Hufkens, A. Kidimbu, B. Toirambe, G. Baert, D. Huygens, Y. Kebede, P. Defourny, J. Bogaert, H. Beeckman, K. Steppe, P. Boeckx, and H. Verbeeck, Conventional tree height-diameter relationships significantly overestimate aboveground carbon stocks in the Central Congo Basin. *Nat. Commun.* **4**, 2269 (2013).

2 Y. Malhi, and J. Grace, Tropical forests and atmospheric carbon dioxide. *Trends Ecol. Evol.* **15**, 332–336 (2000).

3 FAO. *Global Forest Resources Assessment 2010*. (2010). at www.fao.org/docrep/013/i1757e/i1757e.pdf/

4 S. Englhart, V. Keuck, and F. Siegert, Modeling Aboveground Biomass in Tropical Forests Using Multi-Frequency SAR Data — A Comparison of Methods. *IEEE J. Sel. Top. Appl. Earth Obs. Remote Sens.* **5**, 298–306 (2012).

5 CBD. Strategic Plan for Biodiversity 2011–20: Provisional Technical Rationale, possible indicators and suggested milestones for the Aichi Biodiversity Targets. in *Conference of the Parties to the Convention on Biological Diversity* 1–20 (2010).

6 M. Herold, and T. Johns, Linking requirements with capabilities for deforestation monitoring in the context of the UNFCCC-REDD process. *Environ. Res. Lett.* **2**, 045025 (2007).

7 UNFCCC. *Report of the Conference of the Parties on its sixteenth session, held in Cancun from 29 November to 10 December 2010.* Addendum. Part Two: Action taken by the Conference of the Parties at its sixteenth session. (2010) at http://unfccc.int/resource/docs/2010/cop16/eng/07a01.pdf

8 A. Baccini, N.T. Laporte, S.J. Goetz, M. Sun, and H. Dong, A first map of tropical Africa's above-ground biomass derived from satellite imagery. *Environ. Res. Lett.* **3**, 045011 (2008).

9 H. Verbeeck, P. Boeckx, and K. Steppe, Tropical forests: include Congo basin. *Nature* **479**, 179 (2011).

10 C.O. Justice, D.S. Wilkie, Q. Zhang, J. Brunner, and C. Donoghue, Central African forests, carbon and climate change. *Clim. Res.* **17**, 229–246 (2001).

11 P. Mayaux, and F. Achard, "REDD-plus" requirements for the Congo Basin countries. in *Monitoring Forest Carbon Stocks and Fluxes in the Congo Basin* (eds M. Brady, and C. de Wasseige) 4–9 (2010).

12 P. Mayaux, J.F. Pekel, B. Desclée, F. Donnay, A. Lupi, F. Achard, M. Clerici, C. Bodart, A. Brink, R. Nasi, and A. Belward, State and evolution of the African rainforests between 1990 and 2010. *Philos. Trans. R. Soc. Lond. B. Biol. Sci.* **368**, 20120300 (2013).

13 S.S. Saatchi, N.L. Harris, S. Brown, M.A. Lefsky, E.T.A., Mitchard, W. Salas, B.R. Zutta, W. Buermann, S.L. Lewis, S. Hagen, S. Petrova, L. White, M. Silman, and A.C. Morel, Benchmark map of forest carbon stocks in tropical regions across three continents. *Proc. Natl. Acad. Sci. USA* **108**, 9899–904 (2011).

14 R.S. DeFries, R.A. Houghton, M.C. Hansen, C.B. Field, D, Skole, and J.R.G. Townshend, Carbon emissions from tropical deforestation and regrowth based on satellite observations for the 1980s and 1990s. *Proc. Natl. Acad. Sci. USA* **99**, 14256–14261 (2002).

15 G. De Grandi, J.P. Malingreau, and M. Leysen, The ERS-1 Central Africa mosaic: A new perspective in radar remote sensing for the global monitoring of vegetation. *IEEE Trans. Geosci. Remote Sens.* **37**, 1730–1746 (1999).

16 A. Rosenqvist, M. Shimada, B. Chapman, A. Freeman, G. De Grandi, S.S. Saatchi, and Y. Rauste, The global rain forest mapping project-a review. *Int. J. Remote Sens.* **21**, 1375–1387 (2000).

17 P. Mayaux, E. Bartholomé, S. Fritz, and A. Belward, A new land-cover map of Africa for the year 2000. *J. Biogeogr.* **31**, 861–877 (2004).

18 M.C. Hansen, R.S. DeFries, J.R.G. Townshend, M. Carroll, C. Dimiceli, and R.A. Sohlberg, Global percent tree cover at a spatial resolution of 500 meters: First results of the MODIS vegetation continuous fields algorithm. *Earth Interact.* **7**, 1–15 (2003).

19 C.M. DiMiceli, M.L. Carroll, R.A. Sohlberg, C. Huang, M.C. Hansen, and J.R.G. Townshend, *Annual Global Automated MODIS Vegetation Continuous Fields (MOD44B) at 250 m Spatial Resolution for Data Years Beginning Day 65, 2000–10, Collection 5: Percent Tree Cover*, University of Maryland, College Park, MD, USA. (2011) at http://glcf.umd.edu/data/vcf

20 J.O. Sexton, X.-P. Song, M. Feng, P. Noojipady, A. Anand, C. Huang, D.-H. Kim, K.M. Collins, S. Channan, C. DiMiceli, and J.R.G. Townshend, Global, 30-m resolution continuous fields of tree cover: Landsat-based rescaling of MODIS Vegetation Continuous Fields with LIDAR-based estimates of error. *Int. J. Digit. Earth* 1303210312, (2013).

21 M. Hansen, P.V. Potapov, R. Moore, M. Hancher *et al.*, High-resolution global maps of 21st-century forest cover change. *Science* (80) 342, 850–853 (2013).

22 O. Arino, D. Gross, F. Ranera, L. Bourg *et al.*, GlobCover: ESA service for Global Land Cover from MERIS. in IEEE *International Geoscience and Remote Sensing Symposium (IGARSS) 2007* 2412–2415 (2007) doi:10.1109/IGARSS.2007.4423328

23 ESA. *Globcover 2009: products description and validation report.* 1–53 (2010) at http://due.esrin.esa.int/globcover/LandCover2009/GLOBCOVER2009_Validation_Report_2.2.pdf/

24 M. Shimada, T. Itoh, T. Motooka, M. Watanabe, T. Shiraishi, R. Thapa, and R. Lucas, New global forest/non-forest maps from ALOS PALSAR data (2007–10). *Remote Sens. Environ.* **155**, 13–31 (2014).

25 P. Mayaux, T. Richards, and E. Janodet, A vegetation map of Central Africa derived from satellite imagery. *J. Biogeogr.* **26**, 353–366 (1999).

26 FAO. *Global Forest Resources Assessment 2000 Main Report.* pp.321–331 (2000).

27 A. Rosenqvist, C.M. Finlayson, J. Lowry, and D. Taylor, The potential of long-wavelength satellite-borne radar to support implementation of the Ramsar Wetlands Convention. *Aquat. Conserv. Mar. Freshw. Ecosyst.* **17**, 229–244 (2007).

28 E.T. Mitchard, S.S. Saatchi, S.L. Lewis, T.R. Feldpausch, I.H. Woodhouse, B. Sonké, C. Rowland, and P. Meir, Measuring biomass changes due to woody encroachment and deforestation/degradation in a forest–savanna boundary region of central Africa using multi-temporal L-band radar backscatter. *Remote Sens. Environ.* **115**, 2861–2873 (2011).

29 G. De Grandi, P. Mayaux, Y Rauste, A Rosenqvist, M. Simard, and S.S. Saatchi, The Global Rain Forest Mapping Project JERS-1 radar mosaic of tropical Africa: development and product characterization aspects. *IEEE Trans. Geosci. Remote Sens.* **38**, 2218–2233 (2000).

30 M.C. Hansen, D.P. Roy, E.J. Lindquist, B. Adusei, C.O. Justice, and A. Altstatt, A method for integrating MODIS and Landsat data for systematic monitoring of forest cover and change in the *Congo Basin. Remote Sens. Environ.* **112**, 2495–2513 (2008).

31 M. Shimada, Ortho-Rectification and Slope Correction of SAR Data Using DEM and Its Accuracy Evaluation. *IEEE J. Sel. Top. Appl. Earth Obs. Remote Sens.* **3**, 657–671 (2010).

32 C.-C. Chang, and C.-J. Lin, LIBSVM : A Library for Support Vector Machines. *ACM Trans. Intell. Syst. Technol.* **2**, 27:1–27, 1–39 (2011).

33 K. Einzmann, J. Haarpaintner, and Y. Larsen, Forest monitoring in Congo Basin with combined use of SAR C- and L-band. *Proc. IEEE Int. Geosci. Remote Sens. Symp. (IGARSS)*, Munich, Germany 6573–6576 (2012).

34 G. Lopez-Gonzalez, S.L. Lewis, M. Burkitt, and O.L. Phillips, ForestPlots.net: a web application and research tool to manage and analyse tropical forest plot data. *J. Veg. Sci.* **22**, 6573–6576 (2011).

35 R.M. Lucas, J., Armston, and R. Fairfax, An Evaluation of the ALOS PALSAR L-Band Backscatter – Above Ground Biomass Relationship Queensland, Australia : Impacts of Surface Moisture Condition and Vegetation Structure. *IEEE J. Sel. Top. Appl. Earth Obs. Remote Sens.* **3**, 576–593 (2010).

36 T. Le Toan, S. Quegan, M.J.W. Davidson, H. Balzter, P. Paillou, K.P. Papathanassiou, S. Plummer, F. Rocca, S.S. Saatchi, H.H. Shugart, and L.M.H. Ulander, The BIOMASS mission: Mapping global forest biomass to better understand the terrestrial carbon cycle. *Remote Sens. Environ.* **115**, 2850–2860 (2011).

37 ESA. *Sentinel Online User Guides: Applications: Land monitoring.* (2014) at https://sentinel.esa.int/web/sentinel/user-guides/sentinel-1-sar/applications/land-monitoring/

38 A. Rosenqvist,M. Shimada, S. Suzuki, F. Ohgushi, T. Tadono, M. Watanabe, K. Tsuzuku, T. Watanabe, S. Kamijo, and E. Aoki, Operational performance of the ALOS global systematic acquisition strategy and observation plans for ALOS-2 PALSAR-2. *Remote Sens. Environ.* **155**, 3–12 (2014).

39 CONAE. SAOCOM – *Introducción.* (2014). at www.conae.gov.ar/index.php/espanol/misiones-satelitales/saocom/objetivos/

40 W. Wagner, A. Luckman, J. Vietmeier, K.J. Tansey, H. Balzter, C.C. Schmullius, M.W.J. Davidson, D. Gaveau, M. Gluck, T. Le Toan, S. Quegan, A. Shvidenko, A. Wiesmann, A. and J.J. Yu, Large-scale mapping of boreal forest in SIBERIA using ERS tandem coherence and JERS backscatter data. *Remote Sens. Environ.* **85**, 125–144 (2003).

41 K.J. Tansey, A. Luckman, and L. Skinner, Classification of forest volume resources using ERS tandem coherence and JERS backscatter data. *Int. J. Remote Sens.* **25**, 751–768 (2004).

42 C.J. Thiel, C. Thiel, and C.C. Schmullius, Operational Large-Area Forest Monitoring in Siberia Using ALOS PALSAR Summer Intensities and Winter Coherence. *IEEE Trans. Geosci. Remote Sens.* **47**, 3993–4000 (2009).

Part II

Land Cover and Land Cover Change Monitoring

5

Multi-Frequency SAR Applications for Land Cover Classification Within Copernicus Downstream Services

B.F. Spies[1,2], A. Lamb[1] and Heiko Balzter[2]

[1] Airbus Defence and Space Geo-Intelligence, Meridian Business Park, Leicester, UK
[2] University of Leicester, Centre for Landscape and Climate Research, Department of Geography, Leicester, UK

5.1   Background

Copernicus downstream services are a large part of the motivation of the whole Copernicus programme, with the intention to encourage private and public sector users to make use of the available data to build applications that will assist everyone with their daily lives and stimulate economic growth across Europe [1]. A few examples of such downstream services are airTEXT, which shows air pollution, UV and temperature forecasts for the greater London area (www.airTEXT.info) and Weather4D which uses ocean current data from the MyOcean service to show weather patterns in 4D (www. weather4d.com/en). These are some of the many examples proposed by participants in the annual Copernicus Masters competition (www.copernicus-masters.com). Another good example is the Marine Traffic project, which shows the live location of all vessels fitted with an automatic identification system (AIS) transponder. This service broadcasts the information received at ground-based AIS receiving stations live on the web (www.marinetraffic.com). Even more downstream services are being developed with the advent of the launch of the Copernicus dedicated satellites, Sentinel 1 to Sentinel 5.

The first Copernicus Sentinel satellite is a C-band Synthetic Aperture Radar (SAR) satellite, Sentinel 1A, launched in April 2014. Sentinel 1B, which will complete the Sentinel 1 constellation, will follow 18–24 months later [2]. This will take the revisit time from 12 days at the equator using a single satellite to six days at the equator, using Sentinel 1A and 1B in a constellation. At latitudes north of +45° and south of -45°, the average revisit time is every two days with both satellites [2]. The Sentinel 1 mission follows a succession of C-band SAR satellites commissioned by ESA. The first of these was ERS-1, which was launched in 1991, followed by ERS-2 launched in 1995. After the successful ERS missions, Advanced SAR (ASAR) was launched as part of the ENVISAT mission in 2002.

Multi-frequency SAR in this study is defined as using more than one frequency in combination for land cover classification and therefore SAR images from more than one satellite.

Earth Observation for Land and Emergency Monitoring, First Edition. Edited by Heiko Balzter.
© 2017 John Wiley & Sons Ltd. Published 2017 by John Wiley & Sons Ltd.

The benefits of multi-frequency SAR are:

- Increasing the revisit time for any SAR observation, relative to only one or two satellites. This is especially for areas closer to the equator, with longer repeat cycles, due to the polar orbits.
- Offers complementary resolution and swath options.
- Different frequencies respond to different surface roughness and scattering mechanism conditions, so the range of discriminable terrain or land cover features are expanded, relative to a single frequency SAR system.
- It is possible that one frequency may be able to substitute for another for some applications, if one particular satellite went out of action, however this would then be substituting frequencies for the same application, and not in essence a multi-frequency SAR application as defined above.

Even though the current Copernicus SAR configuration only includes the two C-band satellites, Sentinel 1A and 1B, there is the argument that the use of multi-frequency SAR, by combining the Sentinel 1 data with non-ESA SAR satellite data, can provide additional value to using only C-band SAR data. An overview of the Copernicus Services is given in Table 5.1. The context of the services where this report is focusing on is the Land Monitoring and Emergency Management core services.

The main focus of Sentinel 1 is polar monitoring of the Arctic environment and sea-ice zones, marine surveillance, land surface motion risks, land cover and land use monitoring, and global mapping for humanitarian crises management [4]. With the legacy of ESA's previous SAR systems used for ocean monitoring, a large focus of Sentinel 1 will be polar ice monitoring and marine surveillance. With regards to information on the ocean and the arctic environment, SAR is the primary source of information, with its all-weather and day-and-night observation capability [5]. The marine surveillance will cover wave monitoring for extreme wave events, oil spill information and ship detection. The land surface motion risk monitoring will be possible due to the configuration of the Sentinel 1 satellites to do interferometry. This will enable the monitoring of landslides, earthquakes, building and tunnel construction, water and gas extraction and mining activities [5]. For land cover classification (forest, urban areas, agricultural crops etc.), SAR data is not always the primary source of information, not when high resolution optical cloud-free data is available. However, for areas with regular cloud cover, or when urgent images are required for emergency or humanitarian management, SAR images

Table 5.1 Copernicus Services, operational status and website locations [3].

Copernicus Service	Operational Status (June 2015)	Website with access to the service
Land Monitoring	Operational	land.copernicus.eu
Emergency Management	Operational	emergency.copernicus.eu
Marine Monitoring	Operational	marine.copernicus.eu
Atmosphere Monitoring	Operational	atmosphere.copernicus.eu
Security	Pre-operational phase	www.copernicus.eu/main/security
Climate Change	Operational phase	www.copernicus-climate.eu

can be invaluable, as it might be the only data source available. This has been proven many times, especially for water extent mapping after flooding events [6].

All of these examples thus far have only made use of single frequency data (most regularly C-band, but also X- and L-band). Even though there are SAR satellites of different frequencies in orbit and even more planned for the future (see Table 5.3), synergies with other frequencies are not part of the current Copernicus core service activities. Indeed, combining different SAR frequencies for a single application has only been explored sporadically, anywhere, let alone for Copernicus applications [7]. This has been most easily attempted where the different SAR frequencies have been collected concurrently, such as the SIR-C/X-SAR Shuttle mission and from airborne research and development SAR campaigns [8]. It would therefore be an opportunity for the Copernicus downstream services sector to develop such a synergy of multi-frequency images, where application potential is seen to exist. The aim and scope of this chapter is to give use case example of how multi-frequency SAR could be utilized for land cover classification within the context of Copernicus downstream services.

The Copernicus services that are about to go fully operational with the launch of the Sentinel satellites have been many years in development. They originate from a range of projects funded by the European Commission's Sixth Framework Programme (FP6) and Seventh Framework Programme (FP7). The use of SAR in these pre-operational services gives an indication of current use of SAR within Copernicus core services, and of where the potential lies to use more SAR data within the downstream services sector. Table 5.2 shows the projects that lead to the current Copernicus Land monitoring and Emergency management core services.

It can be seen that not many projects up to now used SAR for operational applications apart from the Emergency management core services projects, Respond and SAFER. Within these projects the main use of SAR was for flood mapping. Some examples of flood mapping are given in [12], using TerraSAR-X data in a near real-time rapid mapping scenario. It is shown in [13] how C-band SAR from ENVISAT ASAR is used for flood mapping. The HH polarization is seen as to better discriminate standing water than the other polarization modes. A study comparing different types of satellite imagery for emergency response mapping, lists the main contribution of SAR as deformation mapping, flood mapping and landscape scars after a wildfire [14].

Table 5.2 SAR usage within previous FP6 and PF7 projects leading up to Copernicus Land monitoring and Emergency Management core services.

FP6/FP7 project	Dates	Was SAR used?
Land monitoring		
Geoland (FP6) [9]	2004–07	No
Geoland-2 (FP7) [10]	2008–12	No
Corine Land Cover (EEA) [10]	1990, 2000, 2006	No
Emergency management		
Respond [6]	2005–10	Yes, for flood-mapping
SAFER [10]	2009–11	Yes, for flood-mapping
GARNET-E [11]	2010–12	No

A big limitation to the use of SAR data for operational purposes is the restricted availability of SAR data with regards to revisit time and cost. TerraSAR-X is a commercial satellite and scenes need to be paid for commercial applications in the order of €2000 per scene [15]. The ALOS-2 PALSAR data is much more affordable since November 2014 at €36 per scene [16]. It is foreseen that the Sentinel-1 mission with its frequent revisit time (six-days with Sentinel 1A and 1B) and data that is available free of charge will make SAR data much more accessible and used by a wider user group [2].

The first multi-frequency SAR experiment from orbit was the SIR-C/X-SAR mission flown on the Shuttle Endeavour during two flights, in April and September/October 1994. This mission combined imagers from the United States' SIR-C sensor, using L- and C-bands, each with multiple polarizations (HH, VV, HV), and the German/Italian X-SAR sensor with a single polarization (VV) [17]. Examples of applications include land cover classifications using the intensity images [8], multi-temporal land cover classification [18] and discrimination of rain cells using multi-frequency data [19]. The SIR-C/X-SAR mission was a definite pre-cursor to the many SAR systems available today, such as TerraSAR-X and TanDEM-X, and ENVISAT ASAR which was operational until 2011. More recent multi-frequency SAR applications are for agricultural land use mapping [20] and land cover classification in the arctic coastal ecosystem [21].

There have been several airborne multi-frequency campaigns, such as the DLR's E-SAR, and the latest F-SAR multi-frequency SAR airborne campaign [22]. This enables the simultaneous capturing of P-,L-,C- and X-band images. For a multi-frequency system, this would be the ideal scenario since data is collected simultaneously and with identical specifications (e.g. pixel size), however this is currently limited to airborne instruments. The global reach of satellite missions for a multi-frequency SAR setup is still currently lacking. Therefore, to make use of multi-frequency SAR measurements, one presently need to combine measurements from multiple sensors on different satellites.

5.2 Examples of Multi-Frequency SAR

One example of a formal plan to create a multi-frequency SAR capability is that of the Italian Space Agency (ASI) and the Argentine Space Agency (CONAE – Comision Nacional de Actividades Espaciales). They are working together on a joint programme to form the Italian-Argentine Satellite System for Emergency Management (SIASGE), which will integrate the two satellite constellations of COSMO-Skymed and Satélite Argentino de Observación Con Microondas (SAOCOM) [23]. COSMO-Skymed (Constellation of small Satellites for Mediterranean basin Observation) is a constellation of four X-band satellites [24] and SAOCOM will be a constellation of two L-band satellites [25]. Another innovative way to make use of multi-frequency SAR has been proposed in [26]. Here it is proposed to use data from multi-frequency SAR satellites to supplement the optical thick cloud cover areas in optical imagery, using a Closest Feature Vector (CFV). This is used to fuse the SAR and optical imagery together, specifically in the areas of the optical image, which are affected by clouds. Similar to this study, they used mono-temporal, multi-frequency data.

All the SAR missions that are planned, approved and those currently in orbit, are shown in Table 5.3. Currently, there are C-band, from Radarsat-2 and Sentinel-1, X-band from TerraSAR-X, TanDEM-X and Cosmo-Skymed and L-band from ALOS-2

PALSAR in orbit. Table 5.3 shows the various data opportunities for potential applications of multi-frequency SAR, by combining data from different SAR sensors. It is notable that the spatial resolution keeps improving with the newer satellites, and more satellites are being used in constellations to improve revisit times. Over the next five years, there is the potential to obtain data from L-, C- and X-band SAR sensors, over the same geographical area, and relatively close in time, for multi-frequency SAR applications. The addition of the S-band NOVASAR-S satellite, which will be the UK's first SAR satellite (planned for launch in 2015) and the BIOMASS P-band SAR mission from 2020 onwards will add two more frequencies to the possible SAR images available.

5.3 Methods and Data

Land cover classification is an important monitoring tool used by governments and policy makers in terms of land use planning and observing land cover changes due to various factors including climate change [45–47]. As an example to show how an image can be classified using data from multi-frequency SAR sources, compared to only a single frequency, data for two test sites are considered and compared. The study sites chosen are located across two different biomes in central Africa: a semi-arid region in Tanzania and a forested region in the DRC. Africa was chosen since there are many Copernicus data users interested in Africa. The list of users include the International Charter "Space and Major Disasters" [48], the African-Caribbean-Pacific Observatory (ACP) of the European Commission's Joint Research Centre [49], the Food and Agricultural Organization of the United Nations (FAO) [50], and users and authorities interested in monitoring deforestation [51].

5.3.1 Data Availability and Test Sites

The data was chosen from archived SAR data from TerraSAR-X (X-band), ENVISAT ASAR (C-band) and ALOS PALSAR (L-band). The study sites, as determined by data availability from the archives, are in a semi-arid/savanna area in Tanzania and in a more forested region in the Democratic Republic of the Congo (DRC). The available images of the different SAR sensors for each site and their overlap in time are shown in Figure 5.1, and it can be seen that it is a relatively small area, compared to the full scenes from the individual sensors. For these two sites optical images from the RapidEye sensor were available at nearly the same acquisition time as some of the SAR images.

The timing of acquisition of the SAR and optical images together with the local rainfall data is plotted in Figure 5.2a and Figure 5.2b for the two sites. The rainfall data was obtained from the Famine Early Warning System Network [52]. The rainfall pattern for the Tanzania site is distinctly different to the rainfall at the DRC site. The Tanzania site has a definite wet season with rain up to 200 mm per month at its peak and a definite dry season with almost no rainfall for about three to four months per year. The SAR images were chosen to be as close as possible to each other in acquisition time. The Tanzania images are all acquired in the dry season. The DRC site has a more continuous rainfall distribution throughout the year with 100–200 mm rainfall per month. There seems to be a cycle of about a half-yearly pattern of increasing and decreasing rain. The TerraSAR-X images were deemed

Table 5.3 SAR satellites in orbit, approved and planned for the years 2011 to 2025, for potential multi-frequency SAR applications.

SAR missions [27]	Principal Owner	Polarization	Revisit time (days)	Swath width* (km)	Spatial res.* (m)	2011–2025 (timeline)
P-Band SAR						
BIOMASS [28], [29]	ESA	full	25–45 d	40–60	50	
L-Band SAR						
ALOS PALSAR [30]	JAXA (Japan)	full	46	40–70	7–44	
ALOS-2 PALSAR [31], [32]	JAXA (Japan)	full	14	25–70	3/6/10	
SAOCOM 1A [25]	CONAE (Arg)	full	16	30–350	10–100	
SAOCOM 1B [25]	CONAE (Arg)	full	8, A&B	30–350	10–100	
S-Band SAR						
HJ-1C (Huan Jing) [33], [34]	CRESDA (China)	VV	4	40–100	5–25	
NovaSAR-S [35]	SSTL (UK)	full	1–4	15–20	6	
C-Band SAR						
Envisat (ASAR) [27]	ESA	DualPol	35	56–100	30	
Sentinel 1A [2], [36]	ESA	DualPol	12	80–250	5–20	
Sentinel 1B [2], [36]	ESA	DualPol	6, A&B	80–250	5–20	
Sentinel 1C, … [27]	ESA	DualPol	6	80–250	5–20	
Radarsat-2 [27], [37]	MDA	full	24	20–100	1 × 3	

Mission	Owner	Polarization		Swath width*	Spatial resolution*
Radarsat Constellation Mission (RCM) [38]	MDA/CSA	full	1	5–125	1 × 3
RISAT-1 [39]	ISRO (India)	full	25	30–240	2–6
X-Band SAR					
TerraSAR-X [27], [40]	DLR/Astrium	DualPol	11	10–50	1–3
TanDEM-X [41]	DLR/Astrium	DualPol	11	10–50	1–3
TerraSAR-X2 [42]	DLR/Astrium	full	11	5–24	0.25–3
Cosmo-Skymed-1,2,3 [27], [43]	ASI (Italy)	single/AP	3–36h	10–40	1–5
Cosmo-Skymed-4 [43]	ASI (Italy)	single/AP	2–12h	10–40	1–5
Cosmo-Skymed 2nd Gen A,B [27]	ASI (Italy)	single/AP	2–12h	10–60	1–4
SeoSAR/PAZ [27], [44]	Hisdesat	DualPol	11, 5 (+TSX)	10–30	1–3
SeoSAR/PAZ-2 [27]	Hisdesat	DualPol	?	?	?

Legend: In orbit | Approved | Planned

* Swath width and spatial resolution for high-resolution imaging modes. Wider swaths and lower spatial resolution also available.

Principal Owners:

ASI – Italian Space Agency (Agenzia Spaziale Italiana), CONAE – Argentinian Space Agency (Comision Nacional de Actividades Espaciales), CRESDA – China Center for Resource Satellite Data and Application, CSA – Canadian Space Agency, DLR – German Aerospace Center (Deutsches Zentrum für Luft- und Raumfahrt), ESA – European Space Agency, ISRO – Indian Space Research Organization, JAXA – Japan Aerospace Exploration Agency, MDA – MacDonald, Dettwiler and Associates Ltd., Canada, SSTL – Surrey Satellite Technology Limited, UK.

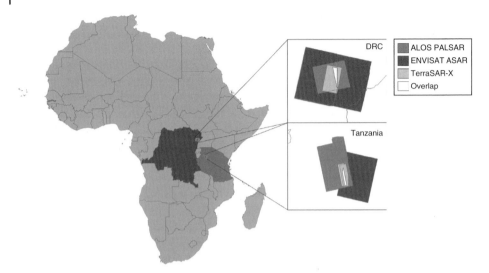

Figure 5.1 Location of the two study sites in the DRC and Tanzania. The overlap between the images from ALOS PALSAR, ENVISAT ASAR and TerraSAR-X are shown.

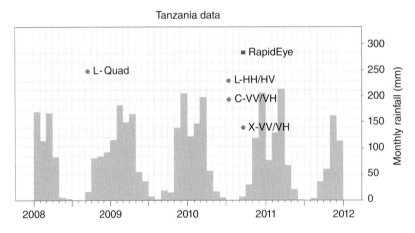

Figure 5.2a The dates of the selected SAR images and optical imagery from RapidEye with corresponding rainfall data, for the Tanzania site.

appropriate to use with the other SAR images, since the rainfall seemed similar, even though it was acquired seven to10 months before the ASAR, PALSAR and RapidEye images. The DRC images are therefore all grouped together as the wet season, with no apparent dry season in this region.

If a specific frequency-polarization option was not available, a suitable image of the same season, but from a different year was chosen, with the assumption that land cover should not change that quickly. This was the case for the PALSAR Quad image in the Tanzania site (taken two years before the others), and the two TerraSAR-X images for the DRC site (taken six to eight months before the others).

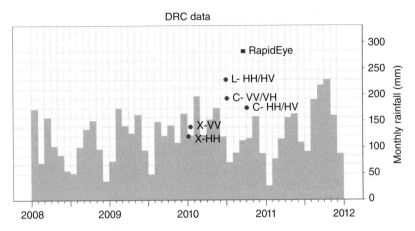

Figure 5.2b The dates of the selected SAR images and optical imagery from RapidEye with corresponding rainfall data, for the DRC site.

5.3.2 Data Preparation

For this experiment to progress, the SAR and optical data needed to be prepared. The different SAR data layers had to be calculated and the optical imagery had to be classified, each to be discussed in the following sections.

5.3.3 SAR Data

The SAR data for each site was obtained via two data request proposals. This was CP1-13326 from ESA for the ASAR, PALSAR and RapidEye data, and proposal LAN1639 for TerraSAR-X data from the DLR. The RapidEye data was obtained from the GMES Safer project.

The SAR data had to be orthorectified, calibrated, multi-looked and converted from linear to dB values. The multi-looking was applied to obtain spatial resolution of 5 m for TerraSAR-X, 10 m for ALOS PALSAR and 15 m for ENVISAT ASAR. The reason for this is to be able to combine all the images into a layer stack, as all are multiples of 5 m spatial resolution. This was performed using Gamma software [53], using a panchromatic Landsat images for the orthorectification since it is close in spatial resolution (15 m) to the SAR images. Additional data layers were derived, namely inter-channel ratio (HH/HV), and channel additions (HH + HV) and subtractions (HH-HV), using scripts from the Geospatial Data Abstraction Library [54]. Finally the mean co-occurrence texture measures of different window sizes were calculated using ENVI radar texture filters. A list of all the resulting SAR layers for the Tanzania site is shown in Table 5.4 and a list of all the resulting SAR layers for the DRC site in Table 5.5. For the L-band and C-band images 7×7 and 9×9 window sizes were computed, and for the X-band 7×7, 9×9, 11×11 and 13×13 window sizes were computed. This is since the X-band images are of a higher spatial resolution than the L-band and C-band images. The inter-channel calculations have been shown to be useful for land cover classification [55].

Table 5.4 List of SAR images and derived images for the Tanzania Dry set of images.

	Tanzania (44 layers)
Image 1	PALSAR_20080915_Quad (13)
variations	HH, HV, VH, VV, VVdivVH, VVminVH, VVplusVH,
	HHdivHV, HHminHV, HHplusHV, HHdivVV, HHminVV, HHplusVV
texture	*None, only texture layers from the second PALSAR images were considered to keep the layer stack reasonably small*
Image 2	PALSAR_20100713_HHHV (9)
variations	HH, HV, HHdivHV, HHminHV, HHplusHV
texture	HH_7×7mean, HH_9×9mean, HV_7×7mean, HV_9×9mean
Image 3	ASAR_20100715_VVVH (9)
variations	VV, VH, VVdivVH, VVminVH, VVplusVH
texture	VH_7×7mean, VH_9×9mean, VV_7×7mean, VV_9×9mean
Image 4	TSX_20100920_VVVH (13)
variations	VH, VV, VVdivVH, VVminVH, VVplusVH
texture	VH_7×7mean, VH_9×9mean, VH_11×11mean, VH_13×13mean
	VV_7×7mean, VV_9×9mean, VV_11×11mean, VV_13×13mean

Table 5.5 List of SAR images and derived images for the DRC set of images.

	DRC (37 layers)
Image 1	PALSAR_20100701_HHHV (9)
variations	HH, HV, HHdivHV, HHminHV, HHplusHV
texture	HH_7×7mean, HH_9×9mean, HV_7×7mean, HV_9×9mean
Image 2	ASAR_20101005_HHHV (9)
variations	HH, HV, HHdivHV, HHminHV, HHplusHV
texture	HH_7×7mean, HH_9×9mean, HV_7×7mean, HV_9×9mean
Image 3	ASAR_20100705_VVVH (9)
variations	VV, VH, VVdivVH, VVminVH, VVplusVH
texture	VV_7×7mean, VV_9×9mean, VH_7×7mean, VH_9×9mean
Image 4	TSX_20100106_HH (5)
variations	HH
texture	HH_7×7mean, HH_9×9mean, HH_11×11mean, HH_13×13mean
Image 5	TSX_20100117_VV (5)
variations	VV
texture	VV_7×7mean, VV_9×9mean, VV_11×11mean, VV_13×13mean

5.3.4   Optical Data for Training and Validation

Since there was no physical ground-truth fieldwork done in this study, the land cover was validated using optical data co-timed with the SAR data. The images from RapidEye were available close to the dates of the main SAR acquisitions. The two optical images for Tanzania and DRC were classified using the five RapidEye spectral bands (blue, green, red, red edge and near infrared), along with a calculated Normalized Difference Vegetation Index (NDVI) and a pseudo albedo layer, calculated from RapidEye's visible bands ('pseudo albedo' = $(1/3) * (R + G + B)$ [56]). To assist in the classification the images were segmented using a multi-resolution segmentation algorithm [57] as implemented in eCognition [58] with a scale parameter of 200, a shape parameter of 0.4 and a compactness parameter of 0.5. This resulted in detailed segments which, upon visual inspection, represented the land cover types accurately in the RapidEye images. An object based classification was preferred over a pixel-based classification, since it leads itself to intuitive manual editing in areas where clouds are prevalent in the optical image. As a second level of classification, the ruleset results were therefore edited where the classification had to be different upon visual inspection. This was specifically surrounding areas with thin clouds. Water features in the DRC image were also misclassified as clouds, due to sun-glare effects, which had to be corrected. The classified optical image for each site was then used as a layer to train the classification algorithm on the SAR layers.

5.3.5   Methodology for Multi-Frequency SAR Comparison

The aim for this study was to compare different frequency combinations to assess the comparison of multi-frequency SAR land cover classification compared to single frequency land cover classification. For both sites a classification from a RapidEye image of the same area and close in time were used as the validation or "pseudo" ground-truth layer to train the classifier using the multi-frequency SAR data.

Figure 5.3 shows the methodology schematically. First of all the optical RapidEye images were classified using the five original spectral bands, along with a calculated NDVI layer and pseudo albedo layer. The land cover classes were visually identified from the RapidEye images as bare, burnt, dense trees, thinner dense trees, sparse vegetation, water and settlement for the DRC site and bare, burnt, dense trees and sparse vegetation for the Tanzania site. They were used as input to classify the RapidEye images into these land cover classes. After some manual refinement, the optical classification layer is used as the response variable for the SAR classification. A layer stack of all the SAR layers together with the derived inter-channel calculations and texture layers were used as the predictor variables. From the combination of the SAR layers and the optical layer samples were selected from each of the land cover classes according to the optical classification. For each land cover type 5000 SAR pixels were randomly selected using the R raster package [59], resulting in 20,000 samples from the four Tanzania land cover classes and 35,000 samples from the seven DRC land cover classes. Table 5.6 shows the number of pixels and the percentage of the samples selected of each land cover type from the RapidEye classification for the DRC and Tanzania sites.

The samples can be split into training and validation sets, with a completely separate test set, kept outside during the model development. However, during this analysis a ten-fold cross validation [60] was performed during the classification stage, instead of a training/validation dataset. This technique provides a more accurate classification that

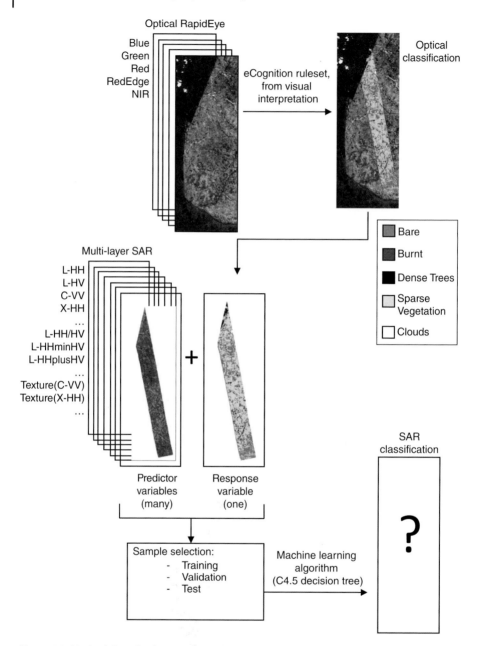

Figure 5.3 Methodology for deriving a SAR classification, using the optical classification result as the response variable in the machine learning algorithm.

is less biased based on a specific sample seed. The land cover samples are used to train a classification algorithm, using a layer stack of SAR images as the predictor variables and the optical classification layer as the response variable. For this comparative study a standard decision tree, namely C4.5 was used to perform the classification [61].

Figure 5.4a,b shows a zoomed-in area of the Tanzania SAR and corresponding optical classification, with Figure 5.4c,d showing a similar view of the DRC images. The data

Table 5.6 Number of pixels, number of samples selected and percentage for each land cover type from the RapidEye classification of the DRC and Tanzania sites.

Land cover	DRC			Tanzania		
	Number of pixels in image	Samples selected	Perc. %	Number of pixels in image	Samples selected	Perc. %
Bare	2,052,327	5,000	0.2	1,944,945	5,000	0.3
Burnt	49,188	5,000	10.2	21,497	5,000	23.3
Dense Trees	4,833,135	5,000	0.1	431,302	5,000	1.2
Thinner Dense Trees	1,145,420	5,000	0.4			
Sparse Vegetation	436,729	5,000	1.1	4,877,657	5,000	0.1
Water	26,476	5,000	18.9			
Settlement	90,342	5,000	5.5			

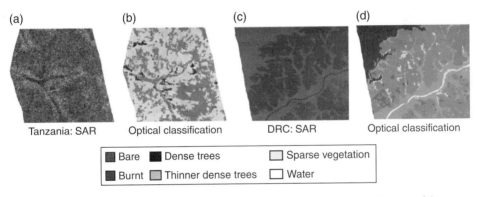

(a) Tanzania: SAR (b) Optical classification (c) DRC: SAR (d) Optical classification

Bare   Dense trees   Sparse vegetation
Burnt   Thinner dense trees   Water

Figure 5.4 (a) Tanzania multi-frequency SAR. (b) Corresponding eCognition classification of the Tanzania RapidEye image. (c) DRC multi-frequency SAR. (d) Corresponding eCognition classification of the DRC RapidEye image.

samples were imported into the Waikato Environment for Knowledge Analysis (Weka) [62]. Weka is an open source program that groups together many machine learning algorithms in a user friendly environment. Within Weka the C4.5 decision tree is implemented in Java and known as J48, which was used as the main classification algorithm for this comparative study. A decision tree is a relatively simple though very effective classifier suitable for linear and non-linear problems [63].

An assumption of a minimum of 100 instances per leaf was applied to all the computed trees. A decision tree was computed for each of the possible frequency combinations, namely: L-band only, C-band only, X-band only, L- and C-band, L- and X-band, C- and X-band and finally L-, C- and X-band. The overall classification accuracy and the class specific classification accuracies are shown in the following section. For both sites a ranking of variable importance was computed using the information gain criteria [63]. This was done to assess a ranking of variable importance between all the different input layers.

5.4 Results

Decision trees were computed for each frequency scenario of L-band only, C-band only, X-band only, L- and C-band, L- and X-band, C- and X-band, and L-, C- and X-band. The overall classification accuracy for each of these scenarios, along with the class-specific classification accuracies are shown in Table 5.8 for the Tanzania site. For the Tanzania site there were four classes, namely *bare, burnt, dense vegetation* and *sparse vegetation*.

Table 5.9 shows the overall classification accuracy along with the class-specific classification accuracies for DRC site. For the DRC site seven land cover classes were identified, namely *bare, burnt, dense trees, thinner dense trees, sparse vegetation, water* and *settlement*. A description of each land cover type in terms of woody cover estimate is shown in Table 5.7.

The class specific accuracies were compiled from the confusion matrix from each classification and correspond to the producer's accuracy for each class. The rank and information gain for all the input variables for the Tanzania site are given in Table 5.10. The different shaded groups of variables combines layers from the same sensor, namely from TerraSAR-X, PALSAR and ASAR. Table 5.10 shows the rank and information gain for all the input variables for the DRC site.

Table 5.7 The land cover types identified in the DRC and Tanzania study sites, with their woodland cover estimate and additional description.

Land cover name	Woodland cover estimate	Description
Dense trees	70–100% canopy cover	Only trees
Thinner dense trees	40–70% canopy cover	Mixed vegetation, trees, grassland, shrubs
Sparse vegetation	10–40% canopy cover	Mixed vegetation, trees, grassland, shrubs
Bare soil	0–10% canopy cover	Bare, with possibly some vegetation
Burnt		Burnt vegetation
Settlements		Settlements
Water		Water

Table 5.8 Overall Accuracy and Class-specific accuracy for the Tanzania site (light shades represent high classification accuracy, and dark shades low classification accuracy).

Frequencies	Overall Accuracy	Bare	Burnt	Dense Trees	SparseVeg
L	50.3	72.2	29.7	64.0	35.4
C	42.8	70.7	19.6	68.6	12.0
X	52.2	78.1	4.9	82.5	43.1
LC	50.1	72.2	30.4	62.3	35.5
LX	54.2	77.5	22.3	80.4	36.4
CX	52.1	78.1	6.8	80.0	43.4
LCX	54.2	77.5	22.3	80.4	36.4

Table 5.9 Overall Accuracy and Class-specific accuracy for the DRC site (light shades represent high classification accuracy, and dark shades low classification accuracy).

Frequencies	Overall Accuracy	Bare	Burnt	Dense Trees	Thinner Dense Trees	Sparse Veg	Water	Settlement
L	47.9	41.0	46.9	83.9	12.6	9.1	69.1	72.4
C	32.2	34.7	33.7	24.0	42.0	3.4	60.7	26.9
X	44.4	20.9	53.9	58.0	44.3	17.3	78.9	37.8
LC	48.9	41.0	46.9	83.9	22.3	10.8	72.0	65.5
LX	53.2	45.5	49.4	80.0	37.1	11.4	80.8	68.2
CX	44.5	18.1	58.0	57.4	41.8	17.7	81.1	37.6
LCX	53.3	45.6	49.5	79.8	37.1	13.0	82.7	65.6

Table 5.10 Variable importance using the Information Gain criteria, for all the input layers for the Tanzania site, using a 10-fold cross validation.

Rank	SAR layer	Information gain
1	TSX_20100920_VVVH_VH_Texture_13 × 13mean	0.48
2	TSX_20100920_VVVH_VH_Texture_11 × 11mean	0.47
3	TSX_20100920_VVVH_VH_Texture_9 × 9mean	0.45
4	TSX_20100920_VVVH_VH_Texture_7 × 7mean	0.42
5	PALSAR_20100713_HHHV_HV_Texture_9 × 9mean	0.41
6	PALSAR_20100713_HHHV_HV_Texture_7 × 7mean	0.41
7	PALSAR_20100713_HHHV_HH_Texture_9 × 9mean	0.35
8.5	PALSAR_20100713_HHHV_HH_Texture_7 × 7mean	0.33
8.5	PALSAR_20100713_HHHV_HHplusHV	0.33
10.1	PALSAR_20080915_Quad_HV	0.32
11.4	PALSAR_20080915_Quad_VH	0.32
11.5	PALSAR_20100713_HHHV_HV	0.32
13	PALSAR_20080915_Quad_HHplusHV	0.30
14	PALSAR_20080915_Quad_VVplusVH	0.29
15	PALSAR_20100713_HHHV_HH	0.26
16	ASAR_20100715_VVVH_VH_Texture_9 × 9mean	0.24
17	ASAR_20100715_VVVH_VH_Texture_7 × 7mean	0.22
18.3	ASAR_20100715_VVVH_VVplusVH	0.21
18.7	TSX_20100920_VVVH_VH	0.21
20	ASAR_20100715_VVVH_VV_Texture_9 × 9mean	0.18
21	PALSAR_20080915_Quad_HHplusVV	0.17
22.4	ASAR_20100715_VVVH_VH	0.16
22.6	ASAR_20100715_VVVH_VV_Texture_7 × 7mean	0.16

(Continued)

Table 5.10 (Continued)

Rank	SAR layer	Information gain
24	PALSAR_20080915_Quad_HH	0.16
25	PALSAR_20080915_Quad_VVminVH	0.16
26.1	PALSAR_20080915_Quad_HHminHV	0.13
26.9	TSX_20100920_VVVH_VVplusVH	0.13
28	ASAR_20100715_VVVH_VV	0.12
29	PALSAR_20080915_Quad_VV	0.12
30	TSX_20100920_VVVH_VVminVH	0.11
31	PALSAR_20080915_Quad_VVdivVH	0.06
32	PALSAR_20080915_Quad_HHdivVV	0.05
33	TSX_20100920_VVVH_VVdivVH	0.05
34	PALSAR_20100713_HHHV_HHminHV	0.04
35	PALSAR_20080915_Quad_HHminVV	0.03
36	PALSAR_20080915_Quad_HHdivHV	0.03
37	TSX_20100920_VVVH_VV_Texture_13×13mean	0.02
38	TSX_20100920_VVVH_VV_Texture_11×11mean	0.01
39.1	PALSAR_20100713_HHHV_HHdivHV	0.01
40.3	TSX_20100920_VVVH_VV_Texture_9×9mean	0.01
40.6	ASAR_20100715_VVVH_VVdivVH	0.01
42	TSX_20100920_VVVH_VV_Texture_7×7mean	0.01
43	ASAR_20100715_VVVH_VVminVH	0.01
44	TSX_20100920_VVVH_VV	0.00

5.5 Discussion

The results for both the Tanzania and DRC sites show an overall trend, that with increased information from multiple frequencies, the classification accuracies also increase. The overall accuracy for this initial result is moderate at only 54% for the Tanzania site and 53% for the DRC site. However, by looking at the class specific accuracies, there are some classes that stand out, such as the *dense vegetation* class in Tanzania with 83% using X-band only and the *dense trees* class in DRC with 84% using L-band.

Table 5.8 shows the classification results for the Tanzania site. The *bare* and *dense trees* has the highest classification accuracy across all the frequency combinations. The *burnt* class has the lowest classification accuracy, with the best being 30% using L-band and C-band, and the *sparse vegetation* class is second last with the best classification at 43% using a combination of C-band and X-band. These results could be interpreted in several ways. One reason the *burnt* class is so low is due to the fact that most of the SAR images were acquired before the RapidEye image, with the ASAR and PALSAR HHHV image two months before and the second PALSAR Quad image during the same month, September, but two years before the RapidEye image. The only SAR image acquired close to and even on the same day as the RapidEye image, is the TerraSAR-X image, acquired on 20 September 2010. However, the X-band image on its own does not seem to classify the *burnt* areas well at all, with only 5% classification accuracy. The timing of the acquisitions is therefore crucial to classify the land cover classes correctly. A similar explanation could be applied to the *sparse vegetation* class, which could be fluctuating between *bare* and *sparse vegetation* over time in the semi-arid environment. The *dense trees* and *bare* classes have consistent high classification accuracy, especially with all the frequency combinations which include X-band. All of these have classification accuracies higher than 77% for *bare* areas, and higher than 80% for *dense trees*.

Table 5.9 shows the classification results for the DRC site. The *dense trees* class has performed the best with up to 84% correctly classified instances. It is a notable observation that all the frequency combinations with high classification accuracy for dense trees contain L-band. Without L-band the dense trees class only reaches 24–58% classification accuracy using C-band or X-band or both. L-band therefore seems to be crucial to identify dense trees in a rain forest region as in the DRC correctly, varying from 80–84% classification accuracy. The *thinner dense trees* class, which can be found on the edges of the dense trees class in the optical image, only achieved 44% classification accuracy using X-band. This might therefore be a class that needs to be merged with another class in the SAR classification. Looking through the confusion matrices, which are not all reproduced here, the *thinner dense trees* are most often misclassified as *dense trees*, which would therefore be a suitable candidate class to merge with. This is true for all the frequencies, and especially with L-band, which misclassifies 53% of the *thinner dense trees* as *dense trees*.

The *water* class performs close to the *dense trees* class, and is the class that is classified well across all the frequency combinations. The observation that water is classified better using X band than L- and C-band should not be misinterpreted here, since it is not only a function of frequency, but also of spatial resolution. The original TerraSAR-X images were 3 m, PALSAR images were 10 m and ASAR images were 15 m spatial resolution. These were all sampled to 5 m spatial resolution, and could therefore be contributing to the misclassification of *water*, since the varying rivers to be classified are only around 60 m wide. Using any combination of frequencies together with X-band produces very good classification accuracies from 79–83%. The *settlement* class performed better than anticipated, since there are many mixed land cover areas surrounding the small village in the DRC scene. This was best classified using L-band on its own, giving 72% classification accuracy.

The *burnt* class has done much better than the *burnt* areas of the Tanzania site, with classification accuracy up to 58%, using a combination of C- and X-band. It is unclear why this is so much better than for the *burnt* areas of the Tanzania class, since the acquisition of the L-band and C-band VVVH SAR images were two months before the RapidEye image. The TerraSAR-X image was acquired nine months before the RapidEye image, and the C-band HHHV image was acquired shortly after the RapidEye image. The biggest contribution in this case seems to come from the X-band, so the reason for the difference between the classification accuracy of the *burnt* areas between the DRC and Tanzania sites is unclear at this stage. Most of the *burnt* instances are misclassified as *bare*, which makes sense, considering the physics of a SAR sensor. The *bare* instances did not classify that well either, with the best being 46% classification accuracy using all three frequencies, or L and X giving 45%.

Table 5.11 Variable importance using the Information Gain criteria, for all the input layers for the DRC site, using a 10-fold cross validation.

Rank	SAR layer	Information gain
1	PALSAR_20100701_HHHV_HV_Texture_9×9mean	0.76
2	PALSAR_20100701_HHHV_HV_Texture_7×7mean	0.73
3	PALSAR_20100701_HHHV_HH_Texture_9×9mean	0.66
4	PALSAR_20100701_HHHV_HHplusHV	0.66
5	PALSAR_20100701_HHHV_HH_Texture_7×7mean	0.64
6	PALSAR_20100701_HHHV_HV	0.62
7	TSX_20100106_HH_HH_Texture_13×13mean	0.55
8	PALSAR_20100701_HHHV_HH	0.55
9	TSX_20100106_HH_HH_Texture_11×11mean	0.52
10	TSX_20100106_HH_HH_Texture_9×9mean	0.49
11	TSX_20100106_HH_HH	0.45
12	TSX_20100106_HH_HH_Texture_7×7mean	0.44
13	TSX_20100117_VV_VV	0.44
14	TSX_20100117_VV_VV_Texture_13×13mean	0.44
15	TSX_20100117_VV_VV_Texture_11×11mean	0.42
16	ASAR_20100705_VVVH_VV_Texture_9×9mean	0.42
17	ASAR_20100705_VVVH_VV_Texture_7×7mean	0.39
18	TSX_20100117_VV_VV_Texture_9×9mean	0.39
19	ASAR_20100705_VVVH_VVplusVH	0.37
20	TSX_20100117_VV_VV_Texture_7×7mean	0.35
21	ASAR_20100705_VVVH_VV	0.35
22	ASAR_20100705_VVVH_VH_Texture_9×9mean	0.29
23	ASAR_20100705_VVVH_VH_Texture_7×7mean	0.27
24	ASAR_20100705_VVVH_VH	0.23
25	PALSAR_20100701_HHHV_HHdivHV	0.14
26	ASAR_20100705_VVVH_VVdivVH	0.13
27	PALSAR_20100701_HHHV_HHminHV	0.13
28	ASAR_20100705_VVVH_VVminVH	0.05
29	ASAR_20101005_HHHV_HH_Texture_9×9mean	0.03
30	ASAR_20101005_HHHV_HH_Texture_7×7mean	0.02
31	ASAR_20101005_HHHV_HHplusHV	0.02
32	ASAR_20101005_HHHV_HH	0.02
33	ASAR_20101005_HHHV_HHdivHV	0.01
34	ASAR_20101005_HHHV_HV_Texture_9×9mean	0.01
35	ASAR_20101005_HHHV_HV_Texture_7×7mean	0.01
36	ASAR_20101005_HHHV_HV	0.01
37	ASAR_20101005_HHHV_HHminHV	0.01

The *sparse vegetation* classes did not do very well at all, with the best being 18% using both C and X-band. The *sparse vegetation* class were mostly confused with *thinner dense trees* and *settlements*. One option would therefore be to merge it with the *thinner dense trees* class, since the settlements are quite localized. The reason that the SAR does not classify the *sparse vegetation* class correctly could be either that it is not easily observable using SAR, or due to the timing of the SAR images compared to the RapidEye image. This will need some further investigation.

The ranking of variable importance according to the *information gain* criteria is given in Table 5.10 for the Tanzania site. Similar to the decision trees, this was also performed using 10-fold cross validation. It clearly stands out that the most influential layers are firstly the texture layers from the TerraSAR-X VH images. Secondly the PALSAR HV and HH texture layers from the 2010 PALSAR image, along with the PALSAR HV and VH layers from the 2008 PALSAR image. Thirdly the ASAR VH texture layers, ASAR VV texture and VH intensity layers. The remainder of the layers are more of a blend between the L-, C- and X-band images. It stands out clearly from the top ranking layers that the texture information adds a lot of value to identify the land cover, compared to the standard intensity layers and band ratios and calculations (e.g. HH/VV, HHminVV and HHplusVV), regardless of the frequency.

The ranking of the SAR layers according to the information gain criteria for the DRC site is given in Table 5.11. The top ranking group of SAR layers for the DRC are all PALSAR HH and HV texture and intensity layers. Then secondly a group of TerraSAR-X texture and intensity layers, and thirdly the ASAR texture layers and remaining layers from all three frequencies. It is interesting to note that the PALSAR layers ranks higher than the TerraSAR-X layers for the DRC site, compared to the Tanzania site which is the other way around. It could be that the texture layers, which are broadly averaging the SAR pixels over bigger areas, and therefore averaging out the speckle, compares better to the slightly larger classification objects from the RapidEye image. It could also be that the *dense trees* of the semi-arid Tanzania region are more distinguishable by the X-band texture than L-band, which performs better to recognise the *dense trees* of the Congo rain forest. This highlights one of the differences between the two study sites. However, this is something to investigate further still.

5.6 Conclusions

This study has shown the additional value of using multi-frequency SAR compared to single-frequency SAR. Even using two frequencies instead of one proved to add a lot of value, especially the combination of L-band and X-band, for both the Tanzania and DRC sites. This is in line with the mentioned SIASGE initiative between COSMO-Skymed and SAOCOM to combine X-band and L-band imagery for emergency response situations.

The main SAR data supply of the future of the Copernicus services will be from the Sentinel missions, which will start with C-band SAR and later optical satellites. This study has shown that for some land cover classes, C-band SAR is not necessarily the optimum frequency to use with a single date acquisition. However, due to the frequent revisit time of Sentinel-1, multi-temporal C-band analysis will probably be more the norm of the day, than multi-frequency SAR. One study that investigated multi-temporal C-band SAR using

ERS-2 and ASAR data found classification accuracies of 90 + % for most classes, over a agricultural test site in Thuringia, Germany [64]. The use of multi-temporal C-band images over an African site could therefore be work to investigate further.

Further research to be conducted is to compare wet and dry season images for the Tanzania and DRC sites. Since no RapidEye imagery was acquired during the wet season, there is a need for optical imagery to be used as a validation layer. It seems that Landsat 8 will most likely be able to fill the gap, although it is at a coarser resolution of 30 m, and acquired in 2013, compared to the very timely acquisition of the RapidEye imagery to the SAR images. In the future Sentinel-2 will provide data at 10 m spatial resolution, which will be of huge benefit to research such as this report, especially with the frequent revisit time of five days. Multi-frequency SAR imagery is also available for more test sites across two forest sites in Cameroon, and two semi-arid sites in Chad and Sudan. These will still be investigated further.

There is definitely a place for multi-frequency SAR applications within the Copernicus downstream services even though the frequencies other than C-band imagery will have to be supplied from third party contributing missions. Multi-frequency SAR has a higher overall classification accuracy than single frequency SAR, where the choice of frequency combinations is class and area specific. Multi-frequency SAR will therefore be beneficial under certain cases, where single frequency SAR does not provide all the needed information.

References

1 Space-Tec-Partners, Assessing the Economic Value of GMES: "European Earth Observation and GMES Downstream Services Market Study," 2012.
2 R. Torres, P. Snoeij, D. Geudtner, D. Bibby, M. Davidson, E. Attema, P. Potin, B. Rommen, N. Floury, M. Brown, I.N. Traver, P. Deghaye, B. Duesmann, B. Rosich, N. Miranda, C. Bruno, M. L'Abbate, R. Croci, A. Pietropaolo, M. Huchler, and F. Rostan, GMES Sentinel-1 mission, *Remote Sens. Environ.*, **120**, 9–24, Feb. 2012.
3 EC, Three new Copernicus services to reach operational status in 2015, *Copernicus Observer*, 2015. Online. Available: http://newsletter.copernicus.eu/issue-09-february-2015/article/three-new-copernicus-services-reach-operational-status-2015. Accessed: 20.11.16.
4 E. Attema, Mission Requirements Document for the European Radar Observatory Sentinel-1, Noordwijk, The Netherlands, 2005.
5 R. Torres, S. Paul, D. Malcolm, B. David, and S. Lokas, THE SENTINEL-1 MISSION AND ITS APPLICATION CAPABILITIES, *IEEE Int. Geosci. Remote Sens. Symp.*, 1703–1706, 2013.
6 ESA, RESPOND Final Report, 2011.
7 V. Turkar, R. Deo, Y.S. Rao, S. Mohan, and A. Das, Classification Accuracy of Multi-Frequency and Multi-Polarization SAR Images for Various Land Covers, *IEEE J. Sel. Top. Appl. Earth Obs. Remote Sens.*, **5**, no. 3, 936–941, Jun. 2012.
8 L.E. Pierce, K. Bergen, M.C. Dobson, and F.T. Ulaby, Land-Cover Classification using SIR-C/X-SAR Data, *IEEE Geosci. Remote Sens. Symp.*, **2**, no. 313, 8–10, 1995.
9 EC, Copernicus, Database of projects, geoland, 2013. Online. Available: http://www.copernicus.eu/pages-principales/projects/project-database/database-of-projects/?idproj=1&what=1&filter=7&page=0&cHash=46db1f94e4354348604d22234dc4ed8b.

10 CSES, Interim Evaluation of the European Earth Monitoring Programme (GMES) and its initial Operations (2011–2013) Final Report, Kent, United Kingdom, 2013.

11 Garnet-E, 2010. Online. Available: www.gmes-garnete.net/.

12 V.H. Cruz, TERRASAR-X RAPID MAPPING FOR FLOOD EVENTS, *Photogramm. – Fernerkundung – Geoinf.*, no. 6, 475, 2010.

13 J. -B. Henry, P. Chastanet, K. Fellah, and Y. -L. Desnos, Envisat multi-polarized ASAR data for flood mapping, *Int. J. Remote Sens.*, **27**, no. 10, 1921–1929, May 2006.

14 K.E. Joyce, S.E. Belliss, S.V. Samsonov, S.J. McNeill, and P.J. Glassey, A review of the status of satellite remote sensing and image processing techniques for mapping natural hazards and disasters, *Prog. Phys. Geogr.*, **33**, no. 2, 183–207, Jun. 2009.

15 Airbus, TerraSAR-X Services International Price List, *Airbus Defence and Space TerraSAR-X Services*, 2014. Online. Available: http://www2.geo-airbusds.com/files/pmedia/public/r463_9_itd-0508-cd-0001-tsx_international_pricelist_en_issue_6.00.pdf. Accessed: 20.11.2016.

16 RESTEC, Announcement for ALOS AVNIR-2 and PALSAR Data Distribution, 2014. Online. Available: www.alos-restec.jp/en/article.php/20141121. Accessed: 20.11.16.

17 F. Stuhr, R. Jordan, and M. Werner, SIR-C/X-SAR : An Advanced Radar, in *IEEE Aerospace Applications Conference*, 1996, **2**, 5–16.

18 L.E. Pierce, K.M. Bergen, M.C. Dobson, and F.T. Ulaby, Multitemporal Land-Cover Classification Using SIR-C/X-SAR Imagery, *Remote Sens. Environ.*, **64**, no. 1, 20–33, 1998.

19 C. Melsheimer, W. Alpers, and M. Gade, Investigation of multifrequency/multipolarization radar signatures of rain cells over the ocean using SIR-C/X-SAR data, *J. Geophys. Res.*, **103**, no. C9, 18867–18884, 1998.

20 K. Zolfaghari, J. Shang, H. McNairn, J. Li, and S. Homyouni, Using Support Vector Machine (SVM) for Agriculture Land Use Mapping with SAR Data: preliminary results from Western Canada, *2013 Second Int. Conf. Agro-Geoinformatics*, 126–130, 2013.

21 S.N. Banks, T. Ullmann, J. Duffe, A. Roth, D.J. King, A. M. Demers, A. Hogg, A. Schmitt, R. Baumhauer, and S. Dech, Multi-frequency analysis of high resolution quad-pol Radarsat-2 and dual-pol TerraSAR-X data for land cover classification in Arctic Coastal Ecosystems, Mackenzie Delta, Beaufort Sea, *Int. Geosci. Remote Sens. Symp.*, 3548–3551, 2012.

22 R. Horn, N. Anton, R. Andreas, F. Jens, and S. Rolf, F-SAR – DLR'S new multifrequency polarimetric airborne SAR, *IEEE Int. Geosci. Remote Sens. Symp.*, **2**, 902–905, 2009.

23 N. Pierdicca, F. Pelliccia, and M. Chini, Thematic mapping at regional scale using SIASGE radar data at X and L band and optical images, *IEEE Geosci. Remote Sens. Symp.*, 1095–1098, 2011.

24 F. Covello, F. Battazza, A. Coletta, E. Lopinto, C. Fiorentino, L. Pietranera, G. Valentini, and S. Zoffoli, COSMO-SkyMed an existing opportunity for observing the Earth, *J. Geodyn.*, **49**, no. 3–4, 171–180, Apr. 2010.

25 CONAE, SAOCOM, 2011. Online. Available: www.conae.gov.ar/eng/satelites/saocom.html.

26 R. Eckardt, C. Berger, C. Thiel, and C. Schmullius, Removal of Optically Thick Clouds from Multi-Spectral Satellite Images Using Multi-Frequency SAR Data, *Remote Sens.*, **5**, no. 6, 2973–3006, Jun. 2013.

27 J. Aschbacher and M.P. Milagro-Pérez, The European Earth monitoring (GMES) programme: Status and perspectives, *Remote Sens. Environ.*, **120**, no. 2012, 3–8, May 2012.

28 ESA, No 13-2013: ESA's next Earth Explorer Satellite, 2013. Online. Available: www.esa.int/For_Media/Press_Releases/ESA_s_next_Earth_Explorer_satellite.

29 T. Le Toan, S. Quegan, M.W.J. Davidson, H. Balzter, P. Paillou, K.P. Papathanassiou, S. Plummer, F. Rocca, S.S. Saatchi, H.H. Shugart, and L. Ulander, The BIOMASS mission: Mapping global forest biomass to better understand the terrestrial carbon cycle, *Remote Sens. Environ.*, **115**, no. 11, 2850–2860, Jun. 2011.

30 JAXA, DAICHI (ALOS) Operation Completion, 2011. Online. Available: www.jaxa.jp/press/2011/05/20110512_daichi_e.html.

31 JAXA, ALOS-2, *eoPortal Directory*, 2013. Online. Available: https://directory.eoportal.org/web/eoportal/satellite-missions/a/alos-2.

32 JAXA, ALOS-2 *Launch Schedule, 2013*. Online. Available: www.jaxa.jp/projects/in_progress_e.html.

33 W. Huang, F. Chen, J. Yang, B. Fu, P. Chen, and C. Zhang, Chinese HJ-1C SAR and Its Wind Mapping Capability, in *The 3rd International Workshop on Advances in SAR Oceanography from Envisat, ERS and ESA third party missions*, 2010, 1–19.

34 R.C. Barbosa, Chinese Long March 2C lofts Huanjing-1C into orbit, 2012. Online. Available: www.nasaspaceflight.com/2012/11/chinese-long-march-2c-huanjing-1c-into-orbit/.

35 SSTL, *Government investment brings low cost radar satellites to market, 2011*. Online. Available: www.sstl.co.uk/News-and-Events?story=1936.

36 ESA, *Sentinel-1, 2013*. Online. Available: www.esa.int/Our_Activities/Observing_the_Earth/Copernicus/Sentinel-1.

37 CSA, *RADARSAT-2, 2011*. Online. Available: www.asc-csa.gc.ca/eng/satellites/radarsat2/.

38 CSA, *RADARSAT Constellation, 2013*. Online. Available: www.asc-csa.gc.ca/eng/satellites/radarsat/.

39 M. Chakraborty, S. Panigrahy, A.S. Rajawat, R. Kumar, *et al.*, Initial results using RISAT-1 C-band SAR data, *Curr. Sci.*, **104**, no. 4, 490–501, 2013.

40 R. Werninghaus and S. Buckreuss, The TerraSAR-X Mission and System Design, *IEEE Trans. Geosci. Remote Sens.*, **48**, no. 2, 606–614, Feb. 2010.

41 M. Zink, TanDEM-X mission status, *IEEE Geosci. Remote Sens. Symp.*, 1896–1899, Jul. 2012.

42 J. Janoth, S. Gantert, W. Koppe, A. Kaptein, and C. Fischer, TERRASAR-X2 – MISSION OVERVIEW, *IEEE Geosci. Remote Sens. Symp.*, 217–220, 2012.

43 F. Covella, F. Battazza, A. Coletta, E. Lopinto, C. Fiorentino, L. Pietranera, G. Valentini, and S. Zoffoli, COSMO-SkyMed an existing opportunity for observing the Earth, *J. Geodyn.*, **49**, no. 3–4, 171–180, Apr. 2010.

44 Hisdesat, *Paz, 2013*. Online. Available: www.hisdesat.es/eng/satelites_observ-paz.html.

45 J.R. Anderson, E.E. Hardy, J.T. Roach, and R.E. Witmer, A Land Use and Land Cover Classification System for Use with Remote Sensor Data, *United States Geol. Surv. Circ.*, no. **671**, 1–28, 1976.

46 M. Bossard, J. Feranec, and J. Otahel, CORINE land cover technical guide – Addendum 2000, Technical Report No 40, Copenhagen, 2000.

47 A.B. Brink and H.D. Eva, Monitoring 25 years of land cover change dynamics in Africa: A sample based remote sensing approach, *Appl. Geogr.*, **29**, no. 4, 501–512, Dec. 2009.

48 A. Mahmood, E. Cubero-Castan, J. Bequignon, L. Lauritson, P. Soma, and G. Platzeck, International Charter 'Space and Major Disasters' Status Report, *Proceedings. 2005 IEEE Int. Geosci. Remote Sens. Symp. 2005. IGARSS '05.*, **6**, 4362–4365, 2005.

49 A. Belward, P. Mayaux, P. Roggeri, and O. Leo, Sustainable Development in Developing Countries : the ACP : African-Caribbean-Pacific Observatory, in *Knowledge-Based Technologies and OR Methodologies for Strategic Decisions of Sustainable Development*, 2009, 209–213.

50 FAO, Food and Agricultural Organization of the United Nations, 2013. Online. Available: www.fao.org/home/en/.

51 H.K. Gibbs, S. Brown, J.O. Niles, and J.A. Foley, Monitoring and estimating tropical forest carbon stocks: making REDD a reality, *Environ. Res. Lett.*, **2**, 045023(1–13), Oct. 2007.

52 FEWSNET, Famine Early Warning Systems Network, 2012. Online. Available: www. fews.net.

53 C. Werner, T. Strozzi, A. Wiesmann, and M. Santoro, GAMMA SAR and Interferometry software, Gumlingen, Switzerland, 2012.

54 GDAL, GDAL – Geospatial Data Abstraction Library. Open Source Geospatial Foundation, 2013.

55 E. Simental, V. Guthrie, and S.B. Blundell, Polarimetry band ratios, decompositions, and statistics for terrain characterization, in *Pecora 16 "Global Priorities in Land Remote Sensing,"* 2005, 1–11.

56 A. Baraldi, V. Puzzolo, P. Blonda, L. Bruzzone, and C. Tarantino, Automatic Spectral Rule-Based Preliminary Mapping of Calibrated Landsat TM and ETM + Images, *IEEE Geosci. Remote Sens.*, **44**, no. 9, 2563–2586, 2006.

57 U.C. Benz, P. Hofmann, G. Willhauck, I. Lingenfelder, and M. Heynen, Multi-resolution, object-oriented fuzzy analysis of remote sensing data for GIS-ready information, *ISPRS J. Photogramm. Remote Sens.*, **58**, no. 3–4, 239–258, Jan. 2004.

58 Trimble, eCognition, 2013. Online. Available: www.ecognition.com/.

59 R.J. Hijmans, Package "raster." 1–216, 2013.

60 R. Kohavi, A study of cross-validation and bootstrap for accuracy estimation and model selection, *Int. Jt. Conf. Artif. Intell.*, 1–7, 1995.

61 J.R. Quinlan, *C4.5: Programs for Machine Learning*. San Mateo, CA: Morgan Kaufmann Publishers Inc., 1993.

62 H. Mark, E. Frank, G. Holmes, B. Pfahringer, P. Reutemann, and I. H. Witten, The WEKA Data Mining Software: An Update, *Spec. Interes. Gr. Knowl. Discov. Data Min. Explor.*, **11**, no. 1, 2009.

63 L. Rokach and O. Maimon, Decision Trees, in *Data Mining and Knowledge Discovery Handbook*, 2nd Editio., O. Maimon and L. Rokach, Eds. Tel-Aviv: Springer, 2010, 149–174.

64 C. Thiel, O. Cartus, R. Eckardt, N. Richter, C. Thiel, and C. Schmullius, Analysis of multi-temporal land observation at c-band, *IEEE Geosci. Remote Sens. Symp.*, **III**, 318–321, 2009.

6

Unsupervised Land Use/Land Cover Change Detection with Integrated Pixel and Object Based Approaches for High Resolution Remote Sensing Imagery

S. Shrestha[1,2], C. Smith[1] and Z. Bochenek[2]

[1] *University of Leicester, Centre for Landscape and Climate Research, Department of Geography, Leicester, UK*
[2] *Remote Sensing Centre, Institute of Geodesy and Cartography, Warsaw, Poland*

6.1   Introduction

Many Change Detection (CD) techniques have been devised since the early days of Earth Observation (EO) satellites. However, the selection of a suitable method or algorithm for a given project is an important, but not an easy task since no approach is optimal and applicable to all cases [1,2]. The selection and preference of a particular method is often influenced by different data sources, image quality, characteristics of the study area, an analyst's knowledge of the change detection methods and the skill in handling remote sensing data [1]. Depending on the involvement of expert knowledge in the process, CD techniques can be broadly divided into unsupervised and supervised approaches. In terms of speed and degree of automation, unsupervised techniques are advantageous as there is no requirement of supplementary expert knowledge. Hence, the use of effective unsupervised CD is often only a choice for applications in which ground truth/prior knowledge is not available or if the method has to be implemented across larger areas, over which collection of ground truth/prior information is not feasible. Many authors have reported that the use of conventional CD methods suitable for low/middle resolution imagery is not suitable for the High Resolution (HR) imagery [3–5]. Therefore, research is now focused on the development of new methods appropriate for HR imagery, which could incorporate contextual information as part of the process to decrease the spectral variability and ultimately increase the accuracy of the CD process.

In the pixel-based approach every pixel in an image is compared with a corresponding pixel in another image for radiometric differences. Change Vector Analysis (CVA) is one of the most commonly used unsupervised pixel-based CD approach [2,6,7]. A newer approach based on the statistical information theory is also very popular [8]. Unsupervised pixel-based CD approach is often supplemented by accounting neighbourhood information at different stages of the process. [9] used Alternate Sequential Filtering (ASF) By Reconstruction for filtering of the magnitude image generated with CVA whereas [10] incorporated contextual information in the later stage of the process

Earth Observation for Land and Emergency Monitoring, First Edition. Edited by Heiko Balzter.
© 2017 John Wiley & Sons Ltd. Published 2017 by John Wiley & Sons Ltd.

by generating a binary mask by exploiting the statistical correlation of pixel values among neighbouring pixels using Markov Random Field (MRF). The author in Ref. [11] claimed that using MRF is computationally expensive and instead proposed the use of the image transformation based on the Principal Component Analysis (PCA) technique. Instead of applying a pixel-based approach, many researches are shifting towards the using newer approach of object-based solutions [12,13]. One of the key advantages of shifting towards image object specific CD approaches is a remedial of the "salt and pepper" effect [3–5,13–15]. The authors in Ref. [12] applied CVA technique using a single scale segmentation whereas those in Refs [7,13] illustrated that a hierarchical multi-scale CVA gives better CD result than considering a single scale segmentation in isolation.

Regardless of the different approaches used, the common element in the above mentioned approaches is that the change/no change information is solely deduced on the basis of the magnitude image comparing spectral bands assuming that significant changes in reflectance values of changed objects are higher than those caused by other factors. However the authors in Ref. [16] asserted that the assumption is only partially valid in HR images due to the difference in atmospheric conditions as well as the difference in sensor geometry during the acquisition of images. Differences in viewing geometry and scene illumination could significantly alter the spectral properties. For example, roof materials exhibit high variability in spectral response due to shadowing effects and interference from many additional small objects such as satellite antennas, photovoltaic panels, glass roofing tops, paintings and dormers in the case of pitched roofs. It is recognized that the shadow problem together with seasonal variation of the vegetation phenology can cause significant alteration of intensity in the magnitude domain, even though no change has occurred from a semantic point of view. In contrast, spectral shapes are comparatively less affected and essentially follow the same pattern. Here, a new method is proposed that instead of solely depending on difference in intensities of pixels for calculation of the magnitude image, the change in shape of the spectral profile of pixels is also considered for a more reliable CD result. In addition, the proposed approach utilizes an image object based on the optimized multi-date segmentation, to allow for the incorporation of adaptive contextual information, and the use of Alternate Sequential Filter (ASF) By Reconstruction for simplification of the difference image and removal of minor misregistration between the two images.

6.2 Datasets

The data set is made up of two multi-spectral and multi-temporal Quick Bird images acquired over an area of Warsaw, Poland in May 2002 (t_1) and March 2007 (t_2). The Gram–Schmidt spectral sharpening algorithm was used for the pan-sharpening process. The co-registration process involved careful selection of ground control point (GCPs), mainly on easily visible unchanged road segments. Second degree polynomial function and nearest neighbour interpolation is applied for the registration process. A relative radiometric correction method was adopted to normalize the images by subtracting the mean value from each spectral band. The set is comprised of images with 2644×2674 pixels where the area is predominately covered by the complex topography associated with the urban environment, including a majority of high-rise buildings.

The main purpose of the CD, in this case, is to identify anthropogenic changes reflecting the development of this area, for example, the appearance and disappearance of buildings and roads. In order to remove the difference in spatial extent of shadows, areas covered by shadows in both images are detected automatically based on an object specific feature "contrast to neighbours" [17] in eCognition software. Similarly, areas covered by vegetation for both time periods were detected automatically based on Normalized Difference Vegetation Index (NDVI). Later on areas covered by vegetation in both images and areas, which are identified as shadows in at least one time period was masked out from a binary CD map produced from the adopted methodology to obtain the final CD map.

6.3 Methods

The proposed CD approach is a combination of a pixel-based approach and an object-based approach, consisting of three major steps (c.f. Figure 6.1). First, the multi-date segmentation with a small scale is used to generate image objects which are subsequently optimized based on image object fusion. Secondly, an object based feature is extracted combining both spectral difference and spectral shape difference. Thirdly, the final CD map is generated where each pixel is associated with change Ω_c class or no change Ω_n class, using a context sensitive thresholding technique. The first step was

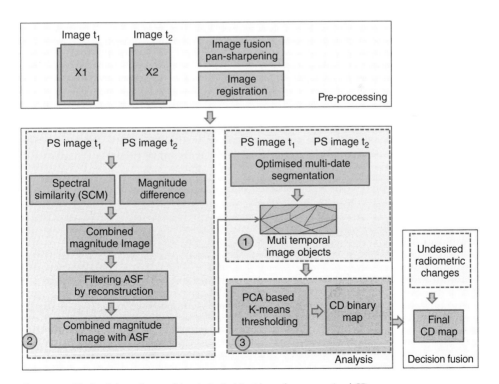

Figure 6.1 Methodology for combined pixel-object based unsupervised CD.

performed with eCognition whereas the second and third steps were performed in MATLAB using in-house developed script.

6.3.1 Optimized Multi-Date Segmentation and Optimization

Image segmentation is the process of partitioning an image into groups of pixels that are spectrally similar and spatially adjacent. A recent categorization of different segmentation techniques from the remote sensing perspective can be found in [18]. There is no simpler way that governs the selection of a suitable segmentation algorithm for a specific domain. Out of many segmentation algorithms, only a few are commercially available in a software package [19]. A comparative analysis of segmentation algorithms by [19,20] asserted the Fractal Net Evolution Approach (FNEA) algorithm in eCognition was comparatively better.

The FNEA segmentation is a region growing approach based on optimization of the heterogeneity of objects such that the average heterogeneity of pixels should be minimized inside an object. A detail description of the algorithm is intentionally absent from this discussion but interested readers are referred to [21,22]. The generation of image objects is performed in one operation from the whole set of temporal spectral bands stacked together (c.f. Figure 6.2). The objects thus generated with multi-date segmentation utilize spatial, spectral and temporal information in combination to produce objects which are spectra-temporally similar in N-dimensional space, where N refers to the number of different spectral bands for the set of bi-temporal images [23]. In the approach adopted, a smaller scale segmentation is used so that objects would represent under-segmentation as oppose to over-segmentation. These are further merged using a spatial aggregation rule, which not only considers spectral properties of neighbouring objects in one image but also considers spectral properties in both bi-temporal images.

In the fusion process, the image object from which fusion process initiates is considered as a "seed object" and all neighbouring image objects of the seed object are potential "candidates objects" for a fusion (c.f. Figure 6.3). After which after each object in the image domain, resulting from the multi-date segmentation, is considered as a seed object in a looping procedure and corresponding neighbouring candidate objects, which are spectrally similar in terms of spectral or textural properties in both of the temporal images, are fused with each other. The problem of how the objects pairs are merged can be handled in number of ways by applying different fitting mode [21]. The similarity of objects can be considered individually for each band of the image as

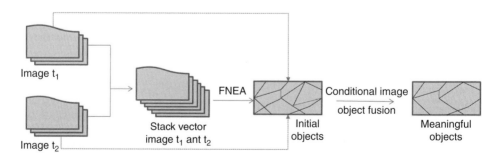

Figure 6.2 Multi-date segmentation and optimization workflow.

(a)

(b)

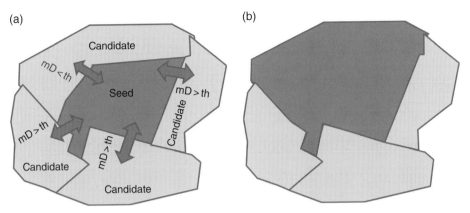

Candidate

$mD_{<th}$

Seed

$mD > th$

$mD > th$

Candidate

Candidate

$mD > th$

Candidate

Candidate

Figure 6.3 Simple illustration of image object fusion (a) objects before fusion (b) objects after fusion.

well as considered in a combined way with multiple conditions. For example, a typical fusion procedure could be to fuse candidate objects with a seed objects which have a large common border with a seed object, where the absolute difference in spectral mean band red or texture is below a specified threshold (th_D) relative to the seed object.

$$mD_{Feature} = \left| M_{Feature}^{Candidates} - M_{Feature}^{Seed} \right|$$ (6.1)

Objects have to satisfy the threshold specified in each feature for fusion when multiple conditions are used. The optimized fusion procedure could be executed multiple times with increasing threshold values or using different features. By using this optimization procedure, a user has more control over the generation of suitable objects, as the user can make decision based on intuition or expert knowledge accumulated from a thorough examination of the spatial or textural characteristic of the data used for the analysis. The selection of suitable features for the fusion procedure can also be directly acquired using the object information provided in eCognition, if necessary. Any object feature, which can be calculated within eCognition, can be used as the basis of the fusion procedure. For example, the user can also control the size of the fusion procedure by applying a restriction fuse objects if the resulting fusion object would be greater than the specified threshold area. The complete optimization procedure is implemented by using a customized algorithm in eCognition that utilize a Graphical User Interface (GUI) which allows the user to specify any number of features they wish, in a graphical way allowing for a more transparent optimization procedure. The customized algorithm utilizes a relatively new algorithm known as image object fusion and concept of Parent Process Object (PPO) [17] available within eCognition.

Two levels of optimization were adopted in the approach. Firstly, initial objects, which have mean absolute difference in spectral value within the specified threshold in multi-spectral bands in both epochs are fused. Secondly, objects that have mean absolute difference in spectral value within a specified threshold in panchromatic bands in both epochs are fused. The threshold can be a small percentage or can be rigorously determined by the visual inspection. After completion of the optimization procedure, objects which have an area less than 10 m^2 are merged with the largest common border surrounding the object in order to remove small objects which give

(a)   (b)

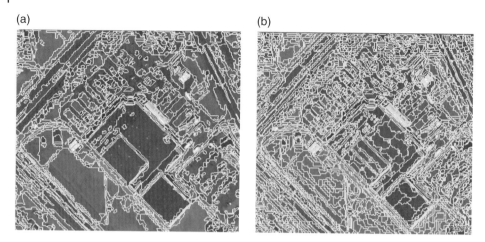

Figure 6.4 Improved knowledge based optimization of objects (a) final object (b) initial object from the multi-date segmentation.

undesired detail, such as satellite antennas, communication receiver, chimney on roofs of the buildings or cars, trucks etc. on the streets.

6.3.2 Object Feature Extraction

The next step is to incorporate spectral information of images with the spatial informa-tion using both a pixel-based calculation and an object-based calculation. Generally, each object generated by segmentation is represented using the mean of the pixels falling within the object boundary [6]. However, representing each region by its vector mean has the drawback that significant spectral information is lost. As the spectral mean of an object can greatly deviate from actual pixel values within the object, it may produce imprecise CD. Therefore, in this approach, the difference vector image or mag-nitude image (X_d) is first calculated pixel-by-pixel using an established CVA concept. In addition, a similarity image (X_s) is computed which represents the similarity score between corresponding pixels to account for similarity between pixel $X^{t_1}(i,j)$ and $X^{t_2}(i,j)$. Afterwards, a filtering technique is used to take pixel-based contextual infor-mation into account, as well as to remove noise from the combined magnitude and similarity image. The resulting image is spatially overlaid over objects generated with the optimized multi-date segmentation, and a combined pixel-object based difference image (X_o) is generated where each object is mapped with the average pixel value falling within the object.

6.3.3 Spectral Similarity Image using Spectral Correlation Mapper (SCM)

Similarity measures have been used extensively in hyper-spectral remote sensing for classification and identification of materials [24]. The use of a similarity score for comparing different spectra has the advantage that it is invariant to small radiometric changes caused by difference in illumination conditions and sensors position. The authors in Ref. [6] used Spectral Angle Mapper (SAM) to represent the direction com-ponent of CVA, which was latter applied to identify the nature of possible changes.

SCM provides better representation of similarity/dissimilarity than SAM as it normalizes the data centralizing itself with respect to the mean. It is also invariant to linear transformation of spectra and therefore suitable for eliminating the effect of shading or illumination difference [25]. SCM calculation of a pixel X between spectra at two different time t_1 and t_2 is given by:

$$SCM = \frac{\sum_{i=1}^{n}\left(X^{t_1}_{i} - \bar{X}^{t_1}\right)\left(X^{t_2}_{i} - \bar{X}^{t_2}\right)}{\sqrt{\sum_{i=1}^{n}\left(X^{t_1}_{i} - \bar{X}^{t_1}\right)^2\left(X^{t_2}_{i} - \bar{X}^{t_2}\right)^2}} \tag{6.2}$$

The output given by SCM is such that $X_s \in [-1,1]$ where -1 denotes that spectra are very different and opposite, 0 denotes they are uncorrelated and 1 denotes they are perfectly identical. Original SCM image resulted using Equation [6.2] is transformed to new SCM image $X_s \in [0,1]$ such that 0 means similarity and 1 means dissimilarity. Subsequently X_d image is normalized within $[0,1]$ to make it compatible with similarity image X_s. Afterwards, image X_d^{norm} and X_s are combined with using Equation(3). The weight of contribution from the image X_s can be controlled with the parameter $C_s \in [0,1]$.

$$X_d^{comb} = \sqrt{\left(1 - C_s\right)X_d^{norm} + C_s X_s} \tag{6.3}$$

6.3.4 Mathematical Morphological Filtering using ASF by Reconstruction

Mathematical Morphology (MM) provides a rich set of powerful tools for incorporating spatial information. In the context of remote sensing MM has been used in image segmentation [26], image filtering [27], image classification [28,29], and building detection [30,31]. MM are best suited for selective extraction or suppression of image structures. The size of the structural element (SE) determines the level of simplification of an image, i.e. it determines the size of the neighbourhood which is used for extracting or suppressing specific structures of the image objects. There are many types of SE such as disk, rectangle, line and circle. Choice of a particular shape and size depends on the specific purpose. Unlike opening (γ) and closing (δ) MM based on reconstruction attempts to restore geometrical properties of the objects that are not completely suppressed by erosion or dilation. By definition, Opening By Reconstruction (OBR) and Closing By Reconstruction (CBR) are useful for removing only a single type of unwanted structure either bright or dark objects respectively. However, when the image to be filtered contains both dark and bright noisy structures, as in the case of the magnitude image generated by CVA, the use of sequential application of OBR-CBR is more appropriate. ASF By Reconstruction [32] is a process of sequential application of OBR and CBR, beginning initially with small SE and proceeding with increasing SE until the given size is reached. Via the sequential increase in the size of the SE, larger regions are processed for reduction of the image complexity in each subsequent step. ASF By Reconstruction has found wide applicability in medical image processing, de-speckling of SAR imagery and in the classification of remote sensing images [9,27,28,32]. For a detailed mathematical description and various application of MM, interested readers are referred to an excellent book [32]. ASF_m that follows the sequence of OBR-CBR was adopted in the current study with increasing disk shaped SE up to size 3.

$$ASF_m\left(f\right)=m_i \ldots m_1\left(f\right), \quad \text{with} \quad m_i = \gamma^i_R \varphi^i_R \tag{6.4}$$

6.3.5 Context Sensitive PCA based K-Means Thresholding

In the combined pixel-object based difference image (X_o), the higher the intensity of a pixel in X_o, the higher is the possibility of that pixel being associated with a change pixel. The authors in Ref. [33] compared many non-fuzzy and fuzzy histogram based thresholding techniques for unsupervised CD and they have argued that when there is no bimodal distribution in the analysed histogram, the result from histogram based techniques may not be accurate. Therefore, a thresholding technique derived from a PCA based K-Means algorithm developed by the author in Ref. [11] was used. The PCA based K-Means algorithm initiates with the extraction of eigenvector space using non-overlapping blocks of the difference image followed by projection of h×h (where h is any odd number e.g. 3×3, 5×5) neighbourhood data of each pixel of the difference image onto the eigenvector space. The vector space thus generated is then separated into two clusters (change/no change) based on the minimum Euclidian distance using the K-Means clustering algorithm. The window size 3 was used for both generating the eigenvector space as well for the projection of the filtered combined image X_d^{comb}. For K-Means clustering, an iteration of 10,000 was used and repeated five times to minimize the effect of the random generation of initial centroids. The final result of K-Means was obtained with the majority vote analysis.

6.4 Results and Discussion

For better understanding and clear visualization of the result, only small part of the images are presented, however for quantitative analysis, the full image extent was used. The advantage of incorporation of additional information with similarity image is apparent by comparing the magnitude image (X_d) and similarity image (X_s) as shown in Figure 6.7. It is to be noted that the difference in value between change/no change areas in the X_d image is subtle. The majority of roads have not changed within the considered time span, however due to the radiometric difference between the two acquisitions they have relatively high values in the X_d image. In contrast, change and no change regions are highly distinguishable and clearly separated in the X_s image. In addition, the X_s image also accentuates residual misregistration even though images were co-registered with utmost care with selection of GCPs.

The X_d^{comb} image was processed using ASF_m to simplify and simultaneously filtering of both dark and bright noisy structures, and to eliminate residual misregistration between images to some extent. The effect of ASF_m is illustrated in Figure 6.6. Small structures, both dark and bright, are removed without affecting other parts of the image.

The result of the CD analysis with the proposed method is shown in Figure 6.7(c). For the purpose of the visual comparison, CD results produced by the standard CVA and

(a)

(b)

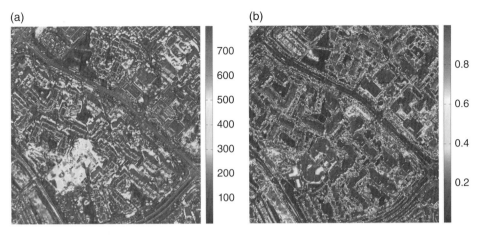

Figure 6.5 Generation of difference images (a) magnitude image (b) similarity image.

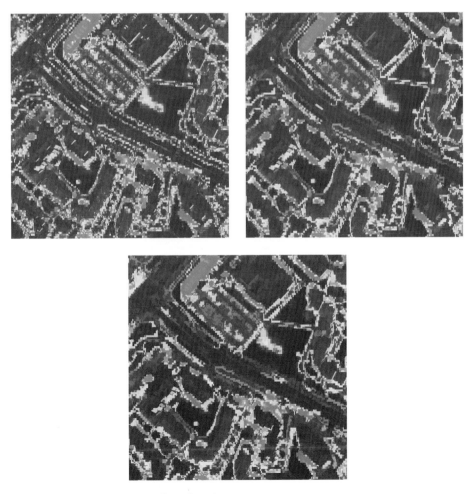

Figure 6.6 Sequential filtering effect of ASF by reconstruction.

(a)   (b)

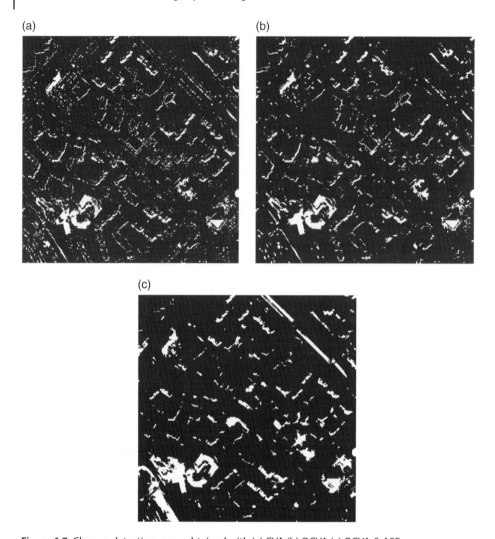

(c)

Figure 6.7 Change-detection maps obtained with (a) CVA (b) OCVA (c) OCVA-S-ASF.

the object based CVA with optimized multi-segmentation (OCVA) are also shown in Figure 6.7(a) and Figure 6.7(b), respectively.

Subjective visual analysis suggests that the result produced by OCVA-S-ASF showed considerable improvement over both other techniques. Standard CVA displayed a salt and pepper appearance, which was reduced to some extent by OCVA. By incorporating additional information, OCVA-S-ASF managed to detect some areas that were undetected by both CVA and OCVA e.g. segments in lower left and upper right (c.f. Figure 6.7). In addition, OCVA-S-ASF resulted in better representation of changed homogeneous areas with respect to the map obtained by CVA. A close up-view of a small portion of the bottom left image is shown for all three techniques in Figure 6.8. For effective comparison, image 2002 and image 2007 are also shown.

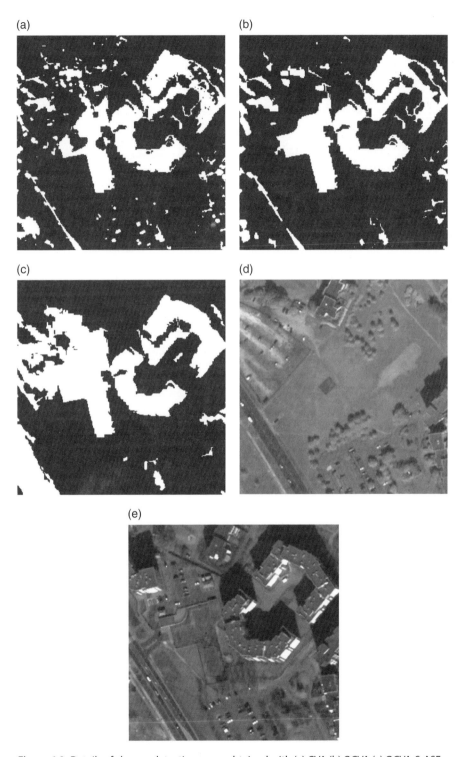

Figure 6.8 Details of change-detection maps obtained with (a) CVA (b) OCVA (c) OCVA-S-ASF (d) image 2002 (e) image 2007.

(a) (b) (c)

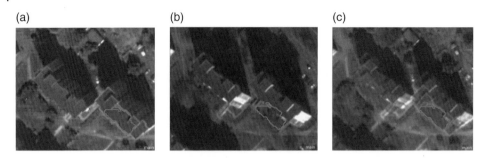

Figure 6.9 Example of the majority of false alarms due to leaning effect of high rise buildings (a) image 2002 (b) image 2007 (c) superimposition of image 2002 and image 2007.

However, due to intrinsic characteristics of the study area there are false alarms (c.f. Figure 6.7) along the periphery of unchanged high-rise buildings (10–18 floors). This is due to considerable shift of building tops due to a leaning effect of buildings (c.f. Figure 6.9) caused by a difference in viewing angles of the sensors. Leaning effect of high-rise buildings is unavoidable in remote sensing images, since removing its effect requires accurate 3D representation of the area (DSM) together with overlapping image areas in order to get information in occluded areas. While that may be feasible for a small area, leaning effect correction over a large areas is time consuming and expensive. Such false alarms are actually changes from radiometric perspective (e.g. building in one instance and vegetation in other instance as shown in Figure 6.9) but from a semantic point of view, there occurred no changes during the time span considered (a building was present in both 2002 and 2007).

The superiority of the method with respect to CVA and OCVA was also verified through a quantitative analysis as presented in Table 6.1 in which the output CD maps were compared to a reference map generated by the visual inspection. The ground truth map thus has 531,704 changed pixels and 6,543,640 unchanged pixels.

As seen in Table 6.1, CVA has the lowest efficiency followed by OCVA. OCVA-S-ASF outperformed both CVA and OCVA which is demonstrated by performance measures such as correct detection, false alarms, missed alarms and total errors. CVA only detected 41.46% of changes correctly whereas false alarms stand at 5.46%. By adopting OCVA, CD improved to higher 44.19% correct detection and reduction of false alarms from 5.46% to 5.11%, however there is no improvement in total errors. With OCVA-S-ASF, correct detection increased significantly from 41.46% with CVA to 84.39% and total errors reduced from 9.86% with CVA to 5.20% only. Nevertheless, reduction in false alarms with OCVA-S-ASF is not that significant in comparison to others techniques.

The result obtained with the proposed method was compared to a single-scale object-based CD approach as described in Bruzzone and Prieto (2000) [10]. A varying segmentation scale was used to generate OCVA of different object sizes without any optimization procedure. For the generation of the magnitude image, the mean values of pixels in each object were used following CVA. No similarity image was incorporated in CD as an additional source of information. Six different scales ranging from 10 to 60 with equal intervals were selected and the results of CD in percentage are shown in Table 6.2.

Table 6.1 Quantitative analysis of change detection errors in pixels and percentage.

Method	Correct detection		False alarms		Missed alarms		Total errors	
	Pixels	%	Pixels	%	Pixels	%	Pixels	%
CVA	220,459	41.46	386,641	5.46	311,245	58.54	697,886	9.86
OCVA	234,973	44.19	361,250	5.11	296,731	55.81	657,981	9.30
OCVA-S-ASF	448,705	84.39	284,608	4.02	82,999	15.61	367,607	5.20

Table 6.2 Results with single level object based change detection in number of pixels.

Scale	Correct detection	False alarms	Missed alarms	Total errors
10	41.69	5.40	58.31	9.78
20	42.66	5.39	57.34	9.70
30	43.45	5.40	56.55	9.65
40	44.28	5.39	55.72	9.57
50	44.67	5.32	55.33	9.48
60	44.89	5.28	55.11	9.42

Since the use of different scales produce objects with different sizes, the CD result using a particular scale is variable. A distinct trend of accuracy increasing with respect to scale of segmentation is clearly visible. The best result, obtained with the highest scale 60, with correct detection of 44.89% and total errors of 9.42%, is similar to one obtained by OCVA. However, compared to the proposed approach, the result is inferior. This illustrates the fact that increasing the scale of segmentation does not have much significance in CD results, but if additional information is incorporated with SCM and ASF reconstruction, the CD results can be significantly improved.

In order to remove false alarms due to leaning effects of high-rise buildings, the CD result obtained from the proposed approach was further improved by using Connected Component (CC) analysis. Using CC analysis, the binary CD was labelled using 4-neighbour connectivity and connected components that have an area less than 200 m^2 are removed from the final CD map. The final CD map covering the full study area is shown in Figure 6.10. However, it is to be noted that if the proposed approach is to be applied in images where high-rise buildings are not dominant, area thresholding with CC analysis may not be required.

6.5 Conclusion

A technique based on a combination of pixel-based approach and an object-based approach has been proposed for unsupervised CVA based CD in HR images. The method is based on multi-date segmentation to generate image objects that are further improved

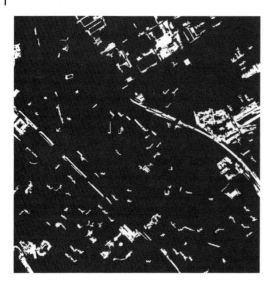

Figure 6.10 The final CD result with CC area thresholding.

by using conditional image object fusion. In addition, the method takes advantage of similarity/dissimilarity between two spectra based on similarity measure SCM to cater radiometric difference due to change in acquisition conditions and the viewing angle of sensors between two epochs. The benefit of using ASF By Reconstruction is noticeable for removing small bright structures and dark structures for simplification of combined magnitude image as well accommodating residual misregistration. The quantitative analysis and qualitative analysis of CD results confirmed the effectiveness of the proposed approach and significantly improved the correct detection of changes as well as reducing overall errors compared to both standard CVA technique and standard object based CVA. The method can be effectively used for monitoring changes from remote sensing images for various applications where change/no change information is sufficient.

References

1   G. Jianyaa, H. Sui, G. Ma and Z. Qiming, A review of multi-temporal remote sensing data change detection. *International Archives of the Photogrammetry, Remote Sensing and Spatial Information Sciences*, **37**, 757–762, (2008).

2   D. Lu, P. Mausel, E. Brondizio and E. Moran, Change detection techniques. *International Journal of Remote Sensing*, **25**, 2365–2401, (2004).

3   T. Blaschke, Towards a framework for change detection based on image objects. In: *Remote Sensing and GIS for Environmental Studies*. Erasmi, S., Cyffka, B. and Kappas, M. (Eds). Goettingen: Goettinger Geographische Abhandlungen, 2005.

4   T. Blaschke, Object based image analysis for remote sensing. *ISPRS Journal of Photogrammetry & Remote Sensing*, **65**, 2–16, (2010).

5   G. Chen, G.J. Hay, L.M.T. Carvalho and A. Wulder, Object-based change detection. *International Journal of Remote Sensing*, **33**, 4434–4457, (2012).

6 F. Bovolo and L. Bruzzone, A theoretical framework for unsupervised change detection based on change vector analysis in the polar domain. *IEEE Transactions on Geoscience and Remote Sensing*, **45**, 218–236, (2007).

7 L. Bruzzone and L. Carlin, A multilevel context-based system for classification of very high spatial resolution images. *IEEE Transactions on Geoscience and Remote Sensing*, **44**, 2587–2600, (2006).

8 O. a. C. Júnior, R.F. Guimarães, A.R. Gillespie, N.C. Silva and R.a.T. Gomes, A new approach to change vector analysis using distance and similarity measures. *Remote Sensing*, **3**, 2473–2493, (2011).

9 M.D. Mura, J.A. Benediktsson, F. Bovolo and L. Bruzzone, An unsupervised technique based on morphological filters for change detection in very high resolution. *IEEE Geoscience and Remote Sensing Letters*, **5**, 433–437, (2008).

10 L. Bruzzone and D.F. Prieto, Automatic analysis of the difference image for unsupervised change detection. *IEEE Transactions on Geoscience and Remote Sensing*, **38**, 1171–1182, (2000).

11 T. Celik, Unsupervised change detection in satellite images using principal component analysis and K-means clustering. *IEEE Geoscience and Remote Sensing Letters*, **6**, 772–776, (2009).

12 L. Bruzzone and D.F. Prieto, An adaptive parcel-based technique for unsupervised change detection. *International Journal of Remote Sensing*, **21**, 817–822, (2000).

13 F. Bovolo, A multilevel parcel-based approach to change detection in very high resolution multitemporal images. *IEEE Geoscience and Remote Sensing Letters*, **6**, 33–37, (2009).

14 T. Blaschke, G.J. Hay, M. Kelly, S. Lang, P. *et al.*, Geographic object-based image analysis – towards a new paradigm. *ISPRS Journal of Photogrammetry and Remote Sensing*, **87**, 180–191, (2014).

15 W. Zhou, A. Troy and M. Grove, Object-based land cover classification and change analysis in the Baltimore metropolitan area using multitemporal high resolution remote sensing data. *Sensors*, **8**, 1613–1636, (2008).

16 S. Marchesi and L. Bruzzone, ICA and kernel ICA for change detection in multispectral remote sensing images. *IEEE International Geoscience and Remote Sensing Symposium*, Cape Town, South Africa, 980–983, 2009.

17 Trimble 2014. *eCognition Developer 9.0, Reference Book*. Munich, Germany: Trimble Germany GmbH, 2014.

18 V. Dey, Y. Zhang and M. Zhong, A review on image segmentation techniques with remote sensing perspective. In: Wagner, W. and Székely, B., (Eds) *ISPRS TC VII Symposium – 100 Years ISPRS*, Vienna, Austria, July 5–7, 2010.

19 G. Meinel and M. Neubert, A comparison of segmentation programs for high resolution remote sensing data. *International Archives of the ISPRS*, **35**, 1097–1105, (2004).

20 P.R. Marpu, M. Neubert, H. Herold and I. Niemeyer, Enhanced evaluation of image segmentation results. *Journal of Spatial Science*, **55**, 55–68, (2010).

21 M. Baatz and A. SchäPe, Published. Multiresolution Segmentation: an optimization approach for high quality multi-scale image segmentation. In: Strobl, J. (Ed.) *Angewandte Geographische Informationsverarbeitung XII*. Beiträge zum AGIT-Symposium Salzburg 2000, Karlsruhe, Herbert Wichmann Verlag, 12–23, 2000.

22 U.C. Benz, P. Hofmann, G. Willhauck, I. Lingenfelder and M. Heynen, Multi-resolution, object-oriented fuzzy analysis of remote sensing data for GIS-ready information. *ISPRS Journal of Photogrammetry & Remote Sensing*, **58**, 239–258, (2004).

23 B. Desclée, P. Bogaert and P. Defourny, Forest change detection by statistical object-based method. *Remote Sensing of Environment*, **102**, 1–11, (2006).

24 Y. Du, C.I. Chang, H. Ren, C.C. Chang, J.O. Jensen and F.M. D'Amico, New hyperspectral discrimination measure for spectral characterization. *Optical Engineering*, **43**, 1777–1786, (2004).

25 A.O. De Carvalho and P.R. Meneses, Spectral Correlation Mapper (SCM): An Improvement on the Spectral Angle Mapper (SAM). In: Green, R.O. (Ed.), *Summaries of the 9th Airborne Earth Science Workshop*, Pasadena, California: JPL Publication, 2000.

26 P. Soille, Constrained connectivity for hierarchical image partitioning and simplification. *IEEE Transactions on Pattern Analysis and Machine Intelligence*, **30**, 1132–1145, (2008).

27 P. Soille, Beyond self-duality in morphological image analysis. *Journal of Visual Communication and Representation*, **23**, 249–257, (2005).

28 M. Volpi, D. Tuia, F. Bovolo, M. Kanevski and L. Bruzzone, Supervised change detection in VHR images using contextual information and support vector machines. *International Journal of Applied Earth Observations and Geoinformation*, **20**, 7785, (2013).

29 J.A. Benediktsson, M. Pesaresi and K. Amason, Classification and feature extraction for remote sensing images from urban areas based on morphological transformations. *IEEE Transactions on Geoscience and Remote Sensing*, **41**, 1940–1949, (2003).

30 X. Huang and L. Zhang, A multidirectional and multiscale morphological index for automatic building extraction from multispectral GeoEye-1 imagery. *Photogrammetric Engineering and Remote Sensing*, **77**, 721–732, (2011).

31 H. Arefi and M. Hahn, *A Morphological Reconstruction Algorithm for Separating Off-Terrain Points from Terrain Points in Laser Scanning Data.* ISPRS WG III/3, III/4, V/3 Workshop "Laser scanning 2005", Enschede, the Netherlands, 2005.

32 P. Soille, *Morphological Image Analysis: Principles and Applications*, Berlin Heidelberg New York, Springer-Verlag, 2004.

33 S. Patra, S. Ghosh and A. Ghosh, Histogram thresholding for unsupervised change detection of remote sensing images. *International Journal of Remote Sensing*, **32**, 37–41, (2011).

7

Earth Observation Land Data Assimilation System (EO-LDAS) Regularization Constraints over Barrax Site

M. Chernetskiy[1], N. Gobron[2], J. Gómez-Dans[3], P. Lewis[3] and C.C. Schmullius[1]

[1] *Friedrich-Schiller-University, Institute of Geography, Department for Earth Observation, Jena, Germany*
[2] *EC Joint Research Centre, Ispra (VA), Italy*
[3] *University College London (UCL), London, UK*

7.1   Background

Vegetation plays a key role in regulating the life of earth's biosphere. Accurate and continuous monitoring of land canopy biophysical parameters such as Leaf Area Index (LAI) is required for understanding the growth of vegetation [1]. These parameters in turn also provide essential input to bio-geochemical cycle modelling, climate modelling, agricultural irrigation management, forecasting of crop production, forest mapping and management and many other fields [2]. Nowadays information for monitoring the state and changes of terrestrial surfaces is provided by a number of optical remote sensing sensors such as Moderate Resolution Imaging Spectroradiometer (MODIS), Satellite Pour l'Observation de la Terre (SPOT), Landsat, etc. The spectral reflectances are used to retrieve different state variable products such as LAI, Chlorophyll concentration (Cab) and radiative fluxes like surface albedo and fraction of photosynthetically active radiation (fAPAR) [3–7].

There are two main methods to obtain these land variables: one is based on empirical parametric relationship with vegetation indices (VI) and another is physically based. The first one uses statistical relationships between biophysical parameters and spectral measurements. Such a method is relatively easy for implementation in the case of comprehensive field measurements. However empirical relationships can be used only with the same configuration of observations, i.e. for a single sensor and type of land cover. The second way is physically based as a canopy radiative transfer model, used through inversion technique. Extraction of vegetation properties is possible only if we can understand how solar radiation interacts with vegetation and environmental factors. This interaction is described by RTM. Bio-geophysical properties of the earth surface can be obtained by inversion of RTM. This approach requires several spectral bidirectional reflectance factor (BRF) values as inputs and/or the knowledge of multi-angular information. The task of an inversion technique is to infer canopy parameters by known spectral BRFs, i.e. input variables of a radiative transfer model adjusted to best explain measured fields of reflectance. Examples of canopy 1-D RT models, among

Earth Observation for Land and Emergency Monitoring, First Edition. Edited by Heiko Balzter.
© 2017 John Wiley & Sons Ltd. Published 2017 by John Wiley & Sons Ltd.

others, are Scattering by Arbitrarily Inclined Leaves (SAIL) [8], semi-discrete model [5], 2-stream based models [9,10] and A two-layer Canopy Reflectance Model (ACRM) [11]. More complex 3-D models describe canopy structure in 3-dimensional space [12,13]. These models are very expensive and have too many input parameters. Usually they are not appropriate for operational inversion of remote sensing derived reflectance.

One problem of RT modelling inversion is that the number of unknown parameters is usually higher than the number of input data, especially for mono-angular sensors [14,15]. This means that a number of free parameters (inputs of a model) has to be smaller than the number of spectral bands of a sensor. This leads to the so-called ill-posed problem, i.e. the number of possible solutions is too high and so there is no a unique solution. In order to solve this problem, various techniques or assumptions can be made such as fixing a few parameters or using a time period of BRFs to increase the number of observations. All these methods characterized by introducing some additional information can be called methods of regularization [16].

The quality of the inversion may be improved by increasing the dimensionality of observations by constraining or in other words regularizing the free variables. Hyperspectral measurements can significantly increase dimensionality, but such kinds of data are not always available. In addition, increasing spectral dimensionality does not always increase the amount of relevant information. One of the ways to improve inversion of variables is to introduce a priori information. Such information may be collected from literature, field measurements, other sensors, or previous experiments [17]. Usually a priori information is used for the fixation of variables or for defining the range of their possible change. Other methods of regularization use additional information from temporal or spatial development of reflectance fields [18,19].

Recently many efforts have been made in order to find possibilities for regularization of the inversion problem as well as for uncertainties estimation [20–22]. Over land surface, Lewis *et al.* (2012) [23] proposed the Earth Observation Land Data Assimilation System (EO-LDAS) which is based on weak constraint data assimilation (DA), meaning that a model is not perfect and has model error, which is regulated by a so-called regularization parameter [24]. The system implements mechanisms for constraining input remote sensing information in order to obtain the optimal estimate of vegetation parameters of the canopy surface. In EO-LDAS, one can constrain the optimal solution by prior information, time or/and spatial regularization.

The main task of the system is the minimization of a cost function J_{post}, which is the sum of the three following cost functions:

$$J_{post} = J_{prior} + J_{obs} + J_{model} \tag{7.1}$$

where J_{prior} provided a priori knowledge of state variables:

$$J_{prior} = -\frac{1}{2}\left(x - x_p\right)^T C_p^{-1}\left(x - x_p\right) \tag{7.2}$$

in which C_p is the covariance matrix which describes uncertainty of the prior state, x is a vector of state variables and x_p the prior estimates.

The function of a prior constraint is to correct the cost function J_{post} by the prior term J_{prior}. This constraint is controlled by the prior state vector x_p and our belief to this state, i.e. the uncertainty of the prior model state C_p.

The second term J_{obs} corresponds to the observations cost function:

$$J_{obs} = -\frac{1}{2}\left(x - H(x)\right)^T C_0^{-1}\left(x - H(x)\right) \tag{7.3}$$

where $H(x)$ corresponds to the radiative transfer model for the scattering of light by vegetation and C_0 the covariance matrix describing the uncertainty in the observations.

Finally the dynamic model cost function J_{model}:

$$J_{model} = -\frac{\gamma^2}{2}x^T\left(D^T D\right)x \tag{7.4}$$

where $\gamma = 1/\delta$ is the regularization parameter which represents the model error and controls the smoothness of output. D is the differential operator in the form of a matrix:

$$D = \begin{bmatrix} 1 & -1 & 0 & \cdots & 0 & 0 \\ 0 & 1 & -1 & \cdots & 0 & 0 \\ \vdots & \ddots & \ddots & \cdots & -1 & 0 \\ 0 & 0 & 0 & \cdots & 1 & -1 \end{bmatrix} \tag{7.5}$$

Time regularization can be applied when time series of remote sensing data are available. It is assumed that the development of biophysical variables can be described by a dynamic model. Time regularization is already proven as a powerful tool for extraction of biophysical information from time series of remote sensing reflectance data [20,25,26] and has been already used in EO-LDAS with Sentinel-2 like data in [23].

Spatial regularization is a relatively new branch in constraining RT canopy model inversion. The main idea of spatial regularization is to use statistics of surrounded pixels in addition to a spectral signature. There are several studies which have demonstrated increased performance of the inversion after applying such spatial constraints [18,19,21,27].

As can be seen, the main purpose of EO-LDAS is the retrieval of biophysical land variables. However, once the state is known after inverting some observations, the system can be used to forward model and predict other observations. In this case availability of hyperspectral measurements such as CHRIS/PROBA data is highly desirable, because it allows validation of a full spectrum in a certain range of wavebands.

Another technique which can be employed in EO-LDAS is using multi-angular information. Conventional remote sensing methods use only one view zenith angle. Meanwhile a multi-angular sensor uses several view angles. This allows better characterization of the structure of vegetation. Knowledge of multi-angular spectral observations has been recognized as essential [28,29]. Since multi-angularity introduces additional information it can be considered as a regularization technique.

The main aim of this contribution is to give an overview of the possibilities offered by the EO-LDAS tool using separately prior information, time regularization and spatial regularization. We also propose the use of multi-angular information in addition to spectral signatures of low spatial/spectral resolution sensors in order to simulate the spectral signatures of higher spatial/spectral resolution sensors. These four exercises will be conducted using space data over the test site Barrax in Spain.

7.2 Methods and Data

7.2.1 Site Description

An agriculture area near Barrax in Spain, was chosen as a test site due to the availability of both field measurements of the European Space Agency (ESA) SPectrabARrax Campaign(SPARC) campaign 2004 (Figure 7.1) [30] and a series of satellite acquisitions (see Table 7.1).

Barrax is a widely studied area, where many field campaigns have taken place over the years 2003, 2004, 2005 and 2009. These campaigns have mostly dealt with the retrieval

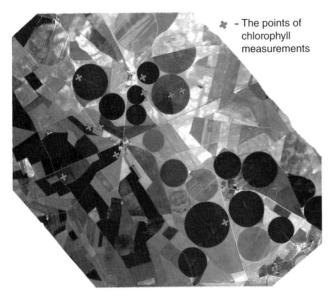

✱ – The points of chlorophyll measurements

Figure 7.1 INTA-AHS 80 Airborne Hyperspectral Scanner (AHS) scene of the Barrax test site for 15.07.2004, 2 m spatial resolution. Measurement points are marked as crosses. Imagery from the SPARC 2004 database.

Table 7.1 Summary of data characteristics.

Sensor	Spatial resolution	Spectral bands	Number of View Zenith Angels (VZA)	Date of acquisition
CHRIS/PROBA	25 m	62 bands from 400 to 1100 nm	5	2004.07.16
ETM+/Landsat	30 m	483, 565, 660, 825, 1650 and 2220 nm	1	2004.07.18
MODIS/Terra	500 m	483, 565, 660, 825, 1240, 1650 and 2220 nm	1	2004.07.16
MISR/Terra	275 m	446, 558, 672 and 867 nm	9	2004.07.16
MERIS/Envisat	300 m	412, 442, 490, 510, 560, 620, 665, 681, 708, 753, 762, 779, 865, 885 and 900 nm	1	2004.07.17

of biophysical parameters, such as LAI, chlorophyll concentration, leaf equivalent water thickness, leaf dry matter and soil spectral measurements. The various works carried out during the campaign mostly related to the retrieval of biophysical parameters by space optical measurements. These latter data have been collected during the campaigns and include two high-resolution hyperspectral sensors CHRIS/PROBA and INTA-AHS.

7.2.2 Remote Sensing Data

The Compact High Resolution Imaging Spectrometer (CHRIS) is an experimental instrument on board the Proba-1 satellite. It can be configured to provide from 19 to 63 spectral bands in range (400–1050 nm) with spatial resolution from 17 to 34 m. CHRIS/ PROBA was constructed for collection of Bidirectional Reflectance Distribution Function (BRDF); therefore it has five view zenith angles from nadir to 55° [31].

One scene of 25 m spatial resolution CHRIS/PROBA was acquired on 16 July 2004, with 62 spectral bands in the range from 400 to 1100 nm at five different viewing zenith angles. The closest nadir geometrical properties are: 8.4°, 283.6°, 20.8°, 325.5° – view zenith angle (VZA), view azimuth angle (VAA), sun zenith angle (SZA) and sun azimuth angle (SAA) respectively.

The Multi-angle Imaging Spectro-Radiometer (MISR) is the operational multi-angular optical sensor, which acquires information globally. We used the MISR top of canopy reflectance at 275 m on 16 July 2004. Only bands with 275 m resolution were used: the four nadir spectral bands and the red bands over eight other cameras.

Now not operational, the MEdium-spectral Resolution Imaging Spectrometer (MERIS) had 15 spectral bands from 390 to 1040 nm and spatial resolution 300 m. We used MERIS data acquired on 17 July 2004.

Additionally we used the surface reflectance from MODIS (MOD09GA) on 16 July 2004 at 500 m spatial resolution in seven spectral bands [32].

The last source of data was acquired on 18 July 2004 by the Landsat ETM+ with spatial resolution of 30 m including six spectral bands. The atmospheric correction was done by the Landsat Ecosystem Disturbance Adaptive Processing System (LEDAPS) software [33]. We used the standard LEDAPS routine. All required input information was taken by LEDAPS from Landsat metadata; ancillary data of (Total Ozone Mapping Spectrometer) TOMS and National Centres for Environmental Prediction (NCEP) Reanalysis.

7.2.3 EO-LDAS

The EO-LDAS [23] is based on weak constraint data assimilation, including a radiative transfer canopy model [5], a leaf spectra model [34] and a spectral soil model [35].

In the following section, we will use prior information (see Table 7.2) [36]. Some values are given in logarithmic scale because EO-LDAS linearizes these parameters by logarithmic transformation.

The only fixed parameter of the canopy model is the leaf angle distribution (LAD) as we will use EO-LDAS over crops fields: it is either fixed as planophile for alfalfa, potato, sugar beet, sunflower and vineyard or to erectophile for corn, garlic and onion.

Reference data are taken from the SPARC field measurements: they represent different crop species and among them both Leaf Area Index and chlorophyll a + b content

Table 7.2 A priori knowledge of state variables in EO-LDAS.

Parameter	Lower limit (non log)	Upper limit (non log)	Prior (non log)
LAI, the single sided leaf area per unit ground area, xlai (log. transform. $e^{(-xlai/2)}$)	0.01(9.2)	0.99 (0.02)	0.05 (5.99)
The canopy height, m, xhc	1.0	5	0.1
The leaf radius/dimension, m, rpl	0.001	0.1	0.01
The concentration of chlorophyll a + b, µg/cm², xkab (log. transform. $e^{(-xkab/100)}$)	0.1 (230.26)	0.99 (1.0)	0.1 (230.26)
The proportion of senescent material, scen	0	1	0.001
Equivalent leaf water, cm, xkw (log. transform. $e^{(-xkw*50)}$)	0.01 (0.092)	0.99 (0.0002)	0.99 (0.0002)
Dry matter, µg/cm², xkm (log. transform. $e^{(-xkm*100)}$)	0.3 (0.012)	0.99 (0.0001)	0.35 (0.01)
The number of leaf layers, xleafn	1.0	2.5	1.5
Soil PC1 (soil brightness), xs1	0.05	0.4	0.05
Soil PC2 (soil wetness), xs2	−0.1	0.1	0.005
Leaf angle distribution, lad	n/a	n/a	fixed

have been measured. The ground-based data is available for 18 fields of different crop types. These fields were masked manually. This allowed excluding mixed pixels at the field borders.

7.2.4 Observational Operator

Observational operator in EO-LDAS is the combination of the 1-D semi-discrete model for the scattering of light by vegetation; a model of leaf optical properties spectra (PROSPECT) and a spectral soil model. The advantage of the semi-discrete model is that it takes into account the true size of the leaves, which allows the "hot spot" effect of parameterization. This effect has quite a big influence on reflectance measured by a satellite sensor. Conventional 1-D models based only on turbid medium assumption are not able to directly take into account this effect. In the semi-discrete model, the two first orders of scattering are calculated by taking into account the size of leaves in three-dimensional space. Multiple scattering is calculated using a turbid medium assumption. In the semi-discrete model the bidirectional reflectance factor (BRF) is given by Equation [7.6]:

$$\rho(\Omega_0,\Omega) = \rho^0(z_0,\Omega_0,\Omega) + \rho^1(z_1,\Omega_0,\Omega) + \rho^M(z_0,\mu_0,\mu) \tag{7.6}$$

where ρ^0 is uncollided term; ρ^1 is first order of scattering; ρ^M is multiple scattering.

PROSPECT model is based on the idea of representing a leaf as a number of absorbing plates which have rough surfaces [34]. The model has two groups of parameters, which are leaf structure parameter and several parameters representing leaf biochemical content.

According to Ref. [35] soil reflectance can be obtained by using the following equation:

$$x = \sum_{i=1}^{K} S_i \varphi_i \qquad (7.7)$$

Where x is reflectance; K is the number of terms – two for this study; S is soil principal components; φ is the Price Empirical Orthogonal Functions (EOF).

In this study EOF are represented by: φ_1 – mean wet soil spectrum, $\varphi2$ – the mean difference between the mean dry and mean wet spectra [37].

7.3 Results

7.3.1 A Prior Constraint in EO-LDAS

The goal of this section is to study the performance of the EO-LDAS *prior constraint* for the retrieval of state variables values. The performance is estimated by comparing the case without and with prior knowledge against ground-based measurements.

In this exercise, the number of CHRIS/PROBA bands is reduced to 17 because of high correlation between hyper-spectral bands [38]; therefore the number of parameters is lower than the number of inputs value.

EO-LDAS without and with prior information is applied over 18 sites where ground-based measurements are available. Inversion was done independently for the three CHRIS/Proba cameras with the fly-by zenith angles 0, +36 and +55 degrees using 17 spectral values. The only fixed canopy parameter is the leaf angle distribution (LAD) which depends on the crop type. Figures 7.2 and 7.3 show the results using logarithmic scale respectively for the chlorophyll content, $e^{-\text{Chl}/100}$ and LAI, $e^{-\text{LAI}/2}$. Top and bottom panels show the results without prior and with prior by comparing retrieval values with in-situ ones (y-axis), respectively. Dotted symbols show the finding values and the error bar the uncertainties associated to the retrieval. These uncertainties are expressed as the standard deviation (σ).

In the case when EO-LDAS without a priori knowledge runs, the Pearson correlation coefficient r^2 ranges from [0.51–0.74] and [0.61–0.72] for LAI and Cab, respectively. The estimated probability of rejecting or accepting the null hypothesis (two sided p value) for all considered cases is less than 0.05. This means that chance of coincidence is low, the correlations are statistically significant and we can use obtained r^2 values. Root mean square error (RMSE) ranges from 15% to 25%. The change of RMSE between data with prior information and data without prior information is small and can be considered as insignificant. However, the uncertainties associated with the results without prior are very high (top panels of Figures 7.2 and 7.3). This means that the probability density functions (PDF) of estimated biophysical parameters having such a broad range of values have to be used with caution.

When prior constraint are used, the range of r^2 values increases to [0.7–0.8] and to [0.6–0.9] for LAI (bottom panel of Figure 7.2) for Cab (bottom panel of Figure 7.3), respectively. It can be seen that additional information content in the form of prior constraint (Table 7.1) is the reason of decreasing of uncertainties by 90–100%. Prior uncertainty for LAI and Cab in logarithmic scale was equal to 1: the uncertainties for

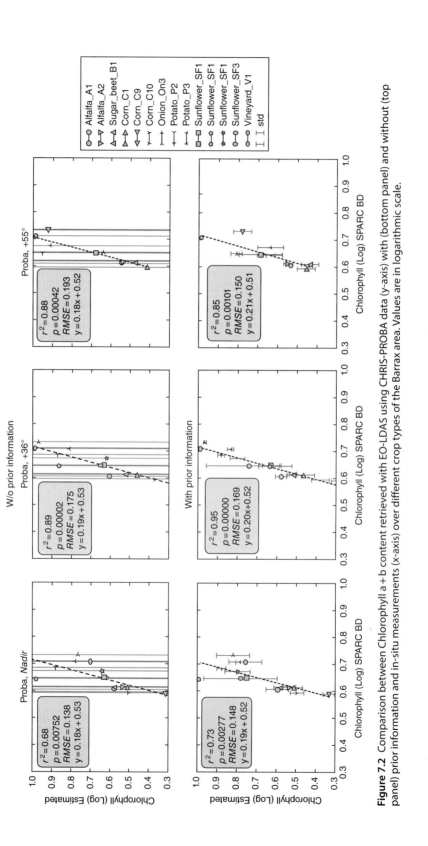

Figure 7.2 Comparison between Chlorophyll a + b content retrieved with EO-LDAS using CHRIS-PROBA data (y-axis) with (bottom panel) and without (top panel) prior information and in-situ measurements (x-axis) over different crop types of the Barrax area. Values are in logarithmic scale.

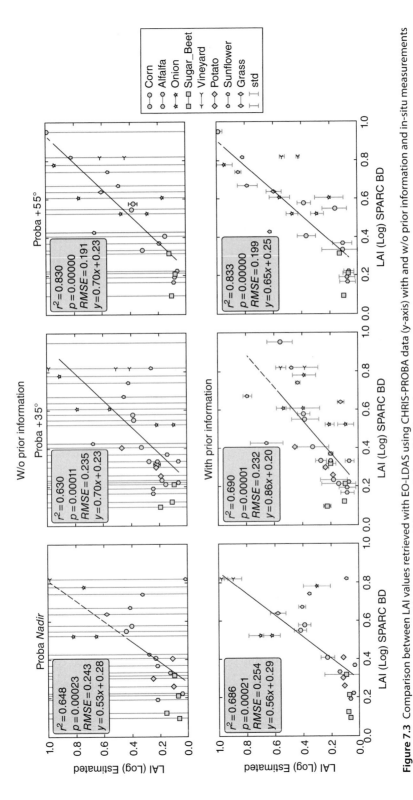

Figure 7.3 Comparison between LAI values retrieved with EO-LDAS using CHRIS-PROBA data (y-axis) with and w/o prior information and in-situ measurements (x-axis) over different crop types of the Barrax area. Values are in logarithmic scale.

retrieved LAI and Cab are less than 0.1. At the same time, the correlation values increase: this can be explained by a better convergence of the cost function. The only exception is the chlorophyll content for camera +55°. However the difference in 0.01 can be considered as an error during optimization process.

7.3.2 Time Regularization with MISR Data

The following exercise explores the time regularization using one year of MISR surface reflectance over the Alfalfa A1 site (39°4′55″N, 2°8′8″W), one of the Barrax fields. The nadir camera (An) of four spectral bands was used. The run was performed for retrieving 10 model parameters (LAD value was fixed to be planophile).

Time regularization imposes temporal smoothness constraint on the surface reflectance data. This constraint is controlled by the regularization parameter γ. However finding optimal γ value can be a challenging and computationally expensive task. One of the possible solutions for obtaining the best γ is cross-validation by removing part of observations and used them for verification of restored temporal dynamic. In this exercise every second observation was removed from 19 observations over 2004.

Figure 7.4 shows a set of solutions for different values of γ.

One can see that when the value of γ is bigger or equal to 200, the dynamic curve of the solution starts to look more stabilized with uncertainties decreased by more than 100% and at $\gamma = 14000$ becomes a flat line with uncertainties close to zero. In order to find the best γ, each set of retrieved parameters values were used in a forward model run to simulate the four MISR bands at the nadir view. The difference between observed surface reflectance data (not used in the optimization) and the simulated one is used as a verification tool. The minimum difference between retrieved reflectance and actual MISR reflectance corresponds to $\gamma = 900$ (Figures 7.4 and 7.5). An important note is that EO-LDAS dynamical "model" doesn't have any information about changing of LAI with time. However with increasing of γ, the LAI curve starts looking like LAI measurements, which include the start of the growing season and the end of the growing season. It demonstrates the ability of EO-LDAS time regularization to constrain the solution.

7.3.3 Spatial Regularization with CHRIS/PROBA Data

Regularization with CHRIS/Proba Only

The spatial constraint is applied with only four CHRIS/Proba spectral bands at 452, 553, 683 and 890 nm. We fix the LAD at planophile or erectophile according to crop type in order to retrieve 10 state variables.

As an example we use the data around the cornfield C9 site (39°4′53″N, 2°7′18″) results of retrieved LAI and chlorophyll content (Cab) values with no space constraint and without prior knowledge are displayed at the top panel of Figure 7.6 and Figure 7.7, respectively. Both retrieved and associated are reported. The ground-based measurement values of LAI are in the range of 2.92 to 3.1 whereas Cab is about $52.94\,mg/cm^2$. The estimations of LAI are from 3 to 4.5 and Cab in the range of $60-78\,mg/cm^2$. The uncertainties are displayed in the right column of the top panel of Figures 7.6 and 7.7: These values are quite high for the whole area of the field and correspond to 300% and 400% on average.

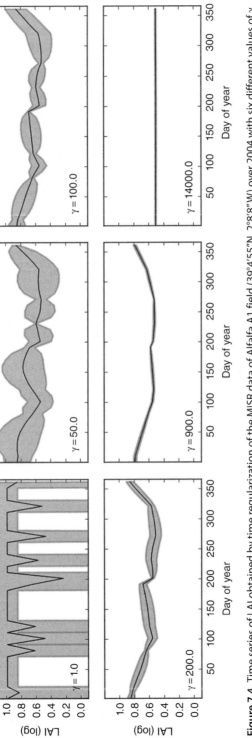

Figure 7.4 Time series of LAI obtained by time regularization of the MISR data of Alfalfa A1 field (39°4′55″N, 2°8′8″W) over 2004 with six different values of γ.

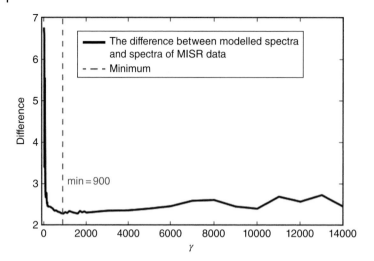

Figure 7.5 Estimation of optimal γ. The difference between time series modelled spectra and spectra of MISR data, which were taken out for cross-validation.

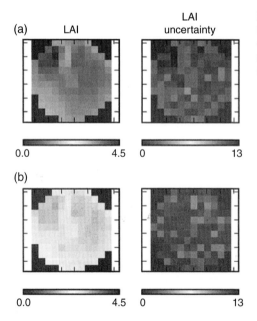

Figure 7.6 Estimation of LAI, CHRIS/PROBA (using 4 bands) over the Corn Field C9 (39°4′53″N, 2°7′18″W). (a) – without priors and spatial regularization (J = Jobs), (b) – with priors and spatial regularization (J = $J_{obs} + J_{prior} + J_{model}$).

When we apply the space constraint we obtain values of LAI in the range of 2.5 to 3.5, and Cab at about 50 to 60 mg/cm² (see bottom panels of the Figures 7.6 and 7.7). It means that the spatial regularization and the prior information provide better solutions, close to the range of the ground-based values. In addition, the uncertainties values decrease significantly to 50% and 30% as illustrated in the bottom panels of Figures 7.6 and 7.7.

Figure 7.8 shows that after using EO-LDAS retrieval with prior information and spatial regularization, uncertainties are reduced for all the state parameters from about 60% to 100%.

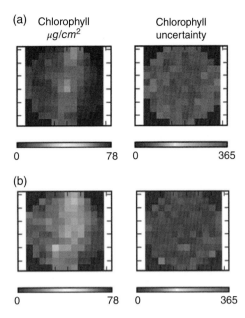

Figure 7.7 Estimation of Chlorophyll a+b content, CHRIS/PROBA (4 bands). Corn, Field C9. (a) – without priors and spatial regularization ($J=J_{obs}$), (b) – with priors and spatial regularization ($J=J_{obs}+J_{prior}+J_{model}$).

(a) Chlorophyll $\mu g/cm^2$

Chlorophyll uncertainty

(b)

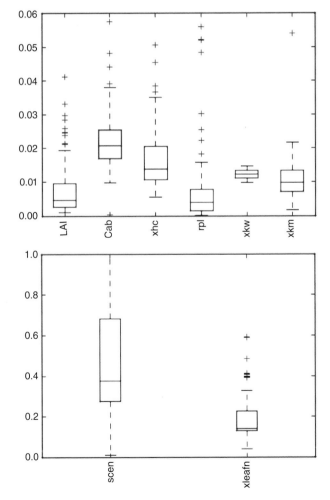

Figure 7.8 Reduction of the uncertainties for eight retrieved canopy parameters: standard deviation with prior and model divided by without prior and model. Parameters: LAI, Cab, canopy height (xhc), leaf radius (rpl), senescent material (scen), leaf water (xkw), dry matter (xkm) and number of leaf layers (xleafn).

Spatial Regularization with CHRIS/Proba and MERIS Data

To test the EO-LDAS efficiency, an additional exercise is run with two optical sensors with different spatial and spectral resolution separately and together. We first use CHRIS/Proba data with only two spectral bands, i.e. the red and near-infrared ones and secondly 15 bands of MERIS/Envisat data. This exercise is performed over the sunflower field (39°4'51"N, 2°6'50"W) with true values of LAI in the range 0.4–0.8 and true values of Cab in the range 43–44 mg/cm^2. The only fixed parameter is LAD, which is set to erectophile. The priors are the same as in the previous example. On the one hand, MERIS does not have enough spatial information for spatial constraining because of the relatively small Barrax field, i.e. size of a MERIS pixel is 300 m, which is comparable to the size of a Barrax field (~300 m). It means no spatial regularization can be made in this case. As a result MERIS-based solution has an overestimation of LAI (2.8–2.9) and underestimation of Cab (5–6 mg/cm^2) (see middle panels of Figures 7.9 and 7.10).

On the other hand, two bands of the high-resolution sensor cannot have enough information for the inversion once 10 model parameters have to be retrieved, despite the use of spatial regularization (upper panels of Figures 7.9 and 7.10). The results show an overestimation of LAI values (0.9–1.22) and an underestimation of Cab ones (9–14 mg/cm^2). In both cases, values of chlorophyll content are quite far away from the ground-based estimates (upper and middle panels of Figure 7.10). However, after solving the problem by following Equation 7.8, i.e. with datasets of the two sensors together,

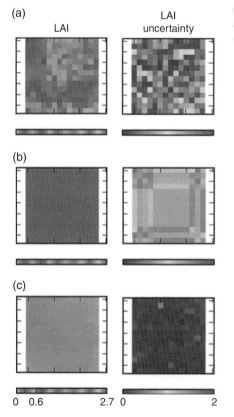

(a)

LAI

LAI uncertainty

(b)

(c)

0   0.6   2.7 0   2

Figure 7.9 Estimation of LAI. Sunflower, Field SF1 (39°4'51"N, 2°6'50"W). (a) – CHRIS/Proba, (b) – MERIS, (c) – CHRIS/Proba + MERIS.

Figure 7.10 Estimation of Chlorophyll content. Sunflower, Field SF1. (a) – CHRIS/Proba, (b) – MERIS, (c) – CHRIS/Proba + MERIS.

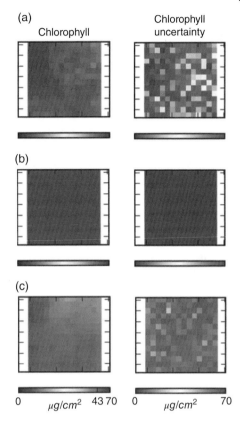

values of chlorophyll content are shifted to the range of the estimate values (Bottom panel of Figure 7.10).

$$J_{\text{posterior}}(x) = J_{\text{space\_model}}(x) + J_{\text{obs\_lores}}(x) + J_{\text{obs\_hires}}(x) + J_{\text{prior}}(x) \qquad (7.8)$$

Thus, using both sensors separately does not retrieve good ground-based estimate values, but when combining them together, the optimal solution is found.

7.3.4 Combination of Low and Medium Resolution Sensors

In this section spectral signatures obtained by EO-LDAS using low and medium resolution sensors, such as ETM+/Landsat, MODIS and MISR, are compared with actual CHRIS/PROBA spectral signatures. Simulated reflectance is calculated for the same geometry of the "nadir PROBA" image for 16.07.2004.

Combination of high and low spatial resolution imagery allows the use of high spatial details of a fine resolution sensor and high temporal frequency of a low-resolution sensor. In addition low-resolution surface reflectance inputs increase information content and may improve quality of filling the time/space gaps in the terms of accuracy and uncertainty.

In the case of Barrax coarse resolution data, such as those of MODIS and MISR, the signal corresponds to mixed vegetation with bare soil. It means that the spectral profile of coarse resolution pixels could mainly correspond to the soil component. The diameter of Barrax fields, i.e. 300 m on average, is comparable to one MODIS or MISR spatial resolution pixel. So if a ground-based point is measured at the border of a field, the spectral profile of a coarse resolution pixel, which corresponds to this ground-based point, can contain mainly soil component.

Considering relative spectral homogeneity of the Barrax fields, a procedure is implemented when higher resolution pixel spectral measurements are compared with the corresponding coarse resolution ones which have the same geographical coordinates and its eight neighbouring pixels. We consider that a coarse resolution pixel fits a corresponding ground-based measurement point if the difference between its spectral bands values and a higher resolution pixel one is minimal. Here we ignore spectral differences of corresponding bands of ETM+, MODIS and MISR. So this procedure is looking for a best match between high-resolution pixel and 9 low resolution pixels. This procedure helps to combine high and low resolution data. In addition, it can correct some errors of geo-referencing.

EO-LDAS takes into account all available space data and its uncertainties: Therefore, the assumption of this exercise is that higher resolution data is more trusted than coarse resolution. Due to this assumption, standard deviation (SD) for all ETM+ bands was set to 0.01. SD for all MODIS and MISR bands was set to 0.02. Therefore, Landsat data are trusted more.

ETM+ and MODIS

Figure 7.11 illustrates the modelled spectra over Alfalfa A1 (39°5′83″N, 2°8′15″W), after applying EO-LDAS using coarse resolution sensors together with CHRIS/PROBA ones. The input bands of multispectral sensor are over-plotted with symbols.

The left panel of Figure 7.11 corresponds to the solution, which was found when only ETM+ data are used. This solution is depicted as a solid grey spectral curve with corresponding uncertainties. The CHRIS/PROBA spectrum is shown as a solid black line. Triangle and dotted symbols correspond to MODIS and ETM+ bands data, respectively. The middle panel shows the result when only MODIS is used in EO-LDAS. The right panel illustrates the solution when ETM+ and MODIS data are combined. The main difference between ETM+ and MODIS solutions (left and middle panels) is that the latter has lower values in the NIR region, which can be explained by lower resolution of MODIS sensor (500 m), which is the reason for a mixture of signals by vegetation and bare soil. Usually reflectance in the NIR region is greater for vegetation than for inorganic materials. The best agreement between modelled spectra with the reference one is found in the latter case when both MODIS and ETM+ are used. We can observe that the red-edge region (680–730 nm) was well retrieved even if the multi-spectral data have no band in this spectral region. In this example, we can see that ETM+ data pull the solution to higher values in the NIR band whereas MODIS data provide better correspondence in red-edge region.

ETM+ and MISR

The MISR data were used in the EO-LDAS by increasing number of the view zenith angles (VZA) from one to nine. The first VZA was the nadir and then we increase the number of VZA one by one from angle at nadir to angles up to 70° (i.e. D cameras). Figure 7.12 shows that MISR-alone solution is similar to MODIS-alone solution (middle panel of Figure 7.11). However it provides slightly better fit in the NIR region because MISR has better spatial resolution (275 m versus 500 m). The results of the solution based on the combination of MISR and ETM+ show that with the increase of number of MISR cameras impact strongly on the results (Figure 7.13). From one to

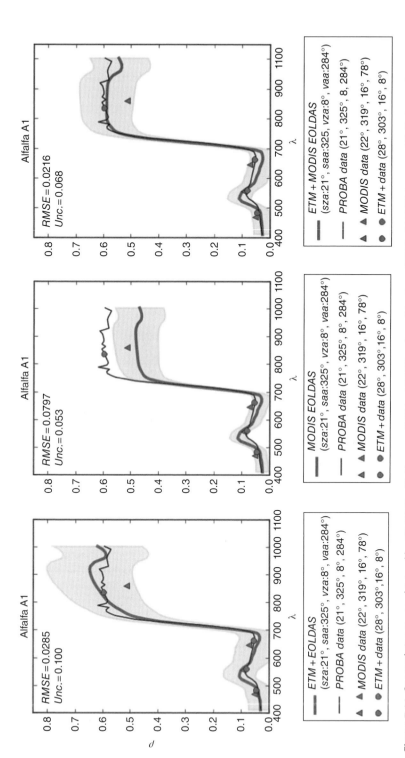

Figure 7.11 Spectral signatures obtained by EO-LDAS with the use of ETM+, MODIS and combination of ETM+/MODIS data over Alfalfa A1 site (39°5′83″N, 2°8′15″W). CHRIS/PROBA view/sun geometry: (sza: 21°, saa: 325°, vza: 8°, vaa: 284°); ETM+: (28°, 303°, 16°, 78°); MODIS: (22°, 319°, 16°, 78°); EO-LDAS reported solutions have the same geometry as CHRIS/PROBA. RMSE values which estimate agreement between Proba and modelled spectra are provided. Unc. values are mean uncertainty of modelled spectra over all bands.

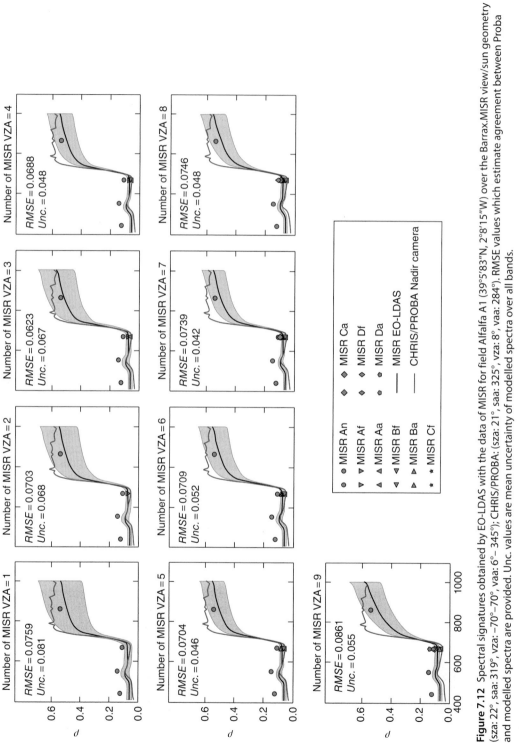

Figure 7.12 Spectral signatures obtained by EO-LDAS with the data of MISR for field Alfalfa A1 (39°5′83″N, 2°8′15″W) over the Barrax.MISR view/sun geometry (sza: 22°, saa: 319°, vza: −70°–70°, vaa: 6°–345°); CHRIS/PROBA: (sza: 21°, saa: 325°, vza: 8°, vaa: 284°). RMSE values which estimate agreement between Proba and modelled spectra are provided. Unc. values are mean uncertainty of modelled spectra over all bands.

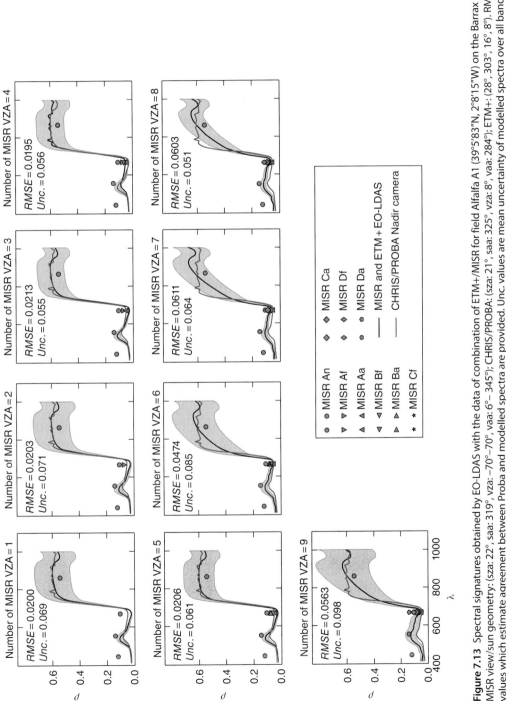

Figure 7.13 Spectral signatures obtained by EO-LDAS with the data of combination of ETM+/MISR for field Alfalfa A1 (39°5′83″N, 2°8′15″W) on the Barrax test site. MISR view/sun geometry: (sza: 22°, saa: 319°, vza: −70°–70°, vaa: 6°–345°); CHRIS/PROBA: (sza: 21°, saa: 325°, vza: 8°, vaa: 284°); ETM+:(28°, 303°, 16°, 8°). RMSE values which estimate agreement between Proba and modelled spectra are provided. Unc. values are mean uncertainty of modelled spectra over all bands.

five MISR cameras help to improve the results, providing even better fit in the terms of RMSE and lower uncertainties (Figure 7.13) than MODIS and ETM+ solution (right panel of Figure 7.11). The reason is that additional cameras give more information in the red region. However increasing the number of view angles from six to nine has a rather negative effect, i.e. MISR starts to "pull" solution from ETM+. This is because by increasing the amount of MISR view angles, we increase the number of MISR-conditioned constraints. These constraints contain additional information about surface, which supplements ETM+ information, but it is still mixed low-resolution pixel.

7.3.5 Discussion

This contribution overviewed the EO-LDAS functionalities over the Barrax agriculture area in Spain. Various constraints were experimented upon: prior information on the canopy parameters, time regularization, spatial regularization and multi-sensor information.

The three first exercises were used for retrieving the biophysical information in terms of state variables such as LAI and Chlorophyll content. The last experiment was performed to simulate hyper-spectral signatures over the canopy surface.

The results of using prior information show that EO-LDAS is able to estimate leaf area index and chlorophyll concentration on a pixel-by-pixel basis: this helps to decrease the number of possible solutions and significantly decrease the output uncertainty and increase correlation between estimated and ground truth data of LAI and chlorophyll. Correlation ranges from 0.63 to 0.95 for all cases. Average increase of the correlation values when prior information is used is 0.05. RMSE varies from 0.138 to 0.254 without significant change between solutions with prior information and without prior information. However, the uncertainties decrease from 90% to 99% for all considered cases.

Feasibility of the EO-LDAS time regularization was already proven using synthetic time series of Sentinel-2 and actual MODIS time-series [39]. In this work we used four spectral bands of MISR at 275 m and the LAI estimation demonstrates a reduction of uncertainties up to 100%, which means a decrease from mean value 42 to 0.015.

The spatial regularization was explored over small homogeneous fields. However, for the purpose of validation we had to assume that each individual field was homogeneous and one–two ground measurements represent a whole field. In this exercise, we did not take into account the edge problem, but in the case of data that are more realistic, we managed it using clustering, before estimation of the parameters, as proposed in Ref. [21]. Another complexity was the estimation of the regularization parameter γ. We solved this problem by using cross-validation as was proposed in Ref. [23]. Despite the small size of the studied area, the spatial regularization was shown to be efficient by decreasing the uncertainties of retrieved estimates by 70–96%.

The goal of the last section was to demonstrate that we could use the EO-LDAS to simulate hyper-spectral signatures of vegetation. We used Landsat/ETM+, Terra/MODIS and Terra/MISR data. It was shown that combining spectral information of MISR or MODIS to ETM+ can improve results by decreasing the difference with CHRIS/PROBA spectral signature.

7.4   Conclusions

The results of this study show that there is a lot of further work to be done for EO-LDAS applications. One issue is the knowledge of input data uncertainties, especially at the top of the canopy level. There is a requirement to investigate how to propagate uncertainties through the pre-processing chain: i.e. calibration, georeferencing, atmospheric correction, dependency on spatial resolution, etc. This will provide a proper balance between the cost function terms: models, radiometric information and prior information and can significantly improve results, i.e. it can improve the balance of cost function terms and this balance strongly depends on the input uncertainties definition. Further improvements in spatial regularization are related to investigation of clustering schemes, which should help in defining proper homogeneous regions. Also it is essential to make new efforts in understanding how multi-angular information can improve the results of EO-LDAS solutions. Synergy of optical RT inversion with Synthetic Aperture Radar (SAR) information on the basis of physical modelling and data assimilation can provide a lot of new possibilities in whole land remote sensing. Another interesting possibility is the coupling of atmosphere and canopy RT models together in EO-LDAS.

Acknowledgements

We gratefully acknowledge financial support of this project through GIONET, funded by the European Commission, Marie Curie Programme Initial Training Network, Grant Agreement number PITN-GA-2010-264509. We also acknowledge the European Space Agency for providing data access by the "EO-Support" system (projects ID 13803 and 13931) and Olivier Morgan (JRC) for his technical support.

References

1 A. Noormets, *Phenology of Ecosystem Processes: Applications in Global Change Research*. Springer, (2009).

2 B. Pinty *et al.*, Retrieving surface parameters for climate models from Moderate Resolution Imaging Spectroradiometer (MODIS)-Multiangle Imaging Spectroradiometer (MISR) albedo products, *J. Geophys. Res.*, **112**, no. D10, p. D10116, (2007).

3 R.B. Myneni *et al.*, Global products of vegetation leaf area and fraction absorbed PAR from year one of MODIS data, *Remote Sens. Environ.*, **83**, no. 1–2, 214–231, (2002).

4 J. Dash *et al.*, The use of MERIS Terrestrial Chlorophyll Index to study spatio-temporal variation in vegetation phenology over India, *Remote Sens. Environ.*, **114**, no. 7, 1388–1402, (2010).

5 N. Gobron *et al.*, A semidiscrete model for the scattering of light by vegetation, *J. Geophys. Res.*, **102**, 9431–9446, (1997).

6 N. Gobron *et al.*, Theoretical limits to the estimation of the leaf area index on the basis of visible and near-infrared remote sensing data, *IEEE Trans. Geosci. Remote Sens.*, **35**, no. 6, 1438–1445, (1997).

7 N.J.J. Bunnik, *The multispectral reflectance of shortwave radiation by agricultural crops in relation with their morphological and optical properties*, Wageningen, Netherlands, (1978).

8 W. Verhoef, Light scattering by leaf layers with application to canopy reflectance modeling: The SAIL model, *Remote Sens. Environ.*, **16**, 125–141, (1984).

9 B. Pinty *et al.*, Simplifying the Interaction of Land Surfaces with Radiation for Relating Remote Sensing Products to Climate Models, *J. Geophys. Res. – Atmos.*, **111**, no. 2, (2006).

10 P.J. Sellers, Canopy reflectance, photosynthesis and transpiration, *Int. J. Remote Sens.*, **6**, no. 8, 1335–1372, (1985).

11 A. Kuusk, A two-layer canopy reflectance model, *J. Quant. Spectrosc. Radiat. Transf.*, **71**, 1–9, (2001).

12 W. Ni *et al.*, An analytical hybrid GORT model for bidirectional reflectance over discontinuous plant canopies, *IEEE Trans. Geosci. Remote Sens.*, **37**, 987–999, (1999).

13 Y.M. Govaerts, *A model of light scattering three-dimensional plant canopies: a Monte Carlo ray tracing approach*, Ispra, Italy, 1996.

14 R. Myneni and J. Ross, *Photon-Vegetation Interactions: Applications in Optical Remote Sensing and Plant Ecology*. Springer Verlag, New York, (1991).

15 D.S. Kimes *et al.*, Inversion methods for physically based models, *Remote Sens. Rev.*, **18**, pp. 381–439, (2000).

16 A.N. Tikhonov and V.Y. Arsenin, *Solutions of Ill-Posed Problems*. New York: Winston, (1977).

17 W. Dorigo *et al.*, Retrieving canopy variables by radiative transfer model inversion an automated regional approach for imaging spectrometer data, in *5th EARSeL Workshop on Imaging Spectroscopy*, (2007).

18 C. Atzberger and K. Richter, Spatially constrained inversion of radiative transfer models for improved LAI mapping from future Sentinel-2 imagery, *Remote Sens. Environ.*, **120**, 208–218, (2012).

19 Y. Wang *et al.*, Regularizing kernel-based BRDF model inversion method for ill-posed land surface parameter retrieval using smoothness constraint, *J. Geophys. Res.*, **113**, no. D13, p. D13101, (2008).

20 T. Quaife and P. Lewis, Temporal Constraints on Linear BRDF Model Parameters, *IEEE Trans. Geosci. Remote Sens.*, **48**, no. 5, 2445–2450, (2010).

21 V.C.E. Laurent *et al.*, A Bayesian object-based approach for estimating vegetation biophysical and biochemical variables from APEX at-sensor radiance data, *Remote Sens. Environ.*, **139**, 6–17, (2013).

22 O. Dubovik *et al.*, Statistically optimized inversion algorithm for enhanced retrieval of aerosol properties from spectral multi-angle polarimetric satellite observations, *Atmos. Meas. Tech.*, **4**, no. 5, 975–1018, (2011).

23 P. Lewis *et al.*, An Earth Observation Land Data Assimilation System (EO-LDAS), *Remote Sens. Environ.*, **120**, 219–235, (2012).

24 D. Zupanski, A general weak constraint applicable to operational 4DVAR data assimilation systems, *Mon. Weather Rev.*, **125**, 2274–2292, (1997).

25 R. Houborg *et al.*, Combining vegetation index and model inversion methods for the extraction of key vegetation biophysical parameters using Terra and Aqua MODIS reflectance data, *Remote Sens. Environ.*, **106**, no. 1, 39–58, (2007).

26 C. Lauvernet *et al.*, Multitemporal-patch ensemble inversion of coupled surface–atmosphere radiative transfer models for land surface characterization, *Remote Sens. Environ.*, **112**, no. 3, 851–861, (2008).

27 C. Atzberger, Object-based retrieval of biophysical canopy variables using artificial neural nets and radiative transfer models, *Remote Sens. Environ.*, **93**, no. 1–2, 53–67, (2004).

28 X. Liu *et al.*, Multi-angular satellite remote sensing and forest inventory data for carbon stock and sink capacity in the Eastern United States forest ecosystems, in *ISPRS Congress Istanbul 2004*, pp. 1–6, (2004).

29 J.-L. Widlowski *et al.*, Canopy structure parameters derived from multi-angular remote sensing data for terrestrial carbon studies, *Clim. Change*, **67**, 403–415, (2004).

30 S. Gandia *et al.*, Retrieval of vegetation biophysical variables from CHRIS/PROBA data in the SPARC campaign, in *Proc. 2nd CHIRS/Proba Workshop, ESA/ESRIN, Frascati, Italy*, no. July, (2004).

31 M.A. Cutter and L.S. Johns, *CHRIS Data format* (Product Document). p. 31, (2005).

32 E.F. Vermote and A. Vermeulen, *Atmospheric correction algorithm: spectral reflectances (MOD09)*. Algorithm Technical Background Document, University of Maryland, Department of Geography, pp. 1–107, (1999).

33 J.G. Masek *et al.*, *LEDAPS Landsat Calibration, Reflectance, Atmospheric Correction Preprocessing Code. Model product.* Available on-line http://daac.ornl.gov. from Oak Ridge National Laboratory Distributed Active Archive Center, Oak Ridge, Tennessee, USA. 10.3334/ORNLDAAC/1080, (2012).

34 S. Jacquemoud and F. Baret, PROSPECT: A model of leaf optical properties spectra, *Remote Sens. Environ.*, **34**, no. 2, 75–91, (1990).

35 J.C. Price, On the information content of soil reflectance spectra, *Remote Sens. Environ.*, **II**, pp. 113–121, (1990).

36 M. Chernetskiy *et al.*, Validation of the Earth Observation Land Data Assimilation System by the field data of ESA SPARC field campaign, in *Proc. ESA Living Planet Symposium 2013, Edinburgh, UK 9–13 September 2013 (ESA SP-722, December 2013)*, vol. **2013**, September, pp. 1–5, (2013).

37 J.L. Gomez-Dans *et al.*, EO-LDAS Validation Report, (2012).

38 F. Vuolo *et al.*, Assessment of LAI retrieval accuracy by inverting a RT model and a simple empirical model with multiangular and hyperspectral CHRIS/PROBA data from SPARC, in *Proc. of the 3rd ESA CHRIS/Proba Workshop, 21–23 March, ESRIN, Frascati, Italy, (ESA SP-593, June 2005)*, (2005).

39 P. Lewis *et al.*, Data assimilation of Sentinel-2 observations: preliminary results from EO-LDAS and outlook, in *Proc. First Sentinel-2 Preparatory Symposium, Frascati, Italy 23–27 April 2012, ESA SP-707, July 2012*, no. July, (2012).

Part III

Coastal Zone and Freshwater Monitoring

8

SAR-Based EO of Salt Marsh Habitats in Support of Integrated Coastal Zone Management

S.J. van Beijma[1], A. Comber[2] and A. Lamb[3]

[1]*Airbus Defence and Space, Geo-Intelligence, Europa House, Southwood Crescent, Farnborough, UK*
[2]*Chair in Spatial Data Analytics, School of Geography, University of Leeds, Leeds, UK*
[3]*Airbus Defence and Space, Geo-Intelligence, Meridian Business Park, Leicester, UK*

8.1   Introduction

In this article a case study is presented on the use of Earth Observation (EO) technologies in support of the mapping of coastal areas. The EO technology under consideration is Synthetic Aperture Radar (SAR); this technology provides important information about land surface structure. The coastal habitats under consideration are salt marsh habitats. These habitats provide botanically and environmentally unique characteristics which are excellent test ground for further development of EO tools based on SAR data.

8.1.1   Coastal Issues

Coastal zones are areas where, put simply, the land meets the sea. They can be very different in appearance: their morphology can vary between cliffs, estuaries, dune systems to low-lying coastal salt marshes or mangrove forests [1]. One of the distinctive characteristics of coastal zones is that form and morphology is constantly subject to change [2], due to the impact of wave activity, tidal currents and sediment supply. This makes development of long-term and sustainable coastal management plans a major challenge. Many coastal habitats are degrading at alarming rates [3–6]. In tropical climates, intertidal mangrove communities have declined by 20% between 1980 and 2005 [3].

Coastal zones with elevation of less than 10 meters cover only 2% of the global land mass, but they host 10% of the World's population [7]. It is predicted that low-lying coastal areas experience the majority of human population growth and economic development [8,9]. Although it is still uncertain to what extent sea level rise will occur in the near future, coastal areas are expected to experience the adverse effects of it [10]. An increased frequency of floods or other coastal hazards as a result of climate changes can have a profound effect on the degradation of coastal zones due to increased erosion rates [11].

In 2002, the European Union (EU) drafted the Integrated Coastal Zone Management (ICZM) Recommendation [12]. There is much debate about what ICZM implies, although in recent years much attention has been given to ecosystem-based coastal management policies [13–15]. A recent example of this approach is the "Building with

Earth Observation for Land and Emergency Monitoring, First Edition. Edited by Heiko Balzter.
© 2017 John Wiley & Sons Ltd. Published 2017 by John Wiley & Sons Ltd.

Nature" project in the Netherlands [16]. This program promotes ecosystem-driven coastal management and aims to show that ecosystem-driven coastal management practices can work. Ecosystem services are goods and services derived from ecosystem functions that benefit human populations directly or indirectly [15,17]. Coastal ecosystems provide a number of ecosystem services in the form of (1) biodiversity support; (2) water quality improvement; (3) flood mitigation; and (4) carbon management [18]. It has been found that healthy coastal ecosystems have a high capability to dissipate wave energy from storms and extreme events like tsunamis [19,20].

8.1.2 Salt Marsh Ecology

Salt marshes fringe many of the world's soft sedimentary coasts exposed to low-energy wave action [9]. Many salt marshes are located along estuarine shores, sheltered by barrier islands, spits, embayments, lagoons and along open shores exposed to low wave energy [21]. Salt marsh communities in the United Kingdom (UK) are most abundant in the south, where most of the UK's soft sedimentary coastlines are located.

In temperate regions salt marshes are characterized by a suite of herbaceous or low woody vascular plants, with *Salicornia* and *Spartina* as the most abundant and best researched species [9]. Salt marshes are influenced by fluxes from the sea, like tides, waves and sediment, as well as climatic variations. Salt marsh communities also respond to seasonal temperature changes and rainfall. The resumption of plant growth occurs in early spring. During summer droughts, growth can be inhibited by increased evapotranspiration which leads to hyper-saline conditions [6].

The spatial distribution of different salt marsh vegetation habitats is not random, but shows a distinctive pattern that can be recognized in many different salt marsh areas around the world [22,23]. These patterns or zonation are highly dependent upon soil salinity and inundation frequency. Soil salinity increases from the lower, sea-facing parts of the salt marsh to the landward upper parts due to shorter sea water inundation time, allowing soil salinity to increase by evapotranspiration [23]. Topography and related flooding period are therefore important, albeit not the only, factors that drive variations of salt marsh vegetation habitats. These vegetation habitats can be classified according to several different classification schemes [22,24]. The European Union has adopted the Habitats Directive in 1992 to improve the conservation of habitats [25]. They proposed the Annex 1 habitat classification scheme, in which intertidal habitats are subdivided similarly to the aforementioned salt marsh zonations. In the UK, the National Vegetation Classification (NVC) is often used [22]. For salt marsh habitats, these classification schemes are based on similar habitat distinctions, which are driven by inundation frequency and associated terrain altitude [6] (Figure 8.1):

- **Pioneer zone:** always submerged except during spring low tides, dominant species *Spartina* spp., *Salicornia* spp., *Aster tripolium*.
- **Lower–mid marsh:** zone covered by most tides, dominant species *Puccinellia maritima* and *Atriplex portulacoides* as well as the previous species.
- **Middle–upper marsh:** zone covered only by spring tides, dominant species *Limonium* spp. and/or *Plantago*, as well as the previous species.
- **Upper marsh:** Zone covered only by highest spring tides, dominant species *Festuca rubra*, *Armeria maritima*, *Elytrigia* spp., as well as the previous species. Vegetation transition from halophytic to adjoining non-halophytic species.

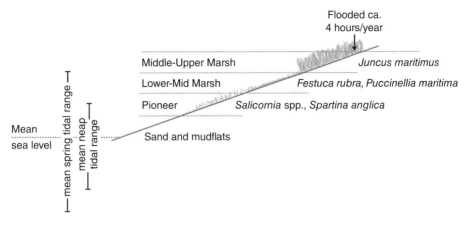

Figure 8.1 Idealized cross-profile of salt marsh vegetation habitat succession, as often found in Northwestern Europe.

Specific management of salt marshes in the UK is not carried out to a formal protocol yet. It is however recognized that salt marshes do provide very useful ecosystem services [6] and they provide excellent buffers against storms and other extreme events [19]. In the Essex estuaries experiments have been carried out in managed realignment of the coastline, due to erosion of fringing salt marsh areas [26]. The program is successful in terms of providing more space for coastal ecosystems, although it was discovered that the biodiversity of recreated salt marsh areas lags behind longer established ones considerably [27].

8.1.3 Remote Sensing Applications for Coastal Zone Management

Research of remote sensing technologies for monitoring and management of coastal zones has been done to a considerable extent. A common research domain is the monitoring of geomorphological changes of the coastline. The repetitive nature of data acquisition of spaceborne platforms provides an excellent opportunity to monitor the changes of coastline position on a regular basis. A recent overview of the remote sensing of coastline is provided by Gens [28], which summarizes different technologies and approaches that are investigated for monitoring of coastal areas.

Airborne optical and LiDAR surveys are routinely used for the mapping of salt marsh habitats and other low-lying coastlines. Airborne operations provide better spatial resolution than spaceborne image acquisition. The UK's Environment Agency (EA) has carried out a number salt marsh vegetation extent surveys over the past three decades, based on aerial photography. Significant differences in estimated salt marsh extent have been recorded over the decades [29]. High-resolution aerial photography and hyperspectral imagery has been used effectively to discriminate different plant species communities in coastal wetlands [30–33]. Further classification improvement was achieved by using multi-temporal data sets using phenological expression difference between wetland vegetation habitats throughout the year [34]. The use of LiDAR for monitoring of intertidal habitats is becoming increasingly popular, as it provides information about vegetation structure as well as geomorphological features [32,35].

SAR has become an alternative acquisition technology for coastal mapping and monitoring. One major advantage of the use of SAR for morphological mapping is the distinct difference of backscatter between land and sea. A number of methodologies have been developed to identify coastline position from SAR imagery by using different edge detection methodologies [36–39]. Another application of SAR capabilities is monitoring of water level fluctuations, based on repeat-pass Interferometric SAR (InSAR) analysis [40]. InSAR methodologies have successfully been used for monitoring of water level changes in wetland areas due to double-bounce effect of radar pulses in areas with flooded vegetation. With increased availability of multi-polarimetric SAR (PolSAR) systems new possibilities for flood monitoring are being researched [41]. The use of SAR for salt marsh vegetation mapping has yielded promising results [42–44]. It was observed that longer SAR wavelengths are more responsive to different vegetation types [44]. Besides mapping of coastal vegetation, SAR has been shown to differentiate between different sediment types (gravel, sand and mud), macrophytes and mussel beds on intertidal flats [45–47], due to its ability to detect differences in surface roughness. Temporal analysis of multi-temporal SAR data indicates that phenological changes of different salt marsh vegetation habitats can be detected and used for improvement of habitat mapping and monitoring [48]. Data products obtained from decomposition of quad PolSAR data [49,50] have shown to improve change detection in wetland environments [51], as well as classification of wetland species [52].

8.2 Case Study: Llanrhidian Marsh

In this section a case study is described in which SAR data sets have been used for the mapping and monitoring of salt marsh vegetation habitats in novel ways. This is achieved by processing and analysis of an airborne quad-polarimetric S-band and X-band SAR data set, acquired over the Llanrhidian salt marshes in July 2010. These data sets provide excellent opportunities to research:

1) Interaction of both S- and X-band SAR with salt marsh vegetation habitats. S-band is very rarely used for SAR applications so far, but is the frequency band proposed for the upcoming NovaSAR-S mission [53,54].
2) Use of polarimetric decomposition SAR descriptors as input variables for unsupervised K-means Wishart classification [55].

This chapter will only highlight the initial part of the analysis, focusing on the respective performance of S-band and X-band. Further results will be published in peer-reviewed journal articles and the PhD thesis. It is aimed that the results from this study will be used to improve the RS toolbox for coastal management and land cover mapping as a whole.

8.2.1 Research Area

The Llanrhidian salt marshes measure over 2000 ha in size and represent almost 5% of all British saltmarsh, located in the Loughor Estuary, west of Swansea in Wales on the northern shore of the Gower peninsula [6,29,56] (Figure 8.2).

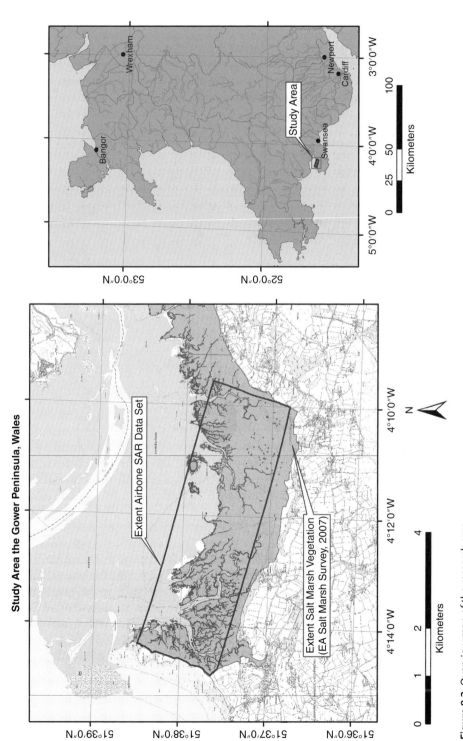

Figure 8.2 Overview map of the research area.

Several ecological and geomorphological studies have been carried out over the past decades [57,58] and it has been found that generally, the salt marsh is in a favourable ecological condition; it is found to be slowly increasing in size [58]. The Llanrhidian Marshes have a number of well-developed geomorphological features: saltmarsh creeks, saltpans, erosive cliffs and bare sand/mudflats. It incorporates all the major vegetation habitats described in the previous section and provides therefore an excellent research area for understanding saltmarsh dynamics.

8.2.2 Data Sets

Quad-polarimetric airborne S- and X-Band SAR data are used for habitat mapping of salt marsh areas. This data set was acquired over the Llanrhidian salt marsh in Wales on 25 July 2010 with the Astrium Airborne Demonstrator. This system operates in X-Band (9.65 MHz) and S-Band (3.2 MHz). An overview of commonly used microwave frequencies in RS is given in Table 8.1.

SAR acquisition in X-Band is well established in both the airborne and space domains (E-SAR, F-SAR, TerraSAR-X and TanDEM-X), but S-Band SAR acquisition is only carried out with the short-lived Russian Almaz-1 satellite; it is the proposed frequency band for the NovaSAR-S satellite. The initial test flights with the system and the calibration routines are described in Ref. [53]. It has been found that lower microwave bands (L-band, P-band) are best suited to map and monitor most vegetation types [59]. Therefore research into the suitability of S-band SAR for vegetation mapping is needed to expand the RS toolbox.

The SAR data is acquired with a single look slant range resolution of 0.69 m in both range and azimuth direction. The system acquires quad-polarimetric data (HH, HV, VV and VH). From the pre-processed scatter matrix a number of polarimetric descriptors were extracted: backscatter intensity channels, Cloude-Pottier (CP) decomposition, Freeman-Durden (FD) decomposition and Van Zyl (VZ) decomposition descriptors [49,50,60]. The rationale behind polarimetric decomposition is that backscattered SAR signal can be deconstructed to extract important information about the structure of the ground target, the scattering mechanism of the return signal as well as the apparent shift in the phase of the signal from the target [61]. The three most commonly used backscatter mechanisms are surface scatter (P_s), double bounce scatter (P_d) and volume or canopy scatter (P_v) (Figure 8.3). Surface scatter describes interaction between

Table 8.1 Microwave bands used in remote sensing.

Band	Frequency (GHz)	Wavelength (cm)
P-band	0.3–1.0	30–100
L-band	1.0–2.0	15–30
S-band	2.0–4.0	7.5–15
C-band	4.0–8.0	3.75–7.5
X-band	8.0–12.5	2.4–3.75
Ku-band	12.5–18.0	1.7–2.4

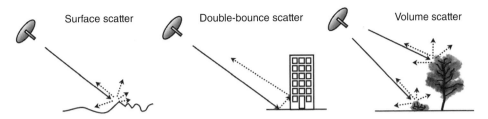

Figure 8.3 Common SAR backscatter mechanisms.

microwaves and bare soils; it is predominant in relatively smooth areas, like sand and mudflats, meadows and fields. Double-bounce scatter occurs whenever distinct vertical structures are present, like buildings and tree stems; it is generally predominant in built-up areas and forested areas. Volume scatter describes interaction between microwaves and smaller branches and leaves and is common over tree and shrub canopies. Backscatter mechanisms are very dependent upon SAR wavelength; surface targets can be dominated by volume scatter on one wavelength, whereas they behave more as surface scatter on other wavelengths [62]. Lower wavelengths have better capability to penetrate canopies than shorter wavelengths and are preferred for estimation of vegetation biomass [63].

The CP decomposition routine applies eigenvector and eigenvalue analysis to describe dominant scattering mechanisms of each target [49]. Three parameters are commonly used for CP decomposition analysis: entropy (H), anisotropy (A) and alpha angle (α). Entropy is a measure of the randomness of a single scatter mechanism. In case of low entropy, a single scatter mechanism is dominant. Alpha angle provides information which scatter mechanism dominates and ranges between 0 and 90 degrees. Low alpha angle signifies dominant odd bounce or surface scattering scatter mechanism, alpha angle values around 45 degrees indicate the dominant volume scatter and alpha angles approaching 90 degrees suggest that the dominant scatter mechanism is double the bounce scatter. Anisotropy characterizes the directional dependence and the relative importance of the secondary scattering mechanism. FD decomposition applies a physical model based on (1) randomly oriented thin cylindrical dipoles; (2) first-order Bragg surface scatter; and (3) double-bounce scattering from a dihedral corner reflector to determine the respective fractions of volume, surface and double-bounce scatter exhibited in each target [50]. The VZ decomposition was proposed by [60] as a modification to the three-component FD decomposition. It is a similar model-based decomposition and decomposes the scatter matrix in the same volume, surface and double-bounce scatter. However, it corrects for the over-estimation of P_v and the possible negative values of P_d and P_s, that can occur with the FD decomposition. Even though several more decomposition methods have been developed, the CP, FD and VZ decompositions are most commonly used and have been successfully applied in (coastal) wetland vegetation mapping and modelling research [44,64,65].

8.2.3 Classification Methods

Image classification was performed to distinguish different salt marsh vegetation habitats. This classification is a combination of classes from both the European Habitat

Directive Annex 1 classification [66] and the British National Vegetation Classification (NVC) [22]. The following classes are distinguished:

1) dry and wet bare sand/mudflats;
2) pioneer *Salicornia* spp.;
3) pioneer *Spartina anglica*;
4) mid-marsh salt marsh meadows (both dominant *Festuca rubra* and *Puccinellia maritima*);
5) upper marsh *Juncus maritimus* swards.

Initially, unsupervised classification based on CP decomposition parameters was carried out. A number of unsupervised classification routines are developed to deal with the SAR speckle effect. Speckle drastically affects the statistics of SAR data. One of the algorithms developed specifically for SAR applications is H/α segmentation, based on the H and α parameters extracted from the CP decomposition [49]. The concept of this segmentation algorithm is the separation of land cover classes in clusters with different H/α relationship. Another way of unsupervised classification makes use of the specific statistical distribution of speckled SAR data. It has been found that measured covariance matrices of SAR imagery follow the complex Wishart distribution [55], which is incorporated in the unsupervised K-means Wishart classifier. This algorithm has been applied successfully in other studies [55,67]. The classifier works by assigning an arbitrary set of initial class centres and classification of the pixels by using Wishart distance. A set of updated class centres is derived from all pixels in each class, and a new class assignment is carried out. This iterative process is repeated until class membership converges. The classifier creates eight classes; for the research area these are clustered according to the most likely land cover class they represent. Two unsupervised habitat classification maps are created, one based on S-band CP variables, the other on X-band variables.

Reference data were collected during an ecological field campaign that was timed seasonally coinciding with the acquisition of the airborne SAR data in July 2013. Random sampling was done in 100 locations by placing a 1×1 metre quadrat on the ground. Within the area bounded by the quadrat, all present vegetation species and their relative abundance were recorded and consequently classified by habitat according to the NVC system. The reference data were used for independent accuracy assessment of the resulting classifications based on S-band and X-band SAR in confusion matrices.

8.2.4 Unsupervised Classification Results

The unsupervised K-means Wishart classifier creates eight classes, which are re-clustered to represent the five land cover classes. The resulting habitat classification maps for S-band and X-band are shown in Figures 8.4 and 8.5, respectively. Accuracy assessments of the two unsupervised classification is given in Table 8.2. The overall accuracy and K-value of the S-band classification are both higher than the unsupervised classification based on X-band SAR. As a rule of thumb $K < 0.4$ is regarded as poor agreement, $K > 0.4$ and $K < 0.7$ as a good agreement and $K > 0.7$ S-band as excellent agreement. It can be concluded that the classification with S-band is just within acceptable limits, although none of the classifications is satisfactory.

Habitat map, S-band SAR, K-means Wishart unsupervised classification

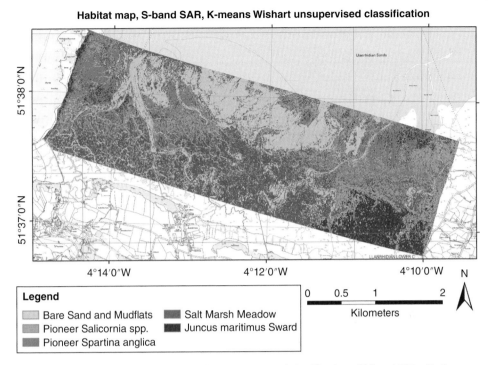

Legend

- Bare Sand and Mudflats
- Pioneer Salicornia spp.
- Pioneer Spartina anglica
- Salt Marsh Meadow
- Juncus maritimus Sward

Figure 8.4 Vegetation habitat map, based on unsupervised classification of S-band SAR with the K-means Wishart classifier.

Habitat map, X-band SAR, K-means Wishart unsupervised classification

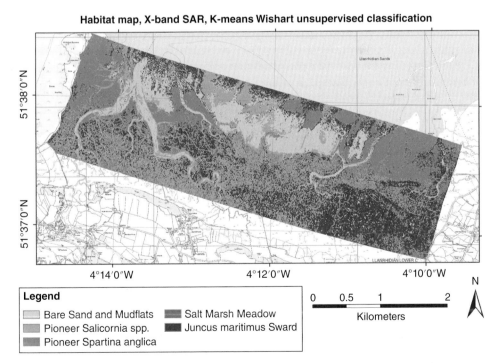

Legend

- Bare Sand and Mudflats
- Pioneer Salicornia spp.
- Pioneer Spartina anglica
- Salt Marsh Meadow
- Juncus maritimus Sward

Figure 8.5 Vegetation habitat map, based on unsupervised classification of X-band SAR with the K-means Wishart classifier.

Table 8.2 Accuracy assessment of unsupervised SAR classifications.

Classification	Overall accuracy (%)	κ-value
Unsupervised S-band SAR	52.33	0.4191
Unsupervised X-band SAR	43.24	0.2967

Visual interpretation of the classification maps shows large differences between the two maps. The habitat classification map based on S-band SAR (Figure 8.4) shows *Juncus maritimus* to be limited to the southeastern part and the western fringe of the research area, whereas it is found much more spread out over the entire research area in the habitat map based on X-band (Figure 8.5).

What is also worth noting is the amount of mis-classifications in the far slant range of the SAR image (the northeastern edge of the research area) in the map based on X-band SAR, with large patches of *Juncus maritimus* classified in this area. This part of the research area is exclusively covered by bare sand and mudflats and some pioneer salt marsh vegetation like *Salicornia* or *Spartina anglica*. The mis-classification can be ascribed to the low signal-to-noise ratio in the X-band image in the far slant range, which, during pre-processing, leads to noisy data that are prone to mis-classification. On the other hand, the habitat map based on X-band SAR shows that X-band is better capable to differentiate between bare soils and vegetated areas, with a more accurate delineation between channels and vegetated areas in the western and central part of the salt marsh. Also, in the S-band habitat map, parts of the salt marsh meadow are mis-classified as bare soil.

8.2.5 Further Work

SAR data are commonly used in conjunction with other RS data sources. In the next stage of this research supervised classification was carried out based on a combination of SAR and several optical and elevation variables. Added to this SAR data sets were other RS data sets:

1) High-resolution aerial photography, acquired in June 2006. This data set contains R, G and B channels with spatial resolution of 0.25×0.25 m.
2) Two Landsat images from the USGS data archive, acquired on 04/06/2010 and 28/04/2011. These were converted to Normalized Difference Vegetation Index (NDVI) products. Spatial resolution of Landsat imagery is 30×30 meters, the NDVI variables were resampled to the working resolution of 2×2 meters,
3) Digital Surface Model (DSM) data with spatial resolution of 2 meters horizontally and 2–3 cm in vertical direction.

In total 30 input variables were created: 12 from S-band SAR, 12 from X-band SAR (three intensity channels, three CP variables, three FD variables and three VZ variables of each SAR frequency band), one DSM layer and five optical bands (R,G,B aerial photography channels and two Landsat NDVI products). A number of supervised classification algorithms can be utilized to perform classification with this aggregate data set. One of the most commonly used is the Maximum Likelihood algorithm. This classification methodology is based on the principle that cells of each class sample in the multidimensional space are normally distributed [68]. More sophisticated classifiers are Support Vector Machines and Random Forest classification. The former classification method is based upon an algorithm that fits hyperplanes (the support vectors) in hyper-dimensional feature spaces for calculation of maximum separability between classes [69]; the latter is based on decision tree classification and it fits many classification trees to a multivariate data set and combines predictions from all the trees [70]. These classifiers also provide information about the performance or importance of the respectively used variables, which can be used to calculate the optimum variable combination.

SAR has been applied successfully to estimate vegetation biomass, primarily in forested areas [59]. It has been found that longer wavelengths are more strongly correlated with vegetation parameters height and volume than shorter wavelengths. Part of the analysis has focussed on the correlation between the available RS variables (SAR, optical and elevation) and percent vegetation cover, vegetation height and gross vegetation volume.

8.3 Discussion and Conclusions

Even though the results of this initial unsupervised classification are not highly accurate, it shows that multi-frequency polarimetric SAR is capable of differentiating between vegetation habitats that differ in three-dimensional appearance. Nonetheless, there are certain limitations to this, as is shown in different responses of S-band and X-band variables to different vegetation types. It appears that X-band SAR differentiates poorly between different salt marsh vegetation types; the X-band microwave signal saturates quickly. S-band SAR differentiates better between vegetation habitats. On the other hand, X-band is more capable than S-band in detecting subtle differences in surface roughness. This is shown in the pioneer zone and the channel networks of the salt marsh, where X-band is more capable than S-band in picking up transitions from non-vegetated sand and mudflats to sparsely vegetated pioneer salt marsh. Research into behaviour of S-band SAR has been limited, due to its limited operational use. With the proposed launch of NovaSAR-S [54] S-band SAR data will become more readily available. Therefore it is advised to expand the research into application of S-band SAR in other application domains like forestry, agriculture or urban studies.

Development of RS applications from SAR data has been very successful in some application areas. Interferometric SAR (InSAR) has been shown to be very useful for detecting changes in terrain and is widely applied in the fields of tectonics, geomorphology engineering, monitoring of land subsidence and glaciers, to name but a few [71]. Applications based on PolSAR are in a less mature stage, although much research is being carried out. PolSAR has proven to be an excellent tool to model three-dimensional structure of ground targets, and is especially suited to estimate biomass [59]. As mentioned before, modelling of scatter mechanisms has provided a set of tools to make this data type more readily interpretable; the development of advanced filtering methods allows suppression of speckle noise while still maintaining detail [72]. There are enough reasons to use PolSAR algorithms and descriptors more commonly for image classification and modelling of surface targets.

In the broad spectrum of coastal zone management, this chapter has described how RS technologies can provide a tool for repeated monitoring of habitats that are important assets for ecosystem-based monitoring, but are difficult to access from land. Although previous salt marsh mapping and monitoring activities have been carried out, it has been shown that inclusion of airborne SAR and other variables can improve mapping and provide more detailed end results. This approach does not need to be limited to salt marsh habitats; inclusion of SAR for monitoring of other coastal habitats or indeed other habitats in general adds valuable extra information that can improve mapping and monitoring considerably.

References

1 Bird, E. *Coastal Geomorphology: An Introduction*, John Wiley & Sons, Chichester, 2008.
2 Haslett, S. K. *Coastal Systems*, Routledge, Abingdon, 2003.
3 FAO 2007. The World's mangroves, 1980–2005. *FAO Forestry Paper*. Rome: Food and Agricultural Organization.
4 Giri, C., Ochieng, E., Tieszen, L. L., Zhu, Z., Singh, A., Loveland, T., Masek, J. & Duke, N., Status and distribution of mangrove forests of the world using earth observation satellite data. *Global Ecology and Biogeography*, **20**, 154–159 (2011).
5 FAO 2003. Status and trends in mangrove area extent worldwide. Rome: Food and Agricultural Organization.
6 Boorman, L. A. 2003. Saltmarsh Review. An overview of coastal saltmarshes, their dynamic and sensitivity characteristics for conservation and management. In: JNCC (ed.). Peterborough: JNCC.
7 McGranahan, G., Balk, D. & Anderson, B., The rising tide: assessing the risks of climate change and human settlements in low elevation coastal zones. *Environment and Urbanization*, **19**, 17–37 (2007).
8 Foresight 2011. Migration and Global Environmental Change. *Final Project Report*. London: The Government Office for Science.
9 Adam, P., Saltmarshes in a time of change. *Environmental Conservation*, **29**, 39–61 (2002).
10 Nicholls, R. J., Wong, P. P., Burkett, V. R., Codignotto, J. O., Hay, J. E., McLean, R. F., Ragoonaden, S. & Woodroffe, C. D. 2007. Coastal systems and low-lying areas. In: M.L. Parry, Canziani, O. F., Palutikof, J. P., Van Der Linden, P. J. & Hanson, C. E. (eds.) *Climate Change 2007: Impacts, Adaptation and Vulnerability. Contribution of Working Group II to the Fourth Assessment Report of the Intergovernmental Panel on Climate Change*. Cambridge, UK: Cambridge University Press.
11 Doody, P., Ferreira, M., Lombardo, S., Lucius, I., Misdorp, R., Niesing, H., Salman, A. & Smallegange, M. 2004. Living with coastal erosion in Europe; Sediment and space for sustainability. *Results from the EUROSION study*. Luxembourg: European Communities.
12 EC 2011. *Integrated Coastal Zone Management, Information and Technology in ICZM, practices around Europe*. In: Czerniak, P., Pickaver, A., Ferreira, M., Steijn, R. & Devilee, E. (eds.). Luxembourg: European Commission.
13 EEA 2006. *The changing faces of Europe's coastal areas*. Copenhagen: European Environmental Agency.

14 Zisenis, M. 2010. *10 messages for 2010; Coastal ecosystems.* In: Agency, E. E. (ed.). Copenhagen: European Environmental Agency.

15 Katsanevakis, S., Stelzenmüller, V., South, A., Sørensen, T. K., *et al.* Ecosystem-based marine spatial management: Review of concepts, policies, tools, and critical issues. *Ocean & Coastal Management*, **54**, 807–820 (2011).

16 Ecoshape. 2011. *Building with Nature* [Online]. Available: www.ecoshape.nl/[Accessed 18/11/2016].

17 Costanza, R., d'Arge, R., De Groot, R., Farber, S., Grasso, M., Hannon, B., Limburg, K., Naeem, S., O'Neill, R. V. & Paruelo, J., The value of the world's ecosystem services and natural capital. *Nature*, **387**, 253–260 (1997).

18 Zedler, J. B. & Kercher, S., Wetland resources: status, trends, ecosystem services, and restorability. *Annu. Rev. Environ. Resour.*, **30**, 39–74 (2005).

19 Möller, I., Quantifying saltmarsh vegetation and its effect on wave height dissipation: Results from a UK East coast saltmarsh. *Estuarine, Coastal and Shelf Science*, **69**, 337–351 (2006).

20 Kamthonkiat, D., Rodfai, C., Saiwanrungkul, A., Koshimura, S. & Matsuoka, M., Geoinformatics in mangrove monitoring: damage and recovery after the 2004 Indian Ocean tsunami in Phang Nga, Thailand. *Nat. Hazards Earth Syst. Sci.*, **11**, 1851–1862 (2011).

21 Adam, P. *Saltmarsh ecology*, Cambridge University Press, Cambridge.1990.

22 Rodwell, J. S. 2000. Salt-marsh communities. In: Rodwell, J. S. (ed.) *British Plant Communities. Volume 5: Maritime Communities and Vegetation of Open Habitats.* Cambridge: Cambridge University Press.

23 Silvestri, S. & Marani, M. 2004. Salt-marsh vegetation and morphology: Basic physiology, modelling and remote sensing observations. In: Fagherazzi, S., Marani, M. & Blum, L. K. (eds.) *The Ecogeomorphology of Tidal Marshes*. Washington, DC: American Geophysical Union.

24 JNCC 2006. Annex 1 Classification schemes. Peterborough, UK: Joint Nature Conservation Committee,.

25 European Commission. 1992. *Council Directive 92/43/EEC of 21 May 1992 on the conservation of natural habitats and of wild fauna and flora* [Online]. Available: http://eur-lex.europa.eu/LexUriServ/LexUriServ.do?uri=CELEX:31992L0043:EN:HTML [Accessed 18/11/2016].

26 Garbutt, R. A., Reading, C. J., Wolters, M., Gray, A. J. & Rothery, P., Monitoring the development of intertidal habitats on former agricultural land after the managed realignment of coastal defences at Tollesbury, Essex, UK. *Marine Pollution Bulletin*, **53**, 155–164 (2006).

27 Garbutt, A. & Wolters, M., The natural regeneration of salt marsh on formerly reclaimed land. *Applied Vegetation Science*, **11**, 335–344 (2008).

28 Gens, R., Remote sensing of coastlines: detection, extraction and monitoring. *International Journal of Remote Sensing*, **31**, 1819–1836 (2010).

29 Environment Agency 2011. *The extent of saltmarsh in England and Wales: 2006–2009.* Bristol: Environment Agency.

30 Lu, D., Mausel, P., Brondízio, E. & Moran, E., Change detection techniques. *International Journal of Remote Sensing*, **25**, 2365–2401 (2004).

31 Tuxen, K., Schile, L., Stralberg, D., Siegel, S., Parker, T., Vasey, M., Callaway, J. & Kelly, M., Mapping changes in tidal wetland vegetation composition and pattern across a salinity gradient using high spatial resolution imagery. *Wetlands Ecology and Management*, **19**, 141–157 (2011).

32 Gilmore, M. S., Wilson, E. H., Barrett, N., Civco, D. L., Prisloe, S., Hurd, J. D. & Chadwick, C., Integrating multi-temporal spectral and structural information to map wetland vegetation in a lower Connecticut River tidal marsh. *Remote Sensing of Environment*, **112**, 4048–4060 (2008).

33 Belluco, E., Camuffo, M., Ferrari, S., Modenese, L., Silvestri, S., Marani, A. & Marani, M., Mapping salt-marsh vegetation by multispectral and hyperspectral remote sensing. *Remote Sensing of Environment*, **105**, 54–67 (2006).

34 Davranche, A., Lefebvre, G. & Poulin, B., Wetland monitoring using classification trees and SPOT-5 seasonal time series. *Remote Sensing of Environment*, **114**, 552–562 (2010).

35 Mason, D. C., Scott, T. R. & Wang, H.-J., Extraction of tidal channel networks from airborne scanning laser altimetry. *ISPRS Journal of Photogrammetry and Remote Sensing*, **61**, 67–83 (2006).

36 Wang, Y. & Allen, T. R., Estuarine shoreline change detection using Japanese ALOS PALSAR HH and JERS-1 L-HH SAR data in the Albemarle-Pamlico Sounds, North Carolina, USA. *International Journal of Remote Sensing*, **29**, 4429–4442 (2008).

37 Yu, Y. & Acton, S. T., Automated delineation of coastline from polarimetric SAR imagery. *International Journal of Remote Sensing*, **25**, 3423–3438 (2004).

38 Shu, Y., Li, J. & Gomes, G., Shoreline Extraction from RADARSAT-2 Intensity Imagery Using a Narrow Band Level Set Segmentation Approach. *Marine Geodesy*, **33**, 187–203 (2010).

39 Twele, A. & Martinis, S. 2009. Flood detection using TerraSAR-X data – Hands-on tutorial. DLR, Wessling.

40 Hong, S.-H., Wdowinski, S. & Kim, S.-W., Evaluation of TerraSAR-X Observations for Wetland InSAR Application. *Geoscience and Remote Sensing, IEEE Transactions on*, **48**, 864–873 (2010).

41 Ramsey, E., Zhong, L., Suzuoki, Y., Rangoonwala, A. & Werle, D., Monitoring Duration and Extent of Storm-Surge and Flooding in Western Coastal Louisiana Marshes With Envisat ASAR Data. *Selected Topics in Applied Earth Observations and Remote Sensing, IEEE Journal of*, **4**, 387–399 (2011).

42 Dehouck, A., Lafon, V., Baghdadi, N. & Marieu, V. 2012. Use of optical and radar data in synergy for mapping intertidal flats and coastal salt-marshes (Arcachon Lagoon, France). *IGARSS 2012*. Munich, Germany.

43 Lee, Y.-K., Park, W., Choi, J.-K., Ryu, J.-H. & Won, J.-S. Assessment of TerraSAR-X for mapping salt marsh. Geoscience and Remote Sensing Symposium (IGARSS), *2011 IEEE International*, 24–29 July 2011, 2330–2333.

44 Clint Slatton, K., Crawford, M. M. & Chang, L.-D., Modeling temporal variations in multipolarized radar scattering from intertidal coastal wetlands. *ISPRS Journal of Photogrammetry and Remote Sensing*, **63**, 559–577 (2008).

45 Gade, M., Stelzer, K. & Kohlus, J. On the Use of Multi-Frequency SAR Imagery for a Surveillance of the Wadden Sea Ecosystem on the German North Sea Coast. 4. *TSX Science Team Meeting, 2011* Wessling. DLR.

46 Adam, E., Mutanga, O. & Rugege, D., Multispectral and hyperspectral remote sensing for identification and mapping of wetland vegetation: a review. *Wetlands Ecology and Management*, **18**, 281–296 (2010).

47 Choe, B.-H., Kim, D.-j., Hwang, J.-H., Oh, Y. & Moon, W. M., Detection of oyster habitat in tidal flats using multi-frequency polarimetric SAR data. *Estuarine, Coastal and Shelf Science*, (2011).

48 Lee, Y.-K., Park, J.-W., Choi, J.-K., Oh, Y. & Won, J.-S., Potential uses of TerraSAR-X for mapping herbaceous halophytes over salt marsh and tidal flats. *Estuarine, Coastal and Shelf Science*, **115**, 366–376 (2012).

49 Cloude, S. R. & Pottier, E., An entropy based classification scheme for land applications of polarimetric SAR. *Geoscience and Remote Sensing, IEEE Transactions on*, **35**, 68–78 (1997).

50 Freeman, A. & Durden, S. L., A three-component scattering model for polarimetric SAR data. *Geoscience and Remote Sensing, IEEE Transactions on*, **36**, 963–973 (1998).

51 Schmitt, A., Brisco, B., Kaya, S. & Murgnaghan, K., Polarimetric Change Detection for Wetlands. (2010). Polarimetric Change Detection for Wetlands. Conference: *Remote Sensing and Hydrology Symposium*, 2010.

52 Brisco, B., Kapfer, M., Hirose, T., Tedford, B. & Liu, J., Evaluation of C-band polarization diversity and polarimetry for wetland mapping. *Canadian Journal of Remote Sensing*, **37**, 82–92 (2011).

53 Natale, A., Bird, R., Whittaker, P., Guida, R., Cohen, M. & Hall, D. Demonstration and analysis of the applications of S-band SAR. *3rd International Asia-Pacific Conference on Synthetic Aperture Radar (APSAR)*, 26–30 Sept. 2011 2011 Seoul. IEEE, 1–4.

54 Bird, R., Whittaker, P., Stern, B., Angli, N., Cohen, M. & Guida, R. NovaSAR-S: A low cost approach to SAR applications. *Synthetic Aperture Radar (APSAR)*, 2013 Asia-Pacific Conference on, 2013. IEEE, 84–87.

55 Lee, J.-S., Grunes, M. R., Ainsworth, T. L., Li-jen, D., Schuler, D. L. & Cloude, S. R., Unsupervised classification using polarimetric decomposition and the complex Wishart classifier. *Geoscience and Remote Sensing, IEEE Transactions on*, **37**, 2249–2258 (1999).

56 May, V. J. 2007. Carmarthen Bay. *Geological Conservation Review*. Peterborough: JNCC.

57 Prosser, M. V. & Wallace, H. L. 1999. *Burry Inlet and Loughor Estuary SSSI, NVC Survey 1998*. Bangor: Countryside Council for Wales.

58 Farleigh, M. R. 2010. *Morphographical Analysis of the Burry Inlet Salt Marshes*. BSc. BSc thesis, University of Glamorgan.

59 Le Toan, T., Beaudoin, A., Riom, J. & Guyon, D., Relating forest biomass to SAR data. *Geoscience and Remote Sensing, IEEE Transactions on*, **30**, 403–411 (1992).

60 Van Zyl, J. J., Arii, M. & Yunjin, K., Model-Based Decomposition of Polarimetric SAR Covariance Matrices Constrained for Nonnegative Eigenvalues. *Geoscience and Remote Sensing, IEEE Transactions on*, **49**, 3452–3459 (2011).

61 Henderson, F. M. & Lewis, A. J., Radar detection of wetland ecosystems: a review. *International Journal of Remote Sensing*, **29**, 5809–5835 (2008).

62 Schmullius, C. C. & Evans, D. L., Review article Synthetic aperture radar (SAR) frequency and polarization requirements for applications in ecology, geology, hydrology, and oceanography: A tabular status quo after SIR-C/X-SAR. *International Journal of Remote Sensing*, **18**, 2713–2722 (1997).

63 Beaudoin, A., Le Toan, T., Goze, S., Nezry, E., Lopes, A., Mougin, E., Hsu, C., Han, H., Kong, J. & Shin, R., Retrieval of forest biomass from SAR data. *International Journal of Remote Sensing*, **15**, 2777–2796 (1994).

64 Ramsey III, E., Rangoonwala, A., Suzuoki, Y. & Jones, C. E., Oil Detection in a Coastal Marsh with Polarimetric Synthetic Aperture Radar (SAR). *Remote Sensing*, **3**, 2630–2662 (2011).

65 Corcoran, J., Knight, J. & Gallant, A., Influence of Multi-Source and Multi-Temporal Remotely Sensed and Ancillary Data on the Accuracy of Random Forest Classification of Wetlands in Northern Minnesota. *Remote Sensing*, **5**, 3212–3238 (2013).

66 Habitats Directive, *Council Directive 92/43/EEC of 21 May 1992 on the conservation of natural habitats and of wild fauna and flora*. Brussels, Belgium, (1992).

67 Reigber, A., Jäger, M., Neumann, M. & Ferro-Famil, L., Classifying polarimetric SAR data by combining expectation methods with spatial context. *International Journal of Remote Sensing*, **31**, 727–744 (2010).

68 Richards, J. A. *Remote sensing digital image analysis: an introduction*, Springer, Berlin.1999.

69 Melgani, F. & Bruzzone, L., Classification of hyperspectral remote sensing images with support vector machines. *Geoscience and Remote Sensing, IEEE Transactions on*, **42**, 1778–1790 (2004).

70 Breiman, L., Random Forests. *Machine Learning*, **45**, 5–32 (2001).

71 Rott, H., Advances in interferometric synthetic aperture radar (InSAR) in earth system science. *Progress in Physical Geography*, **33**, 769–791 (2009).

72 Lee, J.-S., Wen, J.-H., Ainsworth, T. L., Chen, K.-S. & Chen, A. J., Improved sigma filter for speckle filtering of SAR imagery. *Geoscience and Remote Sensing, IEEE Transactions on*, **47**, 202–213 (2009).

9

A Framework for Lakeshore Vegetation Assessment Using Field Spectroscopy and Airborne Hyperspectral Imagery

D. Stratoulias[1,2,3], I. Keramitsoglou[4], P. Burai[3], L. Csaba[3], A. Zlinszky[1], V.R. Tóth[1] and Heiko Balzter[2]

[1] *Balaton Limnological Institute, Centre for Ecological Research, Hungarian Academy of Sciences, Tihany, Hungary*
[2] *University of Leicester, Centre for Landscape and Climate Research, Department of Geography, Leicester, UK*
[3] *Envirosense Hungary Kft., Debrecen, Hungary*
[4] *Institute for Astronomy, Astrophysics, Space Applications and Remote Sensing, National Observatory of Athens, Pendeli, Athens, Greece*

9.1 Background

9.1.1 Wetland Mapping: A Challenging Case

Wetlands are multiple-value systems covering approximately 4–6% of the world's terrestrial area. They host habitats of a dynamic nature and have a high ecosystem value and significance. Wetlands are defined as places where, "water is the primary factor controlling the environment and the related plant and animal life" [1]. Their "tentative minimum" global coverage is estimated at 12.8 million km^2 according to Finlayson *et al.* [2]. They are important because they connect the terrestrial and water biotic and abiotic characteristics of the landscape, and as such they host a high biodiversity of plant and animal species.

Sustainable management of wetlands has been a field for investigation by scientists and practitioners likewise. Traditionally, this is based on field campaigns undertaken by experienced ecologists. Nevertheless this task can become time and cost consuming as well as unmanageable in inaccessible areas and affected by between-surveyors errors [3]. Earth observation has lately an increasing involvement in assessing the conservation status of natural habitats [4,5] with the continuing technological advancements offering imagery with finer specifications. A comprehensive review on the use of remote sensing on wetlands can be found in Ozesmi and Bauer [6], Fitoka and Keramitsoglou [7] and specifically for wetland vegetation at Adam *et al.* [8].

Especially in Europe, several examples of frameworks of ecosystem monitoring have been developed in the past as part of the Copernicus land monitoring services. NATURA 2000 is an EU wide network of nature protection areas established under the 1992 Habitat Directive and is the backbone of EU nature and biodiversity monitoring. The BIOdiversity multi-SOurce Monitoring System: from space to species (BIO_SOS) is an ecological modelling system for monitoring habitats and particularly those exposed

Earth Observation for Land and Emergency Monitoring, First Edition. Edited by Heiko Balzter.
© 2017 John Wiley & Sons Ltd. Published 2017 by John Wiley & Sons Ltd.

to pressure. Multi-scale Services for Monitoring NATURA 2000 Habitats of European Community Interest (MS.MONINA) (2010–13) monitored nature sites of community interest for fostering environmental legislation in Europe and reducing biodiversity loss. Furthermore, one of the targets of EU strategy for 2020 is the better protection and restoration of ecosystems. It proposes the development of framework for degraded ecosystems supported by relevant work for mapping and assessing its state.

Remote sensing is considered a useful tool for such implementations only if the mapping technique is deemed accurate. Wetland mapping based on remotely sensed data presents challenges arising from the inherent complexity in spatial structure of the ecosystems, the broad spectral variability within and among vegetation species and lastly the water level variability over time. In this context, the methodological approach used to assess the status of the ecosystem is important, and this applies especially to lakeshore vegetation, a zone with high biodiversity. Satellite imagery is the choice of most wetland mapping studies because it offers either very fine spatial resolution imagery (WorldView-3 delivers panchromatic data with 0.31 m pixel resolution at nadir angle) or very high spectral resolution data (Hyspiri imaging spectrometer has 212 bands), however there is currently no satellite in orbit offering both capabilities. This can be accomplished at the moment only with hyperspectral sensors adjusted at airplane platforms flying at low altitude.

In this chapter *in-situ* spectroscopic data and airborne hyperspectral imagery were used synergistically to study a reed bed affected by the reed die-back syndrome, the causes of which are still under debate. The aim was to investigate to which extent hyperspectral data are suitable for detecting reed's ecophysiological status and mapping its extent in a lacustrine environment. This case study focuses on the spectral information acquired by macrophytes samples from different distances around Lake Balaton, Hungary, with a specific interest on reed. It presents the capabilities of higher technical specification remote sensing data and can be considered a pilot experiment for future satellite missions.

9.1.2   Reed Die-back in Lake Balaton

Lake Balaton is a large shallow freshwater lake situated in Hungary, Central Europe. According to the last survey conducted by the Hungarian Water Authorities, within the Lake boundary 14.09 km^2 are covered by vegetation, consisting mainly of 11.45 km^2 of *Phragmites australis* (Cav.) Trin. ex Steudel (common reed), 0.93 km^2 of quasi-natural vegetation (e.g. *Typha latifolia, Carex* sp., *Scirpus* sp.) and 1.79 km^2 of other types vegetation, mainly trees [9]. In essence, this is a large wetland area dominated by reed at the vegetated regions. Reed (Figure 9.1) is a tall rhizomatous perennial grass [10] and cosmopolitan plant species frequently encountered in land-water interface zones [11]. Since the beginning of the 1960s a widespread and intensive retreat of reed beds has been observed in Europe [12,13] and is collectively reported in the literature as the "reed die-back syndrome". The phenomenon was first reported 60 years ago [14], cited in [15].

The reed die-back has been a well-studied topic in limnology worldwide. Despite the potential capabilities of earth observation in vegetation studies, remote sensing techniques in Europe have only recently been employed in the framework of reed die-back and macrophyte species composition, with representative examples presented in [16–19] and [20].

Figure 9.1 Typical reed bed at Lake Balaton, as observed at the terrestrial side. The macrophytes consist of a very dense structure of thin stalks with leaves oriented upwards. (Author's personal copy.)

9.2 Methods and Data

9.2.1 Spectral Information from in-situ Measurements

Spectra recorded non-destructively in the field are the most reliable source of information in regard to vegetation's spectral behaviour. Measurements taken at leaf level have the advantage of providing pure leaf spectral signal and with no interference with the atmosphere. Spectra recorded with a field of view over the canopy have little atmospheric interference and can provide position-specific spectra of a ground area analogous to the satellite's pixel resolution.

In order to study the spectral specifications of reed condition and macrophyte species composition, a field campaign was carried out during August 2012 around Lake Balaton. A hand-held ASD portable spectroradiometer (Analytical Spectral Devices Inc., USA) was used to collect leaf and canopy hyperspectral samples. The instrument records radiation in 750 consecutive bands between 325 and 1075 nm. Canopy reflectance was recorded by attaching the device to a pole and placing it perpendicularly at a height of 1m above the vegetation canopy. Leaf reflectance was acquired by attaching a leaf clip and the associated light source to the ASD device through an optical fibre. Macrophyte species samples were collected at the Szigliget bay and samples where *Phragmites* was dominant were collected at the Bozsai bay (Figure 9.2). Access at the waterfront part of the vegetation was accomplished by a boat, while access to the terrestrial part of macrophytes through narrow walkways and transects within the reed stand.

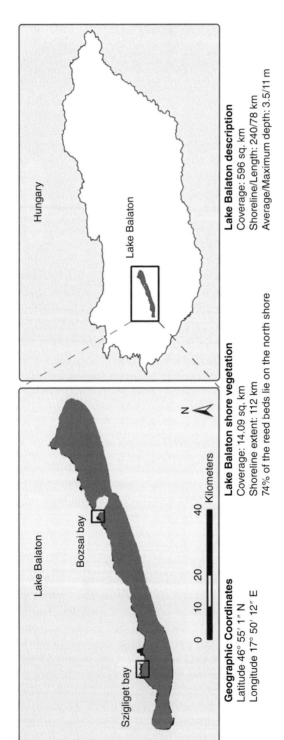

Geographic Coordinates
Latitude 46° 55′ 1″ N
Longitude 17° 50′ 12″ E

Lake Balaton shore vegetation
Coverage: 14.09 sq. km
Shoreline extent: 112 km
74% of the reed beds lie on the north shore

Lake Balaton description
Coverage: 596 sq. km
Shoreline/Length: 240/78 km
Average/Maximum depth: 3.5/11 m

Figure 9.2 Study areas of Bozsai and Szigliget bays and relative position of Lake Balaton in Hungary.

9.2.2 Lake Shore Vegetation Mapping from Airborne Hyperspectral Data

While *in-situ* spectra are the most reliable source of spectroscopic information and provide important information at species and canopy level, field sample collection at large areas such as Lake Balaton is difficult, if not impossible, at frequent intervals. Earth observation data is essentially the only way to monitor a large wetland, synergistically with field measurements. In the second part we present an effort to map the shore vegetation in terms of species distribution and reed categories of interest from airborne hyperspectral imagery. We concentrate on a nature-protected area of Lake Balaton (Bozsai bay) situated at the northwest part of the Tihany peninsula (Figure 9.2). It is quasi-undisturbed from human activity and encompasses a variety of macrophytes, trees and grasslands, however the main ecological focus is placed on the *Phragmites* and a relatively smaller part covered with *Typha latifolia* and *Carex* sp.

9.2.3 Airborne Hyperspectral Imagery

The remotely sensed data were collected in the framework of the European Facility for Airborne Research (EUFAR) programme during a flight campaign undertaken by the Airborne Research and Survey Facility (ARSF, Gloucester, UK.), vested in the Natural Environment Research Council (NERC). AISA Eagle (Spectral Imaging Ltd., Finland) hyperspectral imagery acquired concurrently with 18 cm ground resolution aerial photography from two successive afternoon flight lines 21 August 2010 were used (Table 9.1). The acquisition from an airplane platform assigns a very high spatial and spectral resolution to the hyperspectral data. Due to the cost of the flight campaign, similar datasets are not regularly employed in remote sensing studies and classification of similar images is not well established as in typical multispectral imagery (e.g. Landsat, SPOT).

Table 9.1 Characteristics of the AISA Eagle hyperspectral image acquisition.

Date of acquisition	21 Aug 2010
Time of acquisition	14:08–14:20 UTC
Number of flight lines	2
Spatial resolution	2 m
Spectral resolution	253 bands, 400–1000 nm
Radiometric resolution	12-bit
Full width at half maximum (FWHM)	2.20–2.44
Altitude	1550 m
View angle	Nadir
Field of view	38°
Swath width	992 m
Solar azimuth angle	247.58°
Solar elevation angle	35.23°
Atmospheric conditions	Cloud free atmosphere

Therefore Eagle imagery is a suitable ground for testing techniques that will be applicable in future satellite missions if the technological improvements will continue progressing.

9.2.4 Pre-processing

Imagery collected from airborne sensors differs from that of the satellites in the sense that during image acquisition the airplane is moving through the atmosphere and small fluctuations in the motion of the aircraft can result in large distortions of the raw image. For this reason, simultaneous acquisition of the aircraft's motion variables is necessary in order to place the measurements into a precise geographical reference [21]. The variables include altitude, attitude (yaw, pitch and roll angles), position and velocity of the aircraft and are acquired from integrated Inertial Measuring Unit/Global Navigation Satellite System IMU/GNSS devices. Furthermore, the comparatively short distance between the aircraft and the target introduces distortion in the geometry of the ground pixels as well as glittering at the edge of the image, the magnitude of which depends on the relative position of the sun at the time of image acquisition.

Pre-processing of airborne imagery is a time and effort consuming step and requires specialized software and user skills, however it is critical for the quality of the final product. We followed an approach for correcting the artefacts apparent in the specific images to compensate for radiometric, atmospheric and geometric inaccuracies (Figure 9.3). Cross-track illumination correction (multiplicative model, 3rd polynomial order) was applied to remove the glittering artefact apparent in both images due to the vertical orientation of the sun in regard to the plane direction during the image acquisition. Atmospheric correction was implemented in Fast Line-of-sight Atmospheric Analysis of Spectral Hypercubes (FLAASH), which is part of the ENVI atmospheric correction module [22]. Geometric registration was applied on a 1.5 m × 1.5 m grid with the Airborne Processing Library software [23]. The two images were then subset around the reed bed at the Bozsai bay and mosaicked using a histogram match, based on the image with better clarity. Finally the water pixels were removed by applying a threshold on the spectral index 821 nm/502 nm, ratio which highlights the differences between vegetation and water pixels.

9.2.5 Image Composition

We first studied the spectral signatures of the dominant vegetation at the area of study, mainly reed, die-back reed, *Typha*, grass and trees. Thereafter we identified the spectral regions in which separability between classes is prominent and extracted the bands at 453 nm, 557 nm and 777 nm. We derived the Normalized Difference Vegetation Index (NDVI) [24] and the Photochemical Reflectance Index (PRI) [25] to highlight the vegetation status. Classification of a 253-band image is a computationally intensive

Figure 9.3 Pre-processing flowchart of the airborne hyperspectral imagery.

step; for reducing the spectral dimensionality and thus decreasing the classification computational demand, the first ten bands from the Minimum Noise Fraction (MNF) forward transformation were extracted. A stacking operation of these layers provided a 15-band composite image, which was input in the classification process.

9.2.6   Training and Validation

Vegetation species identification was accomplished by field visits where pure areas were geo-located with a handheld GPS. For the reed-specific categories, close collaboration was established with ecologists experienced with the vegetation of the study area. A set of polygons was selected on the hyperspectral image representing nine classes of interest. It is important to note that we selected the samples homogeneously from the image as the pre-processing has left slivers at the cutline of the mosaic where illumination conditions are different. We used the concurrently acquired aerial photography to assure coherency of the polygons. The set was divided in two groups, one used for training the classifier and the other for validating the output thematic map. The accuracy of the validation exercise was assessed with an error matrix (contingency table) [26].

9.2.7   Classification

We applied supervised classification on the composite image based on the training set. Support Vector Machines (SVM) [27] was chosen as the algorithm of implementation as it is lately employed frequently in remote sensing studies [28] and has been proven to provide comparatively accurate results in land use and land cover classification schemes (e.g. [29–33]). Finally a 3×3 majority filter was applied and a threshold mask for removing areas of no interest such as houses and roads. The cartographic production was carried out in ArcMAP 10.0 (Environmental Systems Research Institute, Inc.).

9.3   Results

9.3.1   Macrophyte Species Leaf Reflectance

Macrophyte leaf reflectance samples were studied at the Szigliget bay for four different vegetation species and categories, mainly terrestrial reed (*Phragmites* not covered with water), aquatic reed (*Phragmites* covered with at least 50 cm of water), *Typha latifolia* and *Carex* sp. (Figure 9.4). The two reed categories show a distinguishing spectral curve comparatively to the other two species as they reflect stronger in the visible region. The difference is especially prominent in the region 560–670 nm, which is regulated by leaf pigments and indicates chlorophyll absorption. In the near-infrared domain reed categories reflect weaker, indicative of the different cell structure. Between terrestrial and aquatic reed, small differences can be observed such as the gradually higher reflectance of the latter below 500 nm. *Carex* sp. and *Typha latifolia* have similarly shaped spectra with *Carex* sp. having 10% relative lower reflectance throughout the spectrum. Macrophyte species have a high degree of separability at leaf reflectance level.

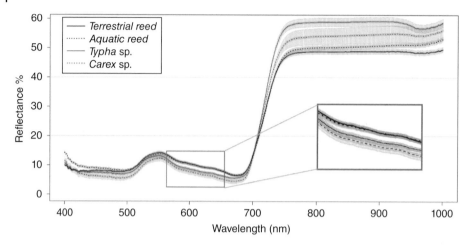

Figure 9.4 Leaf mean reflectance spectra (curves) and 95% confidence intervals (shaded areas) for the macrophyte species collected with an ASD handheld spectroradiometer at the Szigliget bay in Lake Balaton, Hungary. 21, 20, 19 and 10 samples were considered for terrestrial reed, aquatic reed, *Typha* sp. and *Carex* sp. respectively.

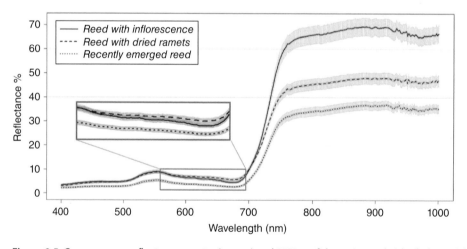

Figure 9.5 Canopy mean reflectance spectra (curves) and 95% confidence intervals (shaded areas) for typical reed samples (reed inflorescence, reed with high dried content and recently emerged reed) collected with an ASD handheld spectroradiometer at the Szigliget bay in Lake Balaton, Hungary.

9.3.2 Phragmites Canopy Reflectance

Phragmites canopy reflectance for different environmental conditions was recorded at the same area with a field of view instrument 1 m above the canopy. *Phragmites* patches with inflorescence (reed inflorescence), *Phragmites* with high dried content and recently emerged *Phragmites* (reed) were sampled (Figure 9.5). Reed shows lower values of reflectance over the spectrum, which is indicative either of the absence of reflecting materials, such as dried content and flowers encountered in the other two classes, either of the better light use efficiency or a combination of the two. Reed patches with higher dried reed content reflects stronger because they contain dead reed stalks from previous years and therefore light utilization is at lower levels. Reed inflorescence has a lower reflectance at the red region and a higher reflectance at the near infrared, which indicates pigment absorption and high photosynthetic activity of the plant at this stage of growth. To add to the complexity of the scene when studying macrophytes, we have to consider that *Phragmites* in most areas cohabits with other macrophyte species, and its presence is not always dominant especially in areas such as the terrestrial edge of the reed bed where drier sediment conditions become favourable to inland vegetation.

9.3.3   Phragmites Reflectance with Regard to Stability

At the Bozsai bay, canopy reflectance was measured for stable and die-back reed samples at the terrestrial and waterfront degraded areas respectively (Figure 9.6). High intra-class variability is observed. The 95% confidence intervals overlap and there are no sharp boundaries where separation between the two classes can be found throughout the spectrum. This fluctuation is attributed to the vegetation material within the field of view as well as the illumination geometry of the scene; the radiance recorded by the instrument with the field of view optics is a combination of reflectance from leaves, stalks, flowers and understory materials. Furthermore, the dense structure of the macrophytes, as well as the upward orientation of the leaves, creates shadows and a generally complex scene, which only resembles the reflectance from the measurements at leaf scale. Variability in spectral response is also introduced by the viewing geometry between the incident light and the observer. Last but not least and from an ecological point of view, the most important factor influencing the macrophytes canopy reflectance is the heterogeneity of the patches, since within the reed stand different patches may contain different species, different genotypes of the same species, and stalks of different age, height, dimension or structure; all these factors cumulatively assign a variability which is common in ecological studies.

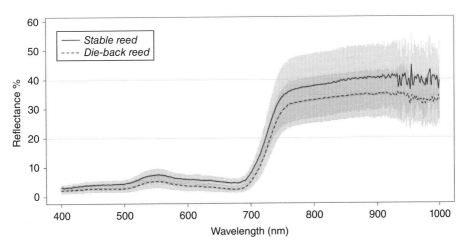

Figure 9.6 Canopy mean reflectance spectra (curves) and 95% confidence intervals (shaded areas) for *Phragmites australis* collected with an ASD handheld spectroradiometer from stable and die-back reed samples at the waterfront area of the Bozsai bay in Lake Balaton, Hungary.

It is important to note that at canopy level die-back reed seems to reflect stronger than the stable reed. This might be attributed to the fact that the photosynthetic activity, and in general the photochemical properties, of the die-back reed are inhibited and the plant absorbs weaker, and hence reflects stronger, than the stable reed.

9.3.4 Airborne Hyperspectral Data Classification

The thematic map (Figure 9.7) presents the main vegetation categories encountered at the study area and the reed sub-classes as extracted from the classification of the hyperspectral airborne imagery. Die-back reed (red colour) is found at the waterward edge of the northeast part of the reed bed. Reed (i.e. *Phragmites*) in the central part of the reed bed was classified as more dominant and botanical survey supports this finding, while at both the terrestrial and waterfront edges of the reed bed and especially at the thin sliver at the west part of the reed bed seems to compete with other macrophytes species. Reed-specific classes are categorized satisfactorily, with some minor misclassification as trees apparent at the intersection of the two airborne images in the centre of the final scene.

Table 9.2 presents the error matrix of the classification. High accuracies are encountered along the main classes, with a very high separability. The overall accuracy achieved was 95.18%, which is considered especially high for vegetation-related studies. Some

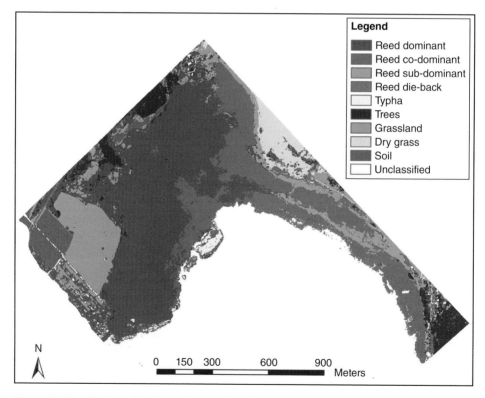

Figure 9.7 Classification of the Bozsai bay in Lake Balaton, Hungary with focus on the macrophyte species and reed synthesis.

Table 9.2 Error matrix of the supervised SVM classification of Bozsai bay. Numbers are in percentages.

Class	Grassland	Soil	Typha	Dry grass	Reed sub-dominant	Reed die-back	Reed dominant	Reed co-dominant	Trees	User's accuracy
Grassland	99.3	0	0	1.32	0	0	0	0	0	99.19
Soil	0	100	0	0	0	0	0	0	0	100
Typha	0	0	90.48	0	0	0	0	0	0	100
Dry grass	0	0	0	98.68	0	0	0	0	0	100
Reed sub-dominant	0	0	0	0	92.97	0	0	1.2	0	80.37
Reed die-back	0	0	0	0	0	93.66	0	0	0.03	99.6
Reed dominant	0.49	0	0.79	0	7.03	0	95.49	11.29	0	85.79
Reed co-dominant	0	0	8.73	0	0	6.34	4.4	87.29	2.78	92.16
Trees	0.21	0	0	0	0	0	0.11	0.23	97.19	99.53
Producer's accuracy	99.3	100	90.48	98.68	92.97	93.66	95.49	87.29	97.19	
Overall accuracy	95.18%									
Kappa coefficient	0.94									

errors exist between the different classes of reed and especially the classes encountered at the edge of the reed bed, i.e. reed die-back, sub-dominant and co-dominant reed. Reed die-back is misclassified only as reed co-dominant, which means that spectrally, and at 1.5 m spatial resolution, is different than the more homogeneous classes of reed. Reed dominant has the highest accuracy across macrophytes, as it is spectrally the most homogeneous macrophytic class in the study area. *Typha* is sometimes misclassified as reed co-dominant. Supplementary classes encountered in the area of interest are classified with very high accuracy (>97%).

9.4 Discussion and Conclusions

In-situ spectral measurements at leaf level provide the most accurate way to study spectrally the physiological status of reed leaves. Imaging spectroscopy at canopy level under adequate solar illumination provides information on the vegetation composition and the geometry of the canopy structure; nevertheless measurements are local-specific due to the high dynamic nature of the lakeshore environment and the effect of the inundation level in the vegetation's phenotypic [34] and subsequently spectral response. In this chapter we presented results, which show separability at leaf and canopy level between species and between canopy measurements containing a composition of species and vegetation states. While field imaging spectroscopy will suffer from some of the drawbacks of field data collection (i.e. time and cost consuming), it can be useful in the framework of reed die-back identification as well as the acquisition of field canopy samples for input in classification directly as training spectra or as reference to training polygons.

Classification of airborne hyperspectral imagery on the other hand might prove to be time and effort consuming. Process chains involve several different software packages and a large amount of processing and disk space in a personal computer. While classification techniques can be automated given the availability of a training set, pre-processing is site- and image-specific and the intervention of the user is essential. Furthermore pre-processing of airborne dataset required specialized knowledge outside the sphere of the ordinary end user. Field data deem to be necessary and important for classification of aquatic vegetation. A classification based solely on the airborne imagery and without the contribution of information on the ecological status of the training samples would provide incomprehensible results.

In this study emergent macrophyte species at the shore of Lake Balaton were classified based on airborne hyperspectral images. A high accuracy of the classes under study is achieved, a fact which can be attributed to the concurrent very high spectral and spatial resolution of the image. Nevertheless the data we employed were collected during a flight campaign with a cost several times higher than that of a satellite image. While several multispectral satellites can offer imagery with similar spatial resolution or even higher (e.g. WorldView-3 multispectral bands have 1.24 m spatial resolution), current hyperspectral satellite sensors do not satisfy this specification. In this study we presented a classification based on individual bands and vegetation indices, hence including only a small number of layers. New sensors, such as the Sentinel-2, are equipped with super-spectral resolution, trying to bridge this gap. It is yet to be seen whether the addition of extra bands in combination with the very high spatial resolution suffices for fine scale aquatic vegetation mapping.

References

1 W. Niering, *Wetlands: The Audubon Society Nature Guides*, Alfred. A. Knopf, New York, 1985.

2 C.M. Finlayson, N.C. Davidson, A.G. Spiers, and N.J. Stevenson, Global wetland inventory – current status and future priorities, *Marine and Freshwater Research*, **50**(8), 717–727 (1999).

3 A. Cherrill and C. McClean, Between-observer variation in the application of a standard method of habitat mapping by environmental consultants in the UK. *Journal of Applied Ecology*, **36**(6), 989–1008 (1999).

4 J.T. Kerr and M. Ostrovsky, From space to species: ecological applications for remote sensing. *Trends in Ecology and Evolution*, **18**(6), 299–305 (2003).

5 T.L. Spanhove, J. Vanden Borre, S. Delalieux, B. Haest and D. Paelinckx, Can remote sensing estimate fine-scale quality indicators of natural habitats? *Ecological Indicators*, **18**, 403–412 (2012).

6 S.L. Ozesmi and M.E. Bauer, Satellite remote sensing of wetlands. *Wetlands Ecology and Management*, **10**(5), 381–402 (2002).

7 E. Fitoka and I. Keramitsoglou (eds.), *Inventory, assessment and monitoring of Mediterranean Wetlands: Mapping wetlands using Earth Observation techniques.* MedWet publication, Sympraxis, Athens, 2008.

8 E. Adam, O. Mutanga and D. Rugege, Multispectral and hyperspectral remote sensing for identification and mapping of wetland vegetation: a review. *Wetlands Ecology and Management*, **18**(3), 281–296 (2010).

9 P. Pomogyi, Remote sensing techniques in applied hydrobotany – Vegetation mapping and classification using aerial photographs (In Hungarian). *Remote Sensing and GIS*, **1**, 5–23, 2013.

10 S.M. Haslam, The development of shoots in *Phragmites communis Trin. Annals of Botany*, **33**(4), 695–709 (1969).

11 G.C. Tucker, The Genera of Arundinoideae (Gramineae) in the Southeastern United-States. *Journal of the Arnold Arboretum*, **71**(2), 145–177 (1990).

12 W.H. Van der Putten, Die-back of *Phragmites australis* in European wetlands: an overview of the European Research Programme on Reed Die-Back and Progression (1993–1994). *Aquatic Botany*, **59**(3–4), 263–275 (1997).

13 H. Brix, The European research project on reed die-back and progression (EUREED). *Limnologica – Ecology and Management of Inland Waters*, **29**(1), 5–10 (1999).

14 H. Hürlimann, *Zur Lebensgeschichte des Schilfs an den Ufern der Schweizer Seen* (In German). Bern, H. Huber, Series: Beitr. Geobot. Landesaufn. Switzerland, **30**, 1–232, 1951.

15 W. Ostendorp, 'Die-back' of reeds in Europe – a critical review of literature. *Aquatic Botany*, **35**(1), S, 5–26 (1989).

16 J. Liira, T. Feldmann, H. Mäemets and U. Peterson, Two decades of macrophyte expansion on the shores of a large shallow northern temperate lake – A retrospective series of satellite images. *Aquatic Botany*, **93**(4), 207–215 (2010).

17 P.D. Hunter, D.J. Gilvear, A.N. Tyler, N.J. Willby, and A. Kelly, Mapping macrophytic vegetation in shallow lakes using the Compact Airborne Spectrographic Imager (CASI). *Aquatic Conservation: Marine and Freshwater Ecosystems*, **20**, 717–727 (2010).

18 A.O. Onojeghuo, and G.A. Blackburn, Optimising the use of hyperspectral and LiDAR data for mapping reed bed habitats. *Remote Sensing of Environment*, **115**(8), 2025–2034 (2011).

19 A. Zlinszky, W. Mücke, H. Lehner, C. Briese and N. Pfeifer, Categorizing wetland vegetation by airborne laser scanning on Lake Balaton and Kis-Balaton, Hungary. *Remote Sensing*, **4**(6), 1617–1650 (2012).

20 P. Villa, A. Laini, M. Bresciani and R. Bolpagni, A remote sensing approach to monitor the conservation status of lacustrine *Phragmites australis* beds. *Wetlands Ecology and Management*, **21**, 399–416 (2013).

21 J. Bange, M. Esposito, D.H. Lenschow, P.R.A. Brown *et al.*, Measurement of aircraft state and thermodynamic and dynamic variables. In: *Airborne Measurements for Environmental Research: Methods and Instruments* (eds. M. Wendisch and J.-L. Brenguier), Wiley-VCH Verlag, GmbH & Co., Germany, 2013.

22 ITT Visual Information Solutions. *Atmospheric Correction Module: QUAC and FLAASH User's Guide, Version 4.7, August, 2009 Edition.* Atmospheric Correction Module. ESRI, ENVI, 2009.

23 M.A. Warren, B.H. Taylor, M.G. Grant and J.D. Shutler, Data processing of remotely sensed airborne hyperspectral data using the Airborne Processing Library (APL): Geocorrection algorithm descriptions and spatial accuracy assessment. *Computers & Geosciences*, **64**, 24–34 (2014).

24 C.J. Tucker, Red and photographic infrared linear combinations for monitoring vegetation. *Remote Sensing of Environment*, **8**, 127–150 (1979).

25 J.A. Gamon, L. Serrano, and J.S. Surfus, The photochemical reflectance index: an optical indicator of photosynthetic radiation use efficiency across species, functional types and nutrient levels. *Oecologia*, **112**, 492–501 (1997).

26 R.G. Congalton, A review of assessing the accuracy of classifications of remotely sensed data. *Remote Sensing of Environment*, **37**(1), 35–46 (1991).

27 V.N. Vapnik, *The Nature of Statistical Learning Theory.* Springer, New York, 1995.

28 G. Mountrakis, J. I. and C. Ogole, Support vector machines in remote sensing: A review. *ISPRS Journal of Photogrammetry and Remote Sensing*, **66**(3), 247–259 (2011).

29 M. Pal and M. Mather, Support vector machines for classification in remote sensing. *International Journal of Remote Sensing*, **26**(5), 1007–1011 (2005).

30 D. Boyd, C. Sanchez-Hernandez and G. Foody, Mapping a specific class for priority habitats monitoring from satellite sensor data. *International Journal of Remote Sensing*, **27**(13), 2631–2644 (2006).

31 I. Keramitsoglou, H. Sarimveis, C.T. Kiranoudis, C. Kontoes, N. Sifakis and E. Fitoka, The performance of pixel window algorithms in the classification of habitats using VHSR imagery, *ISPRS Journal of Photogrammetry and Remote Sensing*, **60**(4), 225–238 (2006).

32 B. Dixon and N. Candade, Multispectral land use classification using neural networks and support vector machines: one or the other, or both? *International Journal of Remote Sensing*, **29**(4), 1185–1206 (2008).

33 M. Dalponte, L. Bruzzone, L. Vescovo and D. Gianelle, The role of spectral resolution and classifier complexity in the analysis of hyperspectral images of forest areas. *Remote Sensing of Environment*, **113**(11), 2345–2355 (2009).

34 V.R. Tóth and K. Szabó, Morphological structural analysis of *Phragmites australis* stands in Lake Balaton. *Annales de Limnologie – International Journal of Limnology*, **48**, 241–251 (2012).

10

Copernicus Framework for Monitoring Lake Balaton Phytoplankton

S.C.J. Palmer[1,2,3], A. Zlinszky[2], Heiko Balzter[3], V. Nicolás-Perea[3] and V.R. Tóth[2]

[1] Department of Biology, University of Prince Edward Island, Charlottetown, Canada
[2] Balaton Limnological Institute, Centre for Ecological Research, Hungarian Academy of Sciences, Tihany, Hungary
[3] University of Leicester, Centre for Landscape and Climate Research, Department of Geography, Leicester, UK

10.1 Introduction

Lakes are important components and regulators of global carbon, nitrogen and phosphorous cycling, are vital within the hydrologic cycle, in addition to providing diverse habitat for rich biodiversity [1–3]. Phytoplankton are the photosynthesizing base of the aquatic food chain, and as such underlie the productivity of lake ecosystems. The quality of lake water can become deteriorated, however, when phytoplankton biomass becomes too high during phenomena known as *eutrophication* and *phytoplankton* (or *algal*) *blooms*. Such events are typically found to occur in response to elevated nutrient levels, from sewage or agricultural runoff, and may result in hypoxic conditions, whereby decomposing phytoplankton biomass decreases oxygen levels in the water with negative effects for many species up the food chain in addition to a range of other negative environmental and economic impacts [4]. Likewise, some species of phytoplankton, such as *cyanobacteria* or *blue-green algae*, are known to produce toxins which may have harmful ecological and health effects and are associated with eutrophic conditions [4]. Phytoplankton dynamics are directly and indirectly affected by temperature changes, are characterized by short generation times, and are thus considered especially sensitive indicators of climate and other environmental change [5]. Experimental and field results generally concur that warming trends will result in increased and intensified eutrophication [3].

Ongoing monitoring has an important role to play in the detection and early-warning of phytoplankton blooms, as well as in attributing and quantifying trends and changes in phytoplankton dynamics. The effects of environmental change on lake systems occur at diverse spatial and temporal scales, which is an important consideration in the design and interpretation of monitoring data. Likewise, natural variability is often crucial to the health and functioning of lake ecosystems, as is the case for most ecosystem types, and may itself be desirable to preserve or recover [6,7] such that the change intended to be monitored is itself is difficult to define. Substantial baseline information is thus a key aspect of monitoring [8]. Traditional monitoring through point measurements has proven limited and resource-intensive at best, and simply impossible at worst.

Earth Observation for Land and Emergency Monitoring, First Edition. Edited by Heiko Balzter.
© 2017 John Wiley & Sons Ltd. Published 2017 by John Wiley & Sons Ltd.

The latter case occurs particularly when lakes are remote, hindering access, regular or otherwise, or when resources necessary to undertake monitoring are limited. Satellite remote sensing has been identified as an alternative, or more realistically, as a complementary source of data, having the potential provide spatially and temporally cohesive information on lake water quality, with regular (on the order of daily, weekly or monthly, depending on the satellite sensor and geographic context of the lake in question) observations of entire lake surfaces [9,10].

The effectiveness of using satellite and airborne imagery to measure and map phytoplankton in lakes has already been demonstrated in a number of cases [11,12], but remains complex, with operationalization of remote sensing-based monitoring presenting additional challenge. This chapter provides an overview of the principles underlying the remote sensing of phytoplankton, discusses the suitability of past and future Global Monitoring for Environment and Security (GMES)/Copernicus sensors aboard ENVISAT, Sentinel-2 and Sentinel-3 platforms, and highlights the potential role of remote sensing in Lake Balaton (Hungary) phytoplankton monitoring activities as a case study. Algorithm calibration and validation using data from the Medium Resolution Imaging Spectrometer (MERIS) and *in situ* measurements are presented and discussed with regards to the impact of algorithm selection on phytoplankton biomass mapping. Early bloom onset detection and monitoring for Lake Balaton is demonstrated, as is the potential insight from long-term monitoring, and the improvements and continuity to MERIS data afforded by the upcoming European Space Agency (ESA) Sentinel missions.

10.2   Remote Sensing of Phytoplankton

10.2.1   Optically Active Water Constituents

The study of water quality parameters through their optical properties can largely be understood through their influence on water colour, given that some parameters of interest (known as *optically-active substances*) reflect or absorb solar radiation at different wavelengths across the visible and near-infrared (VNIR) range of the electromagnetic spectrum. Water containing very little dissolved or suspended matter tends to appear dark blue, absorbing the least amount of light in that spectral range (400–500 nm) [13]. In the context of the optical properties of a water column, three main groups of optically-active substances affect water-leaving radiance through scattering and absorption of VNIR light, in addition to the effect of pure water which is a hypothetical medium consisting of only water molecules and dissolved inorganic matter. These are (1) phytoplankton; (2) suspended particulate matter (SPM); and (3) coloured dissolved organic matter (CDOM), also known as "gelbstoff", "yellow substance", or "gilvin" (Figure 10.1) [14].

The presence and amount of phytoplankton biomass is generally determined through measurements of chlorophyll *a* (Chl*a*), the main pigment common among phytoplankton species. Chl*a* is characterized by strong absorption at blue and red wavelengths as well as between 670 and 690 nm, [15,16], a strong and narrow fluorescence peak at 685 nm [17] and a peak at around 705 due to phytoplankton backscattering. However, thousands of phytoplankton species have been identified, many groups of which are associated with distinguishing auxiliary pigments in addition to Chl*a*, such as chlorophyll *b*,

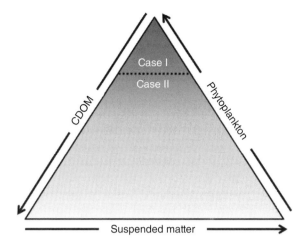

Figure 10.1 Schematic diagram of the Case 1 and Case 2 water types; the domination of phytoplankton and its bi-product, CDOM (Case 1) and all other combinations of the three optically-active substances (adapted from [14]).

phycocyanin, carotenes and xanthophylls, which each have their own characteristic absorption, reflectance and fluorescence signals. Individual phytoplankton species, or groups of species, thus display variations in their spectral response profiles due to characteristic reflectance and absorbance of visible light at distinct wavelengths. Thus, in addition to measuring phytoplankton biomass in general, it can be possible to measure that of individual, targeted species or groups of species given adequate spectral coverage and resolution such as that typically afforded by hyperspectral data [16].

10.2.2 Optically Complex Waters

The retrieval of Chl$a$ concentrations as a proxy for phytoplankton biomass in lake water poses a great challenge, however, due to the optical complexity of the water. All three groups of optically-active constituents listed above may be present in a broad range of concentrations and influence the optical properties of the water independently of each other, to varying degrees and over overlapping spectral ranges. This is in contrast to many cases, generally in open (*pelagic*) ocean settings, where although other components (CDOM, SPM) may be present, their influence on the overall optical properties of the water column is insignificant given their generally low concentrations and the fact that they co-vary linearly with, and in fact are typically by-products of, phytoplankton [14]. Traditionally these types of waters have been classified as "Case 1" whereas the former are known as "Case 2", following the classification introduced by Morel and Prieur (1977) [18] and later updated by Gordon and Morel (1983) [19] and Morel (1988) (Figure 10.1) [20]. Although not universally the case, coastal zone and inland waters are often of the complex, Case 2 type due to the potential for bottom sediment resuspension or suspended matter and CDOM input from adjacent terrestrial environments (rivers or runoff), as well as generally higher and more variable phytoplankton biomass. It should be noted that although this traditional classification system is still commonly used and referred to, its discontinuation has been proposed due to the continuum that the two cases comprise in reality, both in terms of concentrations of parameters (there is no real divide distinguishing so-called Case 1 from so-called Case 2 waters) and because of the temporal variability (i.e., a location may be classified as Case 1 at some times (e.g., high tide) but Case 2 at other times (e.g., low tide)) [21].

In so-called Case 2 waters, such as lakes, the influence of each component on the water-leaving spectral response of a water body overlaps with the others over certain spectral ranges and unique signatures are thus difficult to distinguish; at any given wavelength studied, a linear, one-to-one relationship with any of the components cannot be assumed [14]. The diversity and optical complexity of the so-called Case 2 waters renders the remote sensing of any one parameter, such as phytoplankton, difficult and requires either site-specific or extremely broad retrieval algorithm calibration so as to encompass the range of conditions possible to encounter. Much greater spectral resolution and coverage, and radiometric accuracy and precision (high signal-to-noise ratio) are also required of sensors [14].

10.2.3 Inherent and Apparent Optical Properties

Inherent Optical Properties (IOPs) describe the physical basis of the remote sensing of the water column and its constituents. These are the wavelength-dependent absorption (a) and the Volume Scattering Function (β), from which the total scattering coefficient (b) is determined, of the three optically-active substance types described above and their cumulative effect on the coefficient of beam attenuation (c) (Equation 10.1) [22]. Backscattering (b_b) integrates the volume scattering function (β) over the angles 90–180°. The total absorption ($a_{total}(\lambda)$) and backscattering ($b_{b\ total}(\lambda)$) can be further broken down into the contributions from each of the three constituent types and from the water itself (a_w, $b_{b\ w}$) (Equations 10.2 and 10.3). The IOPs are characteristic of the media, without influence from ambient light conditions [23].

$$c(\lambda) = a(\lambda) + b(\lambda) \tag{10.1}$$

$$a_{total}(\lambda) = a_w(\lambda) + a_{ph}(\lambda) + a_{dg}(\lambda) + a_{CDOM}(\lambda) \tag{10.2}$$

$$b_b\ total(\lambda) = b_{b\ w}(\lambda) + b_{b\ ph}(\lambda) + b_{b\ dg}(\lambda) \tag{10.3}$$

Measurements via remote sensing, however, also include the influence of the light field at the time of measurement, as well as sensor and measurement geometry. Measurements are therefore of Apparent Optical Properties (AOPs), such as remote sensing reflectance (R_{rs}) which is the ratio of radiance upwelling from the surface of the water (L_u) to that downwelling from the sun (E_d) (Equation 10.4). AOPs can be related quantitatively to IOPs via the use of a constant, f, dependant on the measurement light field and β (Equation 10.5), [24] and IOPs are related to concentrations of optically active substances, such as Chl$a$, through different constituent retrieval algorithm types.

$$R_{rs}(\lambda) = L_u(\lambda)/E_d(\lambda) \tag{10.4}$$

$$R_{rs}(\lambda) = f(b_b(\lambda)/a(\lambda) + b_b(\lambda)) \tag{10.5}$$

10.2.4 Constituent Retrieval Algorithms

Given the wavelength dependence of different optically-active substances' influence on the optical properties of the water column, different types of models have been proposed to quantitatively relate remotely sensed AOPs to constituent concentrations, either directly or via quantification of the IOPs. One type makes use of radiative transfer bio-optical modelling of the relationship between optical properties and constituent concentrations, and tends to be more directly transferable between sites as a result of the strong physical basis [12]. Forward model, relating optical properties to concentrations of all three optically-active substance types, is inverted. However, these algorithms are also limited in that they require large datasets of difficult to obtain optical properties and extensive parameterization. This is also the case for neural network type algorithms, which use remotely sensed data as the input layer, and constituent concentrations such as that of Chl*a*, and in some cases IOPs as the output [25–27].

Semi-empirical band arithmetic type models directly fit statistical relationships between reflectance measured at one or a combination of different wavelengths (different bands of a given sensor) and the constituent concentrations [11]. Such algorithms deriving Chl*a* concentrations in lakes typically make use of the Chl*a* absorption, backscattering or fluorescence features in the red and near-infrared range, which is less affected by other optically actively substances compared with the blue and green range typically used to retrieve Chl*a* in open ocean settings [28]. Relationships are determined for one constituent at a time, and coefficients and even band combinations found to successfully retrieve concentrations at one site are likely to completely fail at another given between site variability of optically-active substances (which dominate, at what concentration levels, sediment type, phytoplankton species and groups, etc.). Site-specific calibration and coefficient tuning are required [29]. On the other hand, many band arithmetic type algorithms have been found to very robustly retrieve constituent concentrations for the site for which they were trained, and several band combinations have proven transferable between sites if local calibration of model coefficients is carried out [11,12]. It is important to note that for an algorithm to be expected to perform robustly for a given site, the range of conditions encountered should be included in calibration and validation. Furthermore, given the highly dynamic nature of phytoplankton, it is important that *in situ* measurements for the calibration-validation process correspond closely with image acquisition. In many cases, matchups from the same day as image acquisition, or within a given time frame (e.g., ± three hours) are considered acceptable.

10.3 Medium Resolution Imaging Spectrometer (MERIS)

The Medium Resolution Imaging Spectrometer (MERIS), aboard the European Space Agency (ESA)'s polar-orbiting ENVISAT satellite platform, is a push broom multispectral imaging sensor measuring in 15 spectral bands ranging from the visible (band one; centred at 412.5 nm) to the near-infrared range (band fifteen; centred at 900 nm) [30]. MERIS was primarily envisaged as an ocean observing sensor, with a goal of adding to the understanding of ocean processes and productivity within the climate system context [30]. Nevertheless, sensor characteristics as well as a diverse ground segment component have permitted its extension to various atmospheric and land applications as well, including inland waters, for which it has permitted considerable unprecedented insight. The combination of radiometric, spectral, spatial and temporal resolutions have proven suitable for retrieving meaningful signals from medium to large size optically complex waters. Its band placement allows the various optically active substances suspended or dissolved in the water column to be distinguished spectrally. In the context of oceanic and inland waters measurements, a number of band placement decisions were made in association with the spectral signatures of Chl*a* and other pigments, CDOM and SPM [47] (Table 10.1). Particularly well suited for optically complex waters are the presence of MERIS band 8, centred at 681.25 to capture the Chl*a* absorbance or fluorescence features (depending on the given phytoplankton species composition and concentration range), and band 9, centred at 709 nm and which captures a peak related to high biomass phytoplankton blooms [28,31,32]. Its ability to make use of the red and near infrared spectral range for Chl*a* retrievals reduces the interference by the CDOM signal in the blue-green range.

Given its intended use over water, which tends to be very dark, the radiometric sensitivity was of high priority in the sensor design. A radiometric resolution of less than 0.03 $Wm^{-2}sr^{-1}mm^{-1}$ was identified as necessary to discriminate Chl*a* concentrations at the level of 1 mg m^{-3} [30]. MERIS has a swath width is 1150 km, measured by five overlapping cameras, and operates on two spatial resolutions; full resolution (FR) with a pixel size of 300 m, intended for land and coastal zone applications, and reduced resolution (RR) with a pixel size of 1200 m at nadir, intended for global ocean applications. Two to three day overpass allows for sufficient temporal resolution of data to closely follow phytoplankton bloom dynamics, particularly given the dry, continental climate of Lake

Table 10.1 Spectral positioning and resolution of the 15 MERIS bands, and potential aquatic, vegetation and atmospheric applications. Adapted from ©ESA Earthnet Online 2000–2013 [30].

Band number	Band centre (nm)	Bandwidth (nm)	Potential application
1	412.5	10	Yellow substance, pigments
2	442.5	10	Chl*a* absorption maximum
3	490	10	Chl*a*, other pigments
4	510	10	Suspended sediment, red tides
5	560	10	Chl*a* absorption minimum
6	620	10	Suspended sediment
7	665	10	Chl*a* absorption, fluorescence reference
8	681.25	7.5	Chl*a* fluorescence peak
9	708.75	10	Fluorescence reference, atmospheric correction
10	753.75	7.5	Vegetation, cloud
11	760.625	3.75	Oxygen absorption
12	778.75	15	Atmospheric correction
13	865	20	Vegetation, water vapour reference
14	885	10	Atmospheric correction
15	900	10	Water vapour, land

Balaton, for example, which limits interference by cloud cover. Although ENVISAT stopped operations in April 2012, a ten-year archive of MERIS imagery data from 2002–2012 remains as a valuable legacy.

10.4 Lake Balaton

Lake Balaton, located in the western Transdanubian region of Hungary (104.8 metres above sea level [masl]), is by surface area (597 km^2) the largest lake in Central Europe [33,34] (Figure 10.2). It is very turbid and highly reflective, as a result of its very shallow depth (on average 3.3 m) [35] and wind induced resuspension of its calcareous benthic sediment. The lake is characterized by a trophic gradient from typically mesotrophic (relatively low phytoplankton biomass) in the northeast to eutrophic (relatively high phytoplankton biomass), as a result of the nutrient-rich inflow from the Zala River, in the southwest [36,37]. However spatial and seasonal variability superimposed on this gradient are also observed.

Though in a highly agricultural region, tourism has been estimated to be approximately twelve times more important economically [38]. It has been estimated that more than a million tourists visit the region per year, mainly concentrated in the six to eight week "high" period in July and August [38,39]. The tourism industry associated with Lake Balaton is long lived, already well under development in the 1840s [39]. Such an intensive tourism industry results in consequent pressure on the Lake Balaton ecosystem as a result of increased resource use and waste production. Likewise, the nationally and

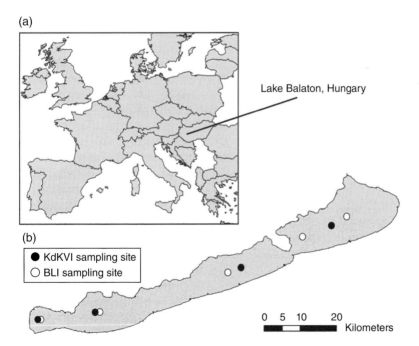

Figure 10.2 The location of Lake Balaton in western Hungary (a). Regular sampling sites for Lake Balaton phytoplankton monitoring by the BLI and the KdKVI (b).

regionally important Lake Balaton tourism industry is heavily dependent on the lake as a natural resource and as such is very sensitive to lake water quality and related ecosystem features and processes.

Over the past decades, water quality has been of ongoing concern. Of particular importance was the severe eutrophication that peaked in the 1970s and 1980s, requiring major remediation and mitigation measures, including the development of a new sewage diversion system and treatment facility, and the construction of the Kis-Balaton Water Protection System (KBWPS) [33,40]. The success of the KBWPS in reducing the nutrient load from the Zala River to Lake Balaton, in combination with sewage diversion and treatment, has been widely acclaimed and quantified [41–43]. Currently, however, speculation surrounds the reversal of KBWPS performance, with sediment saturating in terms of nutrient uptake capacity and nutrient release consequently occurring [44]. This is yet to be confirmed, but will be an important consideration for future monitoring activities. Water quality monitoring undertaken to date has taken the form of sampling two to five points at the centre of the different basins of the lake once to twice monthly by both the Balaton Limnological Institute (BLI) and the Central Transdanubian (Regional) Inspectorate for Environmental Protection, Nature Conservation and Water Management (KdKVI; "*Közép-dunántúli Környezetvédelmi, Természetvedelmi és Vízügyi Felügyeloség*") (Figure 10.2).

10.5 MERIS Phytoplankton Mapping of Lake Balaton

Toward establishing a remote sensing framework for the monitoring of Lake Balaton phytoplankton, an applied review of Chl*a* retrieval algorithms reported to have performed highly for other optically complex waters was undertaken as part of GIONET (*Copernicus* (formerly known as *GMES*) *Initial Operations – Network for Earth Observation Research Training*). These include neural network and semi-empirical, band arithmetic architecture types [45,46]. More than 1 400 MERIS overpasses between January, 2007 and April, 2012 included full or partial coverage of Lake Balaton, and 692 measurements of *in situ* Chl*a* concentrations are available for the same period from the BLI and the KdKVI archives. Considering cloud cover and only same-day (± approximately three hours) coincidence with *in situ* archive data, 68 images and 289 sample points were available for validation match-ups. Pixels flagged at Level 1 as suspect, risk of sun glint, land, coastline, bright or invalid, were excluded from further analysis, leaving the final count of validation match-ups at 211. Although the majority of match-ups occur during the months May to October, during which time cloud cover occurs less frequently and sampling frequency doubles, match-ups represent all seasons and all years of the five year study period. This is an important factor in encompassing to the extent possible the full range of conditions encountered in the lake for algorithm calibration and validation, so as to ensure reliability of image-based Chl*a* maps and future monitoring activities. Full resolution (300 m) imagery improved via AMORGOS geolocation (FSG) was used, and values of Chl*a*, in the case of the neural networks, or radiance, in the case of input for the semi-empirical algorithms, were extracted from a 3 × 3 pixel kernel (approximately 0.8 km^2) centred at each *in situ* matchup.

The selected neural networks have been trained using *in situ* and simulated datasets from elsewhere, and comprise components for atmospheric correction (removal of the atmospheric component from the total signal measured by the satellite) either in a separate neural network module or within the same step as Chl*a* retrieval. These include the Case 2 Regional (C2R) [26], Eutrophic Lake (EUL), Boreal Lake (BL) [27] and Free University of Berlin (FUB) Water processor (FUB-WeW) [25], all of which are implemented within the Basic ERS & ENVISAT (A)ASTER MERIS (BEAM) image processing toolbox [48] and are easily employed by the user. Each of the neural network processors also flag pixels that extend either the training input or output ranges of the atmospheric correction and constituent retrieval modules, and matchups flagged as such were removed from validation analysis. The semi-empirical algorithms, Fluorescence Line Height (FLH) [28,31] and Maximum Chlorophyll Index (MCI) [32], which make use of the relationship between Chl*a* concentration and the height of Chl*a* fluorescence- and phytoplankton backscattering- induced peaks above a baseline respectively (Figure 10.3), are also implemented as processors in BEAM. Semi-empirical algorithms can otherwise be implemented and tuned to Chl*a* concentration using the band-by-band maths tool in BEAM, or similar functions in other image processing packages. Semi-empirical algorithms that have successfully retrieved Chl*a* concentrations on other optically-complex waters over similar concentrations ranges were applied, and included other height-above-baseline (*Maximum Peak-Height* [49]; *Reflectance Line Height* [50,51]) and band ratio [29,51–53] approaches [46]. Of all tested algorithms, the FLH using Level 1b radiance data as input was found to best retrieve Chl*a* concentration for Lake Balaton (Figure 10.4).

As a result, the FLH algorithm is used to demonstrate use of MERIS in the early onset detection and monitoring of phytoplankton blooms. An example of the progression of a summer bloom event, from late July to early September 2010, is presented in Figure 10.5. Blooms typically begin and are most dominant in the southwest, where nutrient-laden water of the Zala River flow into the lake. The rise in Chl*a* concentrations associated with the bloom are identified as early as July 21, but the maximum concentration is

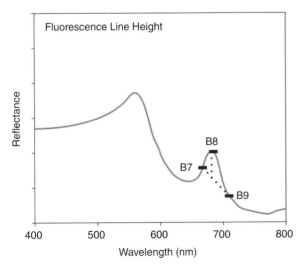

Figure 10.3 Schematic diagram of the FLH algorithm, based on the height of the Chl*a* fluorescence peak near MERIS band 8 (681.5 nm) above a baseline between the two adjacent bands, 7 (664 nm) and 9 (709 nm) (adapted from [28]).

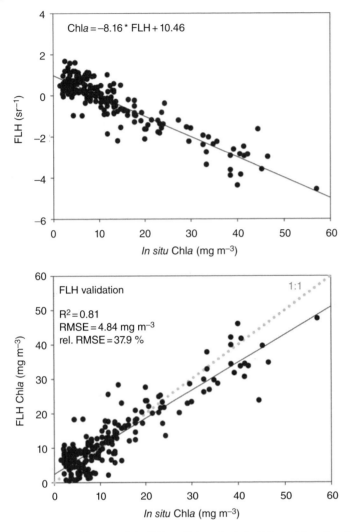

Figure 10.4 Calibration of the FLH coefficients for Lake Balaton (a), and performance evaluation (b) of its Chl*a* retrieval validation ($R^2 = 0.81$, RMSE = 4.84 mg m^3 and relative RMSE = 37.9 %) (adapted from [45]).

reached between August 3 and 8, sustained until after August 14. When bloom progression mapped by MERIS is considered alongside the *in situ* measurements of the BLI (Figure 10.6), 27 MERIS Chl*a* retrievals result, compared with only six BLI measurements. Whereas bloom onset is detected and increasing concentrations monitored as of July 19 using MERIS imagery, the typical BLI sampling interval is such that the bloom is not detected until July 27. Likewise, MERIS data provide addition detail concerning the temporal progression, particularly during the onset period, as well as the spatial features of the bloom not captured by the *in situ* point measurements. Over the full five-year period analyzed and the full ten-year period of the MERIS archive, Palmer et al. [45,47] further demonstrated the ability of the MERIS FLH imagery to consistently capture both

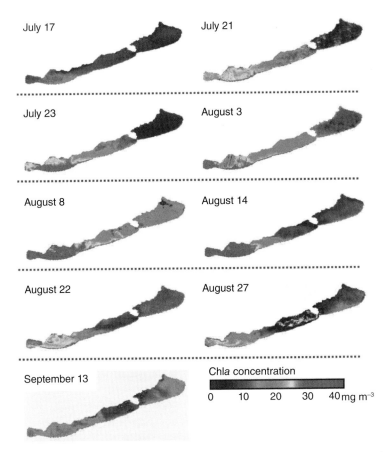

Figure 10.5 Chl*a* maps for Lake Balaton derived from MERIS imagery processed using the FLH algorithm, capturing a summer bloom event beginning in late July 2010. (*See insert for color representation of the figure.*)

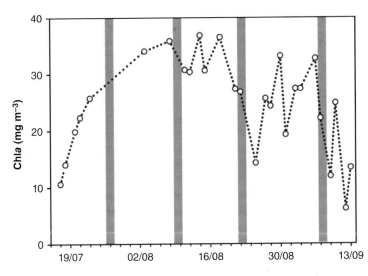

Figure 10.6 Chl*a* concentration retrieved by MERIS-FLH in the southwestern-most basin of Lake Balaton during the 2010 summer bloom (circles and dotted line), contrasted with the timing of BLI sampling (grey bars).

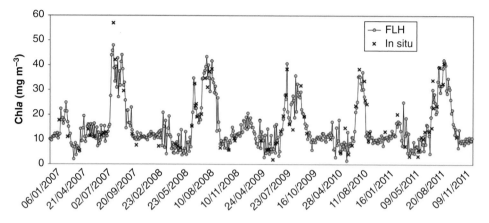

Figure 10.7 Five year time series of MERIS-FLH retrieved Chl*a* in the southwestern-most basin of Lake Balaton, robustly retrieving both bloom and non-bloom periods, and matchup *in situ* measurements (adapted from [45]).

the smaller spring bloom, typically occurring between March and April in Lake Balaton, and the larger summer bloom, between July and September (Figure 10.7), such as that depicted in Figures 10.5 and 10.6.

10.6 Future Outlook

Although MERIS is no longer operating and as such cannot be used for ongoing monitoring activity, its significant data archive remains yet to be fully exploited toward increasing the understanding of lake dynamics and the fine-tuning of biophysical parameter retrieval algorithms. The Ocean and Land Colour Instrument (OLCI) aboard the ESA Sentinel-3 mission has been highly anticipated in the realm of lake remote sensing as it largely provides data continuity to the MERIS sensor data. Another medium resolution imaging spectrometer, OCLI is characterized by similar spectral placement, with several additional bands in the blue, red and NIR spectral ranges, as well as having similar spatial resolution (Figures 10.8, 10.9) [54]. Time series generation bridging the MERIS and OLCI data is facilitated thus and scientific and operational monitoring that began with MERIS can be expected to be resumed with similar algorithms and processing chains, and fine-tuned to make use of the additional bands. Launch of the first of the three Sentinel-3 satellites, with an operational lifespan of seven years, took place on February 16, 2016 [53]. Each has an operational lifespan of seven years [54]. The Multispectral Imager (MSI) to be on board the future series of Sentinel-2 satellites also promises great insight into inland waters. Although its spectral resolution is less well suited to inland waters as compared with OLCI, its spatial resolution of 10, 20 and 60 m will allow imaging of smaller water bodies (Figures 10.8, 10.9). The first Sentinel-2 satellite was launched on June 23, 2015, with a second foreseen for launch in 2017 [54].

Preparatory activities have already been underway for the Sentinels for several years prior to their launch, with inland waters represented on the Sentinel-3 Validation Team

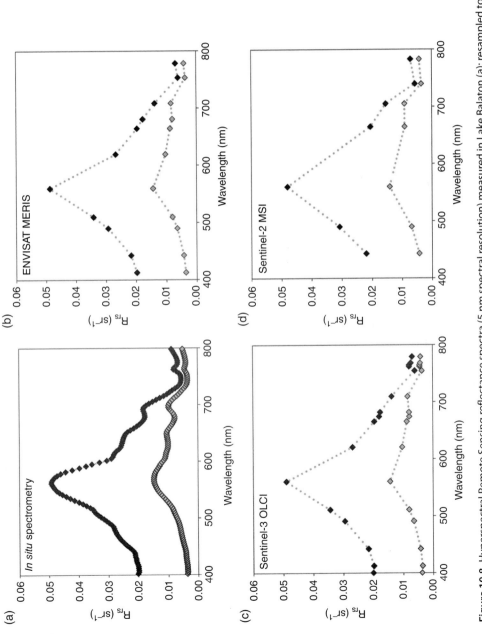

Figure 10.8 Hyperspectral Remote Sensing reflectance spectra (5 nm spectral resolution) measured in Lake Balaton (a); resampled to the spectral bands of ENVISAT MERIS (b); Sentinel-3 Ocean and Land Colour Imager bands (c); and Sentinel-2 MultiSpectral Imager (d).

(a)
(b)
(c)
(d)

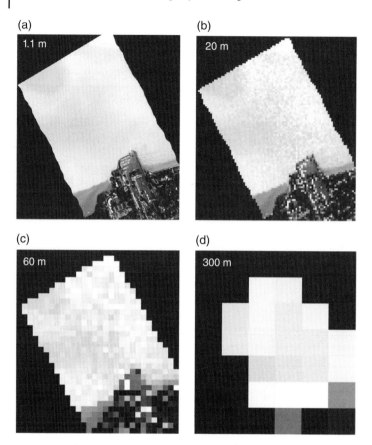

Figure 10.9 AISA airborne hyperspectral imagery (1.1 m spatial resolution) over the southern shoreline of northeastern Lake Balaton (Siofok basin) (a); resampled to the spatial resolutions of Sentinel-2 MultiSpectral Imager (b, c); and Sentinel-3 Ocean and Land Colour Imager bands (d).

which has as an objective "to provide independent validation evidence, experimental data and recommendations to the S3 Mission" [54]. The Global Lakes Sentinel Services (GLaSS; www.glass-project.eu/), European Union 7th Framework Programme project number 313256, is part of the Sentinel-3 Validation Team, and is specifically dedicated to preparing for the wealth and large volume of data for inland waters to arise from the Sentinel satellites, including the development and adaptation of water quality algorithms, integration of Sentinel-2 and -3 data, automated processing and data mining, toward increased accessibility of data generally. Other projects, GloboLakes (funded by the UK National Environment Research Council; www.globolakes.ac.uk/) and Diversity-2 (funded by ESA; www.diversity2.info/) also seek to identify and develop common methods applicable to all lakes, or broad classes of lakes globally. Although currently based predominantly on archive MERIS and matchup *in situ* data, the long term perspective entails transferability to and exploitation of Sentinel data. It has been confirmed that "free and open access to Sentinel satellite data will be granted for the Copernicus operational phase" [55]. This is expected to result in multifold spinoff benefits in terms of research productivity and employment generation, and assures the facility of future Lake Balaton satellite-based monitoring.

A processing chain using MERIS imagery, the selected FLH Chl*a* retrieval algorithm and freely-available image processing software, expected to accommodate future Sentinel imagery, has been developed. Transfer of the methodology to researchers stakeholders of the BLI is foreseen to not only encourage continuity and extension of the use of satellite imagery for Chl*a* mapping by these institutes, but will also provide an improved understanding of both the benefits and limitations of satellite products, allowing advantage to be taken more fully of products and portals resulting from larger projects that include Lake Balaton, such as GloboLakes and Diversity-2, as well as ChloroGIN (www.chlorogin.org/). In this way the future priorities highlighted by the Group on Earth Observations (GEO) related to capacity building [56] are also addressed at the local scale. Future Sentinel-based monitoring will contribute to requirements of the European Commission Water Framework Directive pertaining to phytoplankton biomass [57]. Based on activities undertaken to date, recommendations for slight adjustments to *in situ* sampling so as to optimize complementarity with satellite products have been made to appropriate authorities, particularly to coincide timing of sampling with image acquisition for ongoing validation.

10.7 Conclusions

Concentrations of Chl*a* in Lake Balaton were accurately retrieved using the FLH algorithm applied to MERIS satellite imagery, delineating spatial and temporal patterns that would otherwise go undetected. The MERIS archive remains an invaluable source of data to be exploited toward improved understanding of phytoplankton dynamics during the ten years of data acquisition. Although no longer active, lessons from and progress made through the availability of MERIS data, during and after its lifetime, allow us to understand and prepare for the future possibilities of the ESA Sentinel missions. A processing chain, including the selected Chl*a* retrieval algorithm, transferable from MERIS to Sentinel-3 OLCI, has been developed to encourage continuity of local remote sensing activities by Lake Balaton stakeholders.

Acknowledgements

The authors are grateful for the financial support of this project through GIONET, funded by the European Commission, Marie Curie Programme Initial Training Network, Grant Agreement PITN-GA-2010-264509. H. Balzter was supported by the Royal Society Wolfson Research Merit Award, 2011/R3. A. Zlinszky was supported by the project TÁMOP-4.2.2.A-11/1/KONV-2012-0064.

References

1 L. Duker and L. Borre, Biodiversity conservation of the world's lakes: A preliminary framework for identifying priorities, *LakeNet Report Series*, Number 2. Monitor International, Annapolis, Maryland USA (2001).

2 C.E. Williamson, J.E. Saros, W.F. Vincent and J.P. Smol, Lakes and reservoirs as sentinels, integrators, and regulators of climate change, *Limnol. Oceanogr.* **54**(6 part 2), 2273–2282 (2009).

3 B. Moss, Cogs in the endless machine: Lakes, climate change and nutrient cycles: A review, *Sci. Total Environ.*, **434**, 130–142 (2012).

4 V.H. Smith, Eutrophication of freshwater and coastal marine ecosystems a global problem, *Environ. Sci. Pollut. R.*, **10**(2), 126–139 (2003).

5 R. Adrian, C.M. O'Reilly, H. Zagarese, S.B. Baines *et al.*, Lakes as sentinels of climate change, *Limnol. Oceanogr.*, **54**, 2283–2297 (2009).

6 P.B. Landres, P. Morgan and F.J. Swanson, Overview of the use of natural variability concepts in managing ecological systems, *Ecol. Appl.*, **9**(4), 1179–1188 (1999).

7 M. Erlandsson, J. Folster, H. Laudon, G.A. Weyhenmeyer and K. Bishop, Natural variability in lake pH on seasonal, interannual and decadal time scales: Implications for assessment of human impact, *Environ. Sci. Technol.*, **42**(15), 5594–5599 (2008).

8 G.M. Lovett, D.A. Burns, C.T. Driscoll, J.C. Jenkins, M.J. Mitchell, L. Rustad, J.B. Shanley, G.E. Likens and R. Haeuber, Who needs environmental monitoring? *Front. Ecol. Environ.*, **5**(5), 253–260 (2007).

9 H.B. Glasgow, J.M. Burkholder, R.E. Reed, A.J. Lewitus and J.E. Kleinman, Real-time remote monitoring of water quality: a review of current applications, and advancements in sensor, telemetry, and computing technologies, *J. Exp. Mar. Biol. Ecol.*, **300**, 409–448 (2004).

10 Q. Chen, Y. Zhang, A. Ekroos and M. Hallikainen, The role of remote sensing technology in the EU water framework directive (WFD), *Environ. Sci. Policy*, **7**, 267–276 (2007).

11 M.W. Matthews, A current review of empirical procedures of remote sensing in inland and near-coastal transitional waters, *Int. J. Remote Sens.*, **32**(21), 6855–6899 (2011).

12 D. Odermatt D, A. Gitelson, V.E. Brando and M. Schaepman, Review of constituent retrieval in optically deep and complex waters from satellite imagery, *Remote Sens. Environ.*, **118**, 116–126 (2012).

13 H. Buiteveld, J.H.M. Hakvoort and M. Donze, Optical properties of pure water. In: *Proceedings of SPIE, Ocean Optics XII*, 13–15 June 1994, Bergen, Norway, J.S. Jaffe (Ed.), 174–183 (1994).

14 IOCCG, Remote sensing of ocean colour in coastal, and other optically-complex waters, S. Sathyendranath, (Ed.), *Reports of the International Ocean-Colour Coordinating Group*, No. 3, IOCCG, Dartmouth, Canada (2000).

15 E. Sváb, A.N. Tyler, T. Preston, M. Présing and K.V. Balogh, Characterizing the spectral reflectance of algae in lake waters with high suspended sediment concentrations. *Int. J. Remote Sens.*, **26**(5), 919–928 (2005).

16 P.D. Hunter, A.N. Tyler, M. Présing, A.W. Kovács and T. Preston, Spectral discrimination of phytoplankton colour groups: The effect of suspended particulate matter and sensorspectral resolution, *Remote Sens. Environ.*, **112**, 1527–1544 (2008).

17 R. Barbini, F. Colao, R. Fantoni, C. Micheli, A. Palucci and S. Ribezzo, Design and application of a lidar fluorosensor system for remote monitoring of phytoplankton, *ICES J. Mar. Sci.*, **55**, 793–802 (1998).

18 A. Morel and L. Prieur, Analysis of variations in ocean color, *Limnol. Oceanogr.* **22**, 709–722 (1977).

19 H.R. Gordon and A. Morel, Remote assessment of ocean color for interpretation of satellite visible imagery. A review, *Lecture Notes on Coastal and Estuarine Studies*, R. T. Barber, N. K. Mooers, M. J. Bowman and B. Zeitzschel (Eds.), Springer-Verlag, New York, 114 pp. (1983).

20 A. Morel, Optical modeling of the upper ocean in relation to its biogeneous matter content (Case 1 waters), *J. Geophys. Res.*, **93**(C9), 10749–10768 (1988).

21 C.D. Mobley, D. Stramski, W.P. Bissett and E. Boss, Optical modeling of ocean waters: is the Case 1–Case 2 classification still useful? *Oceanography*, **17**, 61–67 (2004).

22 IOCCG, Remote sensing of inherent optical properties: Fundamentals, tests of algorithms, and applications, Z.P. Lee (Ed.), *Reports of the International Ocean-Colour Coordinating Group*, No. 5, IOCCG, Dartmouth, Canada (2006).

23 J.T.O. Kirk, Volume scattering function, average cosines, and the underwater light field, *Limnol. Oceanogr.*, **36**(3), 455–467 (1991).

24 H.R. Gordon, O.B. Brown and M.M. Jacobs, Computed relationships between the inherent and apparent optical properties of a flat homogeneous ocean, *Appl. Opt.*, **14**(2), 417–427 (1975).

25 T. Schroeder, I. Behnert, M. Schaale, J. Fischer and R. Doerffer, Atmospheric correction algorithm for MERIS above case-2 waters, *Int. J. Remote Sens.*, **28**(7), 1469–1486 (2007).

26 R. Doerffer and H. Schiller, The MERIS Case 2 water algorithm, *Int. J. Remote Sens.*, **28**(3–4), 517–535 (2007).

27 R. Doerffer and H. Schiller, MERIS lake water algorithm for BEAM–MERIS algorithm theoretical basis document, V1.0, 10 June 2008. GKSS Research Center, Geesthacht, Germany (2008).

28 J.F.R. Gower, R. Doerffer and G.A. Borstad, Interpretation of the 685 nm peak in water-leaving radiance spectra in terms of fluorescence, absorption and scattering, and its observation by MERIS, *Int. J. Remote Sens.*, **20**(9), 1771–1786 (1999).

29 K. Kallio, T. Kutser, T. Hannonen, S. Koponen, J. Pulliainen, J. Vepsalainen and T. Pyhalahti, Retrieval of water quality from airborne imaging spectrometry of various lake types in different seasons, *Sci. Total Environ.*, **268**, 59–77 (2001).

30 ESA Earthnet Online 2013a, *MERIS Design*. Available from: <https://earth.esa.int/web/guest/missions/esa-operational-eo-missions/envisat/instruments/meris/design>.

31 J.F.R. Gower, L. Brown and G.A. Borstad, Observation of chlorophyll fluorescence in west coast waters of Canada using the MODIS satellite sensor, *Can. J. Remote Sens.*, **30**(1), 17–25 (2004).

32 J.F.R. Gower, S. King, G.A. Borstad and L. Brown, Detection of intense plankton blooms using the 709 nm band of the MERIS imaging spectrometer, *Int. J. Remote Sens.*, **26**(9), 2005–2012 (2005).

33 S. Herodek, L. Lackó and Á. Virág, Lake Balaton research and management, K. Misley (Ed.), *United Nations Environment Program and Hungarian Ministry of Environment and Water Management co-publication* (1988).

34 G. Szábo, B. Khayer, A. Rusznyák, I. Tátrai, G. Dévai, K. Márialigeti and A. Borsodi, Seasonal and spatial variability of sediment bacterial communities inhabiting the large shallow Lake Balaton, *Hydrobiologia*, **663**, 217–232 (2011).

35 A. Zlinszky and G. Timár, Historic maps as a data source for socio-hydrology: a case study of the Lake Balaton wetland system, Hungary, *Hydrol. Earth Syst. Sc.*, **17**(11), 4589–4606 (2013).

36 M. Présing, T. Preston, A. Takátsy, P. Speőber, A.W. Kovács, L. Vörös, Gy. Kenesi and I. Kóbor, Phytoplankton nitrogen demand and the significance of internal and external nitrogen sources in a large shallow lake (Lake Balaton, Hungary), *Hydrobiologia*, **599**, 87–95 (2008).

37 A. Mózes, M. Présing and L. Vörös, Seasonal dynamics of Picocyanobacteria and Picoeukaryotes in a large shallow lake (Lake Balaton, Hungary), *Int. Rev. Hydrobiol.*, **91**(1), 38–50 (2006).

38 Lake Balaton Development Coordination Agency (LBDCA). *Lake Balaton Region.* 10th International Living Lakes Conference proceedings (2005).

39 L. Puczkó and T. Rátz, Tourist and resident perceptions of the physical impacts of tourism at Lake Balaton, Hungary: Issues for sustainable tourism management, *J. Sustain. Tour.*, **8**(6), 458–478 (2000).

40 V. Istvánovics, A. Clement, L. Somlyódy, A. Specziár, L. G-Tóth and J. Padisak, Updating water quality targets for shallow Lake Balaton (Hungary), recovering from eutrophication, *Hydrobiologia*, **581**, 305–318 (2007).

41 P. Pomogyi, Nutrient retention by the Kis-Balaton Water Protection System, *Hydrobiologia*, **251**, 309–320 (1993).

42 I. Tátrai, K. Mátyás, J. Korponai, G. Paulovits and P. Pomogyi, The role of the Kis-Balaton Water Protection System in the control of water quality of Lake Balaton, *Ecol. Eng.*, **16**(1), 73–78 (2000).

43 I. Tátrai, Á.I. György, K. Mátyás, J. Korponai, P. Pomogyi, Á. Vári, V. Józsa and G. Boros, Intrinsic processes causing periodic changes in stability in a shallow biomanipulated lake, *Mar. Freshwater Res.*, **62**, 197–204 (2011).

44 H. Horváth, K. Mátyás, Gy. Süle, and M. Présing, Contribution of nitrogen fixation to the external nitrogen load of a water quality control reservoir (Kis-Balaton Water Protection System, Hungary), *Hydrobiologia*, **702**, 255–265 (2013).

45 S.C.J. Palmer, P.D. Hunter, T. Lankester, S. Hubbard, E. Spryakos, A.N. Tyler, M. Présing, H. Horváth, A. Lamb, H. Balzter, and V. Tóth, Validation of Envisat MERIS algorithms for chlorophyll retrieval in a large, turbid and optically-complex shallow lake, *Remote Sens. Environ.*, **157**, 158–169 (2015).

46 S.C.J. Palmer, Remote sensing of spatiotemporal phytoplankton dynamics of the optically complex Lake Balaton, Doctoral dissertation, University of Leicester Department of Geography (2015).

47   S.C.J. Palmer, D. Odermatt, P.D. Hunter, C. Brockmann, M. Presing, H. Balzter, and V.R. Tóth, Satellite remote sensing of phytoplankton phenology in Lake Balaton using 10years of MERIS observations, *Remote Sens. Environ.*, **158**, 441–452 (2015).

48 F. Fomferra and C. Brockmann, BEAM – The ENVISAT MERIS and AATSR Toolbox. Proceedings of the MERIS (A) ATSR Workshop (ESA SP-597). 26–30 September 2005 ESRIN, Frascati, Italy. Lacoste, H. (Ed.) (2005).

49 M.W. Matthews, S. Bernard and L. Robertson, An algorithm for detecting trophic status (chlorophyll-a), cyanobacterial-dominance, surface scums and floating vegetation in inland and coastal waters, *Remote Sens. Environ.*, **124**, 637–652 (2012).

50 J.F. Schalles, A.A. Gitelson, Y.Z. Yacobi, and A.E. Kroenke, Estimation of chlorophyll a from time series measurements of high spectral resolution reflectance in a eutrophic lake, *J. Phycol.*, **34**, 383–390 (1998).

51 M.W. Matthews, S. Bernard and K. Winter, Remote sensing of cyanobacteria-dominant algal blooms and water quality parameters in Zeeloevlei, a small hypertrophic lake, using MERIS, *Remote Sens. Environ.*, **114**, 2070–2087 (2010).

52 S. Koponen, J. Attila, J. Pulliainen, K. Kallio, T. Pyhalahti, A. Lindfors, K. Rasmus and M. Hallikainen, A case study of airborne and satellite remote sensing of a spring bloom event in the Gulf of Finland, *Cont. Shelf Res.*, **27**(2), 228–244 (2007).

53 W.J. Moses, A.A. Gitelson, S. Berdnikov and V. Povazhnyy, Estimation of chlorophyll-a concentration in case II waters using MODIS and MERIS data – successes and challenges, *Environ. Res. Lett.*, **4**, 045005 (2009).

54 European Space Agency (ESA) 2017, Copernicus: Observing the Earth. Available from: <http://www.esa.int/Our_Activities/Observing_the_Earth/Copernicus/Overview4>. Accessed 11 January 2017.

55 ESA Earthnet Online 2013d, *Free access to Copernicus Sentinel satellite data*. Available from: <https://earth.esa.int/web/guest/missions/mission-news/-/article/free-access-to-copernicus-sentinel-satellite-data>. Accessed 19 November 2016.

56 GEO, 2013, *Capacity Building: Strategic Target*. Available from: <www.earthobservations.org/geoss_ta_cb_tar.shtml>. Accessed 19 November 2016.

57 EC Guidance Document No 10, *River and lakes – Typology, reference conditions and classification systems, Common Implementation Strategy for the Water Framework Directive (2000/60/EC)*, Working Group 2.3 – REFCOND, Luxembourg: Office from Official Publications of the European Communities, 87 pp. (2003).

Part IV

Land Deformation Mapping and Humanitarian Crisis Response

11

InSAR Techniques for Land Deformation Monitoring

P. Kourkouli[1,2], U. Wegmüller[1], A. Wiesmann[1] and K. Tansey[2]

[1] *GAMMA Remote Sensing AG, Gümligen, Switzerland*
[2] *University of Leicester, Centre for Landscape and Climate Research, Department of Geography, Leicester, UK*

11.1   Introduction

Copernicus [1] – previously known as GMES (Global Monitoring for Environment and Security) – is a European programme for monitoring the Earth, assembling data from different sources, such as Earth Observation satellites and stationary sensors. The users of Copernicus services are primarily policymakers and public authorities, who need the information to develop environmental regulations and policies, or to take critical decisions for emergency reasons, such as a natural disaster or a humanitarian crisis. Concerning ground motion monitoring, there are pan-European Copernicus services providing ground motion hazard information such as the GMES emergency service as well as the Terrafirma and PanGeo projects [2,3]. The goal of those services is to provide support in the mitigation of risk, identify geohazards and improve the safety of the European citizens.

Land deformation (downlift or uplift) is the result of a variety of displacement mechanisms caused by natural processes or/and human activities. Numerous problems might be caused by ground displacement, such as elevation changes, damage to constructions such as buildings, roads, railroads, dams and bridges. Consequently, the effects of land deformation and particularly subsidence cannot be underestimated. Specifically, the environmental and economic impacts of land subsidence phenomena can vary from minor to severe depending on the existent land use of the affected area and the subsidence magnitude and coverage [4,5]. In addition, in many cases the loss of life is certainly the most significant impact of land deformation, which is mainly caused by natural disasters, e.g. a seismic event. For these reasons, ground motion monitoring is essential for avoiding and potentially preventing such hazardous effects.

Earth Observation for Land and Emergency Monitoring, First Edition. Edited by Heiko Balzter.
© 2017 John Wiley & Sons Ltd. Published 2017 by John Wiley & Sons Ltd.

11.2 Background

11.2.1 Synthetic Aperture Radar Systems

Synthetic Aperture Radar (SAR) was first proposed by Carl Wiley [6] in 1954 and demonstrated by Graham [7] in 1974. The main characteristic of SAR, which makes it unique is its imaging capability. It is a side looking all-weather system having the ability to capture scenes at both day and night, even with the existence of cloud coverage. More than 15 SAR satellites have been operating until today and 10 new SAR sensors are planned to be launched within the next five years [8]. It is remarkable that in the last years, a high number of researchers used SAR sensors in order to monitor a huge variety of dynamic processes of the Earth (Figure 11.1), both on a global and local scale, providing high accuracy. Over the last decades, a large range of SAR techniques and applications based on the SAR sensors has been observed. In 1978, the first civilian spaceborne SAR system established into orbit by NASA/JPL was the polar orbiting satellite SEASAT which operated for about six weeks [9]. During the 1990s, the availability of data from the European satellites ERS-1 (ESA, 1991), ERS-2 (ESA, 1995), the Japanese JERS-1 [10,11] and the Canadian RADARSAT-1 [12] researchers showed a particular interest in SAR sensors.

The success of the previous missions was followed by the development and launch of new SAR satellites, firstly with ENVISAT ASAR (Europe) and ALOS PALSAR (Japan) and afterwards with RADARSAT-2 (Canada) [13]. Over the last few years their capability has been considerably improved with the development of new sensors with very high resolution and shorter orbit cycle such as Cosmo-Skymed (Italy) [14] and TerraSAR-X (Germany) [15]. In April 2014, a new satellite named Sentinel-1 was set into orbit. Due to these sensors, SAR based techniques such as SAR Interferometry are becoming more and more quantitative geodetic tools for deformation monitoring, rather than simple qualitative tools [16]. Table 11.1 presents the most well-known SAR satellite missions used for providing observations for surface motion.

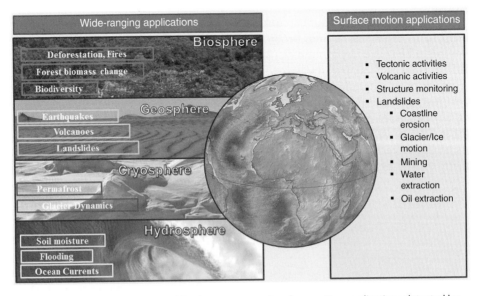

Figure 11.1 Wide-ranging variety of Earth processes and surface motion applications detected by SAR sensors (modified by Moreira *et al.* [8]).

Table 11.1 Important SAR missions are presented with their band and revisit time.

Satellite	Band	Revisit cycle (days)	'91	'92	'93	'94	'95	'96	'97	'98	'99	'00	'01	'02	'03	'04	'05	'06	'07	'08	'09	'10	'11	'12	'13	'14
ERS-1	C	35																								
JERS-1	L	44																								
RADARSAT-1	C	24																								
ERS-2	C	35																								
ENVISAT	C	35																								
ALOS	L	46																								
RADARSAT-2	C	24																								
TerraSAR-X	×	11																								
Cosmo-SkyMed	×	16																								
Sentinel-1	C	12																								

11.2.2 Methods Used for Ground Motion Monitoring

A variety of different methodologies are used for ground motion monitoring. Each methodology is capable of measuring land deformation at a specific area over a certain time period. This means that the ability for each methodology differs at different scales. A number of the most noteworthy and widely used methods (spaceborne, airborne and terrestrial), demonstrating both their strengths and weaknesses, are shown in Table 11.2.

Some methods can measure subsidence only if they are applied on specific processes. For instance, extensometers and GPS estimate ground motion at small spatial scales, while levelling surveys are able to capture measurements over a broader area. Thus, each technique can be used independently, but this does not mean that they cannot be used together, since every method detects ground displacements at different scales.

Despite their ability to measure ground displacement accurately, airborne and terrestrial methodologies are cost and time effective. Over 20 years, SAR Interferometry (InSAR) has become a very powerful tool for measuring subsidence in a large variety of applications, which we present in the following section. Ground motion measurements can be detected in large-scale areas at varied wavelengths. New SAR satellites, however (e.g. TerraSAR-X, Cosmo-Skymed) with high resolution have now made possible measurements of deformation at a distinct level, i.e. at a meter scale (e.g. structures).

11.2.3 InSAR and Applications for Ground Motion Measurements

InSAR techniques can be used over a very broad range of applications. In this section, we briefly present a few of them.

- **Structure monitoring:** Potential knowledge of the instability of a construction is a significant factor for preventing a future damage. Possible structures could be large constructions such as bridges or dams. Moreover, archaeological sites and monuments are of great interest [17,18].
- **Landslide hazards:** InSAR measurements used in synergy with in situ information related to the displacement mechanisms, are useful and powerful tools for monitoring areas characterized by landslides or slope instability [19–21]. The potential of the continuous monitoring of an area and the introduction of the interferometric result into a Geographic Information System (GIS) have increased research related to landslides. With the availability of such kinds of deformation products, scientists have the ability to generate maps showing the risk zones of a specific landslide phenomenon and therefore to assess a potential warning [22].
- **Tectonic and volcanic activities:** InSAR methodologies were applied successfully to applications such as earthquakes in order to measure tectonic displacements. Numerous applications concerning earthquakes were presented, with the most representative example, the Bam earthquake (Iran), extensively studied [23,24]. In addition, InSAR was used in volcanic processes for estimating deformation associated with eruptions or magma accumulation. A characteristic example is the case study of Mount Etna [25].
- **Coastline erosion:** Quantitative understanding of subsidence is important for predicting relative sea-level rise, storm-surge flooding, and successful wetland restoration [26].

Apart from the aforementioned applications, InSAR can be used in other fields like glaciology for measuring glacier movement; subsidence due to mining [27], subsidence caused by underground water, or oil extraction.

Table 11.2 Methodologies used for land deformation monitoring. Strengths and limitations.

	Methods	Strengths	Limitations
	Different methodologies used for ground motion monitoring		
Terrestrial Monitoring	*InSAR*	• Very precisely monitor 2D displacements • All weather system • Short time interval (minutes) • Relatively high spatial density of observations	• Big equipment • Limited field of view (some tenths of degrees in both the horizontal and vertical domain) • Short range (<5 km) • Single fixed look direction – problematic shadowing
	LiDAR	• High-resolution images • Quick surveys and relatively inexpensive • Commercially available technology	• Relatively small apertures of 1–10 square kilometers • Very localized deformation and subsidence processes
	Optical Leveling	• Straight-forward approach for measuring height changes • Very precise height measurements over short distances (1 mm over 1 km)	• Time consuming • Labor intensive • Small scale monitoring
	Borehole Extensometers	• Vertical accuracy (3 mm) • Measurements at specific location • Continuous measurements	• Multiple devices for wide areas • Possible to measure displacements only between the bottom and the top of the borehole
Airborne Monitoring	*InSAR*	• Observations can be scheduled at any time • Full 3D deformation field • High spatial and temporal resolution imaging	• Very high cost for building-maintaining-operating a SAR aircraft platform • Not yet commercially available
	LiDAR	• High-density maps • Surveys can be scheduled and flown any time (within weather limitations) • Measurements in all terrain and vegetation conditions	• 5–10 cm accuracy (vertical component) • Surveys can only be flown in good weather
Spaceborne Monitoring	*GPS*	• mm scale displacements • Estimates of the water content of the troposphere • Continuous temporal monitoring • Sensitive to horizontal displacements	• Intense fieldwork • Measure at only discrete locations • Expensive to install • Spatially dense network
	InSAR	• Low-cost • Large and small scale observations • Short revisiting time period - high frequency monitoring • LOS measurements – sensitive to vertical displacements • Large archive of SAR data • Historical deformation measurements (since 1992)	• Low spatial coverage in rural/natural surfaces • Not sensitive to the horizontal direction

11.2.4 Differential Synthetic Aperture Radar Interferometry

In the early 1990s, DInSAR was developed. DInSAR's principle is the utilization of two or more complex radar scenes acquired over the same region from slightly difference positions and at different times (repeat-pass interferometry) (Figure 11.2).

It exploits the phase difference between two radar acquisitions. In ideal conditions, if the scattering characteristics are equal in both acquisitions, then the interferometric phase is expressed as [28]:

$$\Delta\varphi = \varphi_1 - \varphi_2 = -\frac{4\pi}{\lambda}(r_1 - r_2) = -\frac{4\pi}{\lambda}\Delta R \tag{11.1}$$

Where r_1 and r_2 are the geometric distances and φ_1 and φ_2 are the contributions of the scattering phases in the two images. If between the two acquisitions the earth's surface is deformed, a map of the ground displacement along the radar Line of Sight (LOS) with millimetre accuracy between those two dates can be generated. This map is called interferogram, which is the complex image formed by cross-multiplying two radar images. The interferogram contains the topographic term and an external DEM needed in order to subtract topography from the differential interferogram.

The interferometric phase is estimated for every acquisition. The interferometric phase (φ_{int}) is expressed as the sum of five terms, the topographic phase (φ_{topo}), the displacement phase (φ_{disp}), the atmospheric phase (φ_{atm}) and the phase noise (φ_{noise}) [29]:

$$\varphi_{int} = \varphi_{topo} + \varphi_{disp(LOS)} + \varphi_{atm} + \varphi_{noise} \tag{11.2}$$

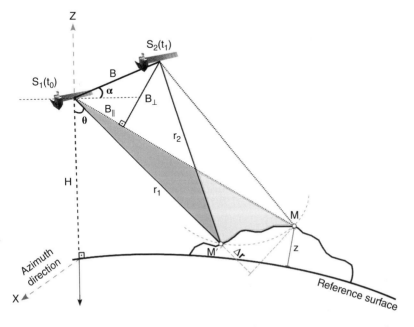

Figure 11.2 Interferometric SAR configuration. Two SAR antennas, S_1 in time t_0 and S_2 in time t_1 are separated by a baseline B, with a perpendicular baseline B_p and a parallel baseline B_n. The two antennas observe the surface at range r_1 and r_2 respectively. θ is the look angle of the reference sensor S_1. Knowing the range r_1 and the height H of the satellite we can determine the height z of point M, which needs to be estimated. (Modified after [28].)

For more than two decades, DInSAR demonstrated its strong capability to detect land deformation over large-scale areas in a wide range of earth science fields. Numerous scientific publications confirmed DInSAR's strong contribution in a variety of applications, natural or anthropogenic [20,26,30–34]. Hence, it is widely recognized that the technique can successfully measure ground displacements with high resolution occurring both in large and small-scale areas. Although DInSAR is established as a powerful technique for detecting ground displacements, there are several limitations. The most significant are the spatial and temporal decorrelation and the atmospheric distortions. To address these issues multi-temporal interferometric techniques have been developed. Hence, the scientific community took advantage of the long temporal series of data and started to analyse the temporal domain of the ground displacements. The multi-temporal interferometric approaches are grouped into two main categories: (a) the Persistent Scatterer (PS) methodologies, such as PSInSAR [35], Interferometric Point Target Analysis (IPTA) [36] and Spatial Temporal Unwrapping Network (STUN) [37] and (b) the Small Baseline Subsets (SBAS) [38] approaches. Recently, new efforts have been made by researchers using an incorporation of the above categories aiming at the improvement of the spatial coverage of the deformation.

11.2.5 Persistent Scatterer Interferometry

In the late 1990s, Ferretti *et al.* [35] introduced the so-called Permanent Scatterer Interferometry (PSInSAR). In general, the PSI methodology is a point-wise technique permitting measurements with millimetre accuracy of individual features or in other words persistent scatterers (PS) such as structures, rocks and artificial reflectors. The PSI's principle is that the measured PS should remain coherent for the entire time period. This means that the signal phase of a point target should remain coherent for every acquisition of the entire data stack. Features that represent a good PS can be man-made objects such as buildings, pipes, dams, bridges, poles or natural objects such as rocks. For this reason, PS density is relatively high in built-up areas, whereas in natural terrains the PS density is relatively low or in many cases no PSs can be detected. This is a limitation that is also observed in DInSAR methodology when it is applied over natural environments and/or over a long timespan where decorrelation occurs due to the loss of coherence. However, PSI has mitigated some DInSAR's limitations such as temporal decorrelation and atmospheric artefacts.

Table 11.3 presents briefly the strengths and weaknesses of the PSI methodology [39–41]. Starting with the strengths, firstly, the methodology provides a wide area for land monitoring (e.g. ERS and ENVISAT acquisitions cover an area of 100×100 km) and with high spatial resolution. In particular, due to the release of new SAR sensors such as Cosmo-Skymed and TerraSAR-X, spatial resolution reaches up to 1 m. These two characteristics permit observation of ground motion both on a regional and local scale, e.g. with many points on one single building [42]. A significant strength of PSI is its sensitivity of very small land movements, with velocities up to 1 mm yr^{-1}. In addition, there is the availability of the big archive of SAR data, for instance the ERS has provided data since 1991. This is an important aspect, as it enables the historical ground motion monitoring of a specific study area covering a period of almost 25 years. Furthermore, PSI compared with other techniques (e.g. DGPS, levelling) is a time and cost effective methodology. It is considered as a complementary technique with DGPS. Due to the fact that GPS data are very accurate in the horizontal direction, and PSI data are more accurate in the vertical direction, they can be used in a complementary way.

Table 11.3 Strengths and limitations of the PSI.

Persistent Scatterer Interferometry	
Strengths	**Weaknesses**
• Big data archives	• Steep slopes prevent the SAR signal for reaching those areas which are characterized by geometric decorrelation (shadow) in the LOS direction
• High accuracy of the estimated displacement rate ($<1\,\mathrm{mm\,yr}^{-1}$)	
• Estimated movements are measured in the Ling of Sight (LOS) and they are sensitive in the vertical direction – more precise than GPS measurements (sensitive on the horizontal direction)	• Satellite revisit cycle remains short for some applications
	• Low coherence over natural terrain makes difficult the potential to achieve any scatterer
• PS density is more sufficient comparing with other conventional techniques (in urban areas >300 points/km^2)	• High number of SAR acquisitions is acquired for a more robust analysis
• Time series – historical deformation for each scatterer with very high accuracy	• Fast movements (approx. > 4–$5\,\mathrm{cm\,yr}^{-1}$) cannot be detected
• Artificial reflectors can be installed in areas where no scatterer exist	• Problems with non-uniform movements
• Complementarity with other techniques (GPS, levelling, etc.)	• Uncertainties in data continuity through dependence on satellites and acquisition plans
• Low cost compared with in-situ surveys	

Besides the strengths, PSI method has weaknesses and limitations. The ground motion monitoring depends on the satellite's revisit time of the satellite (e.g. 35 days for ERS and ENVISAT, 46 days for ALOS, 11 days for TerraSAR-X). Additionally, it is suggested that a large number of SAR scenes be used for the analysis (≥ 20 acquisitions), as the result usually becomes more robust and reliable since it permits the mitigation of atmospheric artefacts [43]. Another limitation is that PSI can detect only the available scatterers that exist in the study area. These scatterers are point targets that remain coherent for the entire time period. Moreover, the PS density is high over urban environments such as cities and villages. For example, using C-band data like ERS and ENVISAT, the PS density reaches up to 1000 PS/km^2, whereas in high-resolution data such as X-band (TerraSAR-X) the achieved density is much higher [44].

However, the PS density is very low in natural terrains characterized by the existence of vegetation or/and forest, which leads to the loss of coherence. Another weakness of the PSI technique is its capability of measuring deformation only towards the LOS direction. Furthermore, PSI can estimate fast movements ($\sim <4$–$5\,\mathrm{cm\,yr}^{-1}$), which means that some deformation phenomena cannot be detected. Nevertheless, during the last years, terrestrial SAR interferometry has detected such movements due to the very short time interval (minutes).

11.2.6 Small Baseline Methodologies

An extension of the DInSAR technique proposed by Berardino *et al.* [38] is the so-called Small Baseline Subset (SBAS) methodology. SBAS utilizes multiple small baseline acquisition subsets, obtaining time series deformation on distributed targets (DS). Using multiple SAR scenes with a short spatial baseline, we can mitigate the spatial decorrelation and topographic errors [45]. The advantage of this approach is that by using small temporal and spatial baselines, coherence values increase, causing the improvement of spatial coverage. SBAS uses the singular value decomposition (SVD) method to derive time series land displacement measurements. It was principally developed to monitor land deformation which occurs on a relatively large spatial scale ($\sim 100 \times 100$ m pixel dimensions) [38]. Therefore, it is not appropriate for detecting deformations at a very local level (e.g. structures).

As it was mentioned previously, lately, scientists have made efforts to develop new algorithms by combining both DInSAR and PSI techniques [46,47]. In this chapter we present an example of our attempt to combine both DInSAR and PSI methodologies.

11.3 Methods and Data

11.3.1 A Case Study of a Combined DInSAR/PSI Methodology

DInSAR and PSI are both well-established and broadly used methodologies usually applied individually or in a complementary way, e.g. providing some differential interferograms together with a PSI result. Hence, to benefit their complementarity we developed an approach using components from both techniques, aiming at the increment of the displacement information. Our attempt deals with the combination of point-like targets and distributed scatterers. For instance, a building surrounded by low vegetation is a very good dominant scatterer itself, whereas its surrounding pixels, i.e. distributed scatterers, cannot be observed with the PSI only method as they are not dominant targets.

To compare the result, we divided the analysis into two main processing steps: the PSI only and the combined DInSAR/PSI processing. Firstly, a single reference PSI approach was applied on the dataset. A data stack of 20 point differential interferograms was calculated. A point target list based on the spectral characteristics and temporal variability of the point targets as well as the point interferograms was generated. A least-squares regression was performed on the point phases to estimate point heights and deformation rates. Filtering was applied on the residual phases, both spatial and temporal, in order to eliminate contributions in the atmosphere and non-linear deformation terms. Final unwrapped phases were converted into displacement measurements (LOS).

To obtain the corresponding multi-look differential phases a data stack of 20 2d differential interferograms was calculated. To achieve a high number of pixels and consequently increasing the spatial coverage of the result, strong multi-looking for every interferogram was applied. For the integration of the differential interferograms into the PSI processing, the multi-looked (ML) phases were extracted into point locations. Then, both ML differential phases and the initial point data interferograms were combined into one point data stack, as well as the point list with the 2d ML point locations. We proceeded with the combined point data stack of differential interferograms and the combined point list to the later analysis using exactly the same procedure as in the PSI processing. Finally, we compared this result with the PSI only outcome.

11.3.2 Materials and Test Case Study

The Advanced Land Observing Satellite (ALOS), which was launched in 2006 is a Japanese Satellite used for cartography and earth observation purposes. ALOS carries the Phased Array type L-band Synthetic Aperture Radar (PALSAR) [48] instrument, an active microwave sensor used for land monitoring. The main reason that we decided to use L-band was the wavelength (~24 cm), which permits a deeper penetration of the canopy. This permits higher coherence values and consequently more information can be derived. For testing the applicability of our combined approach, we decided to analyse a stack of 20 ALOS PALSAR acquisitions captured over a small region in Jordan (Middle East). The data were acquired in descending orbit, covering the period between August of 2007 and February 2011. Here, we focus on a small region of the study area, which covers an extent of approximately 10 × 10 km. The region is a natural terrain, as it comprises a mixture of land cover types such as desert, low vegetation and rocky parts.

11.4 Results and Discussions

The combined approach applied over a natural terrain shows a good potential (Figure 11.3). For both point-like targets and multi-looked distributed targets, the processing worked well as the results show reasonable spatial coverage, the same deformation pattern and are fully consistent. Although the PSI only result achieved an adequate number of points, the combined approach improved the spatial coverage of the ground measurements significantly, showing a higher pixel density and consequently better coverage of

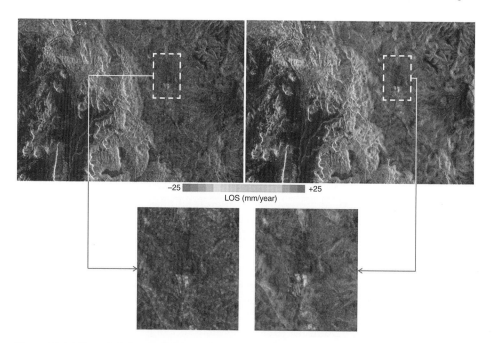

Figure 11.3 PSI result (left pane) compared with the combined outcome (right pane). Dashed box indicates the enlarged small section.

information. Specifically, within a grid of 200 m cell size, in some parts of the area the values of deformation increased up to 120 pixels more than in the PSI technique. The greatest improvement is noticeable in the eastern part of the area where the result increased between 40 and 120 pixels. On the other hand, the central part of the region as well as the major part of the western area improved the least, between 1 to 50 pixels. This happens due to geometric distortions such as layover and shadowing and additionally due to the low coherence levels, i.e. ($\gamma < 0.5$). To further improve the coverage, the result could potentially be improved by applying a multi-reference stack technique using small temporal baselines resulting higher values of coherence. In addition, a characteristic example is demonstrated in the enlarged region shown in Figure 11.3. In this area, which covers an extent of approximately 3×1 km, a few scatterers of the PSI result show significantly high subsidence velocities, whereas in the combined result this information has improved (50–120 pixels). This permits a more reliable result as it makes the interpretation more complete and robust.

11.5   Conclusions

InSAR methodologies are presented in this chapter focusing – in the major part – on the two well-established approaches, DInSAR and PSI. A review of both techniques has been carried out, indicating mainly their drawbacks. The authors conclude that a key limitation is the poor spatial coverage of the deformation result when both methodologies are applied individually. However, significant research has been done using an incorporation of both methodologies in order to overcome such limitations. As an example, we demonstrated a result generated from our effort using a combination of both DInSAR and PSI methods. The improvement of the spatial coverage, which increases the information content, makes the result more valuable for interpretation, e.g. by authorities such as civil protection agencies. These new approaches and the launch of new SAR sensors enable the improvement of deformation mapping in case studies in which previously it had been difficult to obtain information.

Acknowledgments

P. Kourkouli gratefully acknowledges the Marie-Curie fellowship funded by the Marie Curie People Actions under the European Seventh Framework Programme (GIONET), Grant Agreement PITN-GA-2010- 264509. ALOS PALSAR dataset © Jaxa.

References

1 Copernicus: *The European Earth Observation Programme*, <www.copernicus.eu>.
2 *A pan-European Ground Motion Hazard information service*, <www.terrafirma.eu.com/index.htm>.
3 *PanGeo service*, <www.pangeoproject.eu>.
4 V.F. Viets, *Environmental and Economic Effects of Subsidence*, Lawrence Berkeley Natl. Lab. (2010).

5 H.Z. Abidin, I. Gumilar, and H. Andreas, *Study on Causes and Impacts of Land Subsidence in Bandung Basin, Indonesia.* FIG Working Week 2011. Bridging the Gap between Cultures. Marrakech, Morocco, 18–22 May 2011.

6 C.A. Wiley, Pulsed Doppler Radar Methods and Apparatus, 1954. *US Patent 3,196,436.*

7 L.C. Graham, Synthetic interferometer radar for topographic mapping, *Proc. IEEE* **62**, (1974).

8 A. Moreira, P. Prats-Iraola, M. Younis, G. Krieger, I. Hajnsek, A Tutorial on Synthetic Aperture Radar, *IEEE Geosci. Remote Sens. Mag.* (2013).

9 D.B. Lame and G.H. Born, SEASAT measurement system evaluation: Achievements and limitations, *J. Geophys. Res. Ocean.* **87**, 3175–3178 (1982).

10 K. Nemoto, Y. Nishino, H. Ono, M. Mizutamuri, K. Nishikawa, and K. Tanaka, Japanese Earth Resources Satellite – 1 Synthetic Aperture Radar, *Proc. IEEE* **79**, 800–809 (1991).

11 T. Nishidai, Early results from "Fuyo-1" Japan's Earth resources satellite (JERS-1), *Int. J. Remote Sens.* **14** (9), 1825–1823 (1993).

12 S. Raney, R.K. Luscombe, A.P. Langham, E.J. and Ahmed, RADARSAT, *IEEE*, (1991).

13 K. Morena, L. James, K. and Beck, An Introduction to the RADARSAT-2 mission, *Can. J. Remote Sens.* **30**(3), 221–234 (2004).

14 M. De Luca, G.F. Marano, G. Piemontese, M. Versini, B. Caltagirone, F. Casonato, G. Coletta, and A. De Carlo, Interoperability, expandability and multi mission-sensor COSMO-SkyMed capabilities, in *Int. Geo-science Remote Sens. Symp.* (Barcelona, Spain, 2007).

15 A. Eineder, M., Breit, H. Fritz, T. Schuettler, and B. Roth, TerraSAR-X SAR products and processing algorithms, in *Int. Geosci. Remote Sens. Symp.* (Seoul, Korea, 2005).

16 M. Crosetto, B. Crippa, and E. Biescas, Early detection and in-depth analysis of deformation phenomena by radar interferometry, *Eng. Geol.* **79**, 81–91 (2005).

17 I. Parcharidis, M. Foumelis, K. Pavlopoulos, and P. Kourkouli, Ground deformation monitoring in cultural heritage areas by time series interferometry: the case of ancient Olympia site (western Greece), in *Fringe, European Space Agency, Special Publications.* ESA SP, (2010).

18 D. Tapete, R. Fanti, R. Cecchi, P. Petrangeli, and N. Casagli, Satellite radar interferometry for monitoring and early-stage warning of structural instability in archaeological sites, *J. Geophys. Eng.* **9**, S10–S25 (2012).

19 J. Colesanti, C. and Wasowski, Satellite SAR interferometry for wide-area slope hazard detection and site-specific monitoring of slow landslides, in *Int. Landslide Symp.*, (Rio de Janeiro, Brasil, 2004).

20 T. Strozzi, P. Farina, A. Corsini, C. Ambrosi, M. Thüring, J. Zilger, A. Wiesmann, U. Wegmüller, and C. Werner, Survey and monitoring of landslide displacements by means of L-band satellite SAR interferometry, *Landslides* **2**, 193–201 (2005).

21 C. Delacourt, D. Raucoules, S. Le Mouélic, C. Carnec, D. Feurer, P. Allemand, and M. Cruchet, Observation of a Large Landslide on La Reunion Island Using Differential Sar Interferometry (JERS and Radarsat) and Correlation of Optical (Spot5 and Aerial) Images., *Sensors (Basel).* **9**, 616–630 (2009).

22 S. Bianchini, F. Cigna, G. Righini, C. Proietti, and N. Casagli, Landslide Hot Spot Mapping by means of Persistent Scatterer Interferometry, *Environ. Earth Sci.* **67**, 1155–1172 (2012).

23 E.J. Fielding, Surface ruptures and building damage of the 2003 Bam, Iran, earthquake mapped by satellite synthetic aperture radar interferometric correlation, *J. Geophys. Res.* **110**, B03302 (2005).

24 S. Stramondo, M. Moro, F. Doumaz, and F.R. Cinti, The 26 December 2003, Bam, Iran earthquake: surface displacement from Envisat ASAR interferometry, *Int. J. Remote Sens.* **26**, 1027–1034 (2005).

25 A.A.D. Massonnet and P. Briole, Deflation of Mount Etna monitored by spaceborne radar interferometry, *Nature* **375**, 567–570 (1995).

26 I. Parcharidis, P. Kourkouli, E. Karymbalis, M. Foumelis, and V. Karathanassi, Time Series Synthetic Aperture Radar Interferometry for Ground Deformation Monitoring over a Small Scale Tectonically Active Deltaic Environment (Mornos, Central Greece), *J. Coast. Res.* **287**, 325–338 (2013).

27 Z. Perski and D. Jura, ERS SAR Interferometry for Land Subsidence Detection in Coal Mining Areas, 25–29 (1999).

28 R.F. Hanssen, *Radar Interferometry: Data Interpretation and Error Analysis* (Kluwer Academic Publishers, 2001).

29 C. Werner, U. Wegmüller, A. Wiesmann, and T. Strozzi, *Interferometric Point Target Analysis with JERS-1 L-band SAR Data*, 25–27 (2003).

30 P. Teatini, L. Tosi, T. Strozzi, L. Carbognin, U. Wegmuller, and F. Rizzetto, Mapping regional land displacements in the Venice coastland by an integrated monitoring system, *Remote Sens. Environ.* **98**, 403–413 (2005).

31 J. Biggs, E. Bergman, B. Emmerson, G.J. Funning, J. Jackson, B. Parsons, and T.J. Wright, Fault identification for buried strike-slip earthquakes using InSAR: The 1994 and 2004 Al Hoceima, Morocco earthquakes, *Geophys. J. Int.* **166**, 1347–1362 (2006).

32 P. López-Quiroz, M.-P. Doin, F. Tupin, P. Briole, and J.-M. Nicolas, Time series analysis of Mexico City subsidence constrained by radar interferometry, *J. Appl. Geophys.* **69**, 1–15 (Elsevier B.V., 2009).

33 D. Perissin, Z. Wang, and H. Lin, Shanghai subway tunnels and highways monitoring through Cosmo-SkyMed Persistent Scatterers, *ISPRS J. Photogramm. Remote Sens.* **73**, 58–67 (International Society for Photogrammetry and Remote Sensing, Inc. (ISPRS), 2012).

34 S. Gernhardt and R. Bamler, Deformation monitoring of single buildings using meter-resolution SAR data in PSI, *ISPRS J. Photogramm. Remote Sens.* **73**, 68–79 (International Society for Photogrammetry and Remote Sensing, Inc. (ISPRS), 2012).

35 A. Ferretti, C. Prati, and F. Rocca, Permanent scatterers in SAR interferometry, in *Geosci. Remote Sens. Symp. 1999. IGARSS '99 Proceedings. IEEE 1999 Int.* **3** (1999).

36 C. Werner, U. Wegmuller, T. Strozzi, and A. Wiesmann, Interferometric Point Target Analysis for Deformation Mapping, in *IGARSS*, (Toulouse, France, 2003).

37 B.M. Kampes, *Radar Interferometry: Persistent Scatterer Technique* (Springer, 2006).

38 P. Berardino, G. Fornaro, R. Lanari, S. Member, and E. Sansosti, A new algorithm for surface deformation monitoring based on small baseline differential SAR interferograms, *IEEE Trans. Geosci. Remote Sens.* **40**, 2375–2383 (2002).

39 J. Ferretti, A. Prati, C. Rocca, and F. Wasowski, Satellite interferometry for monitoring ground deformations in the urban environment, in *IAEG*, (2006).

40 B. Crosetto, M. Agudo, M. Monserrat, O. Pucci, Inter-comparison of persistent scatterer interferometry results, in *Eur. Sp. Agency, (Special Publ. ESA SP (649 SP)*, (2008).

41 T. Perski, Z. Hanssen, R. Wojcik, A. Wojciechowski, InSAR analyses of terrain deformation near the Wieliczka Salt Mine, Poland, *Eng. Geol.* **106**, 58–67 (2009).

42 N. Adam and R. Bamler, Coherent stacking with TerraSAR-X imagery in urban areas, *2009 Jt. Urban Remote Sens. Event*, 1–6 (IEEE, 2009).

43 R. Bamler, B. Kampes, N. Adam, and S. Suchandt, Assessment of Slow Deformations and Rapid Motions by Radar Interferometry. In *"Photogrammetric Week 05"*. Dieter Fritsch, Ed. Wichmann Verlag, Heidelberg, 2005.

44 M. Crosetto, O. Monserrat, A. Jungner, and B. Crippa, *Persistent Scatterer Interferometry: Potential and Limits*. Available at: www.ipi.uni-hannover.de/fileadmin/institut/pdf/isprs-Hannover2009/Crosetto-136.pdf (2011).

45 P. Shanker, F. Casu, H. A. Zebker, and R. Lanari, Comparison of Persistent Scatterers and Small Baseline Time-Series InSAR Results: A Case Study of the San Francisco Bay Area, *IEEE Geosci. Remote Sens. Lett.* **8**, 592–596 (2011).

46 A. Ferretti, A. Fumagalli, F. Novali, C. Prati, F. Rocca, A. Rucci, A. Permanent, S. Sar, and I. Psinsar, A New Algorithm for Processing Interferometric Data-Stacks: SqueeSAR, *IEEE Trans. Geosci. Remote Sens.* **49**, 3460–3470 (2011).

47 A. Hooper, A multi-temporal InSAR method incorporating both persistent scatterer and small baseline approaches, *Geophys. Res. Lett.* **35**, L16302 (2008).

48 A. Rosenqvist, M. Shimada, and M. Watanabe, ALOS PALSAR: Technical outline and mission concepts, 1–7 (2004).

12

Mapping Land Surface Displacements in the Swiss Alps with Radar Interferometry

J. Papke[1,2], T. Strozzi[1] and N.J. Tate[2]

[1] *GAMMA Remote Sensing AG, Gümligen, Switzerland*
[2] *University of Leicester, Department of Geography, Leicester, UK*

12.1 Introduction

In mountain landscapes, such as in the Swiss Alps, land surface elevation changes and land surface displacements take place over time. This is due to the nature of the mountain environment where a variety of geomorphological processes form the landscape [1].

With current state of the art measurement technology and mapping techniques, the changing relief and the velocity of geomorphological processes can be observed at different temporal and spatial resolutions. Earth observation data acquired with remote sensing techniques from satellites and airborne platforms [2] and with ground-based techniques [3] are a valuable source of information in mountain geomorphology.

In this study, the use of satellite Synthetic Aperture Radar (SAR) interferometry [4,5] and ground-based radar interferometry [6] is discussed for the mapping of mass movements in a selected Alpine test site in Grächen and St. Niklaus in the canton Valais in Switzerland. A short introduction is given to land surface displacements in the Alps and the use of radar interferometry for their mapping. Subsequently, surface velocity measurements of different mass movements in the study site are presented that have been observed with TerraSAR-X satellite SAR interferometry and in a field campaign with a ground-based radar interferometer.

12.2 Background

12.2.1 Land Surface Displacements in the Alps

Climate, with heating and cooling, thaw and freeze, rain and snow, and gravity are the main forces in high mountains acting on ice, bedrock, rocks, sediments and soils and driving the transport of earth material downslope. The rate and the type of geomorphological processes are additionally affected by the composition, structure and form of the mountains [7]. In the low and intermediate zones of mountain slopes, land surface

Table 12.1 Typical land surface displacements in the Alps and their time-spatial characteristics.

Type	Velocity	Spatial extent
Landslide	mm/yr – m/yr	$m^2 - km^2$
Rock glacier	cm/yr – m/yr	$m^2 - km^2$
Glacier	m/yr	$m^2 - km^2$

displacement can have the character of extremely slow, very slow and slow moving landslides [8] with velocities in the order of a few millimetres to several decimetres per year. In addition also gravitational and fluvial processes take place and transport material along stream channels. Above the forest line in the periglacial mountain environment, mountain permafrost [9] and rock glaciers [10,11] characterize the landscape. Active rock glaciers are ice-debris mixtures transporting large amounts of rock debris while creeping downslope with surface speeds in the order of centimetres to a few metres per year. The mapping of their kinematics and surface geometry [12,13] together with geophysical measurements of their internal composition [14] are helpful to provide a better understanding of their dynamics over time. Due to the abundance of ice inside the rock glacier body, the sensitivity to warming air and ground temperatures is of high scientific interest, as well as the role of liquid water inside the rock glacier [15]. When rock glaciers are situated on steep slopes, they can be a source for debris flows, rock fall and landslides.

In glaciated mountain environments, the dynamics of glaciers and their changes over time, especially the retreat of Alpine valley glaciers over the past decades, are a focus for research [16]. The retreat of the glacier ice volume in elevation and extent can lead to instability of the slopes surrounding the glacier and thus triggering new mass movements or changing the activity rates of existing landslides nearby [17,18]. All these different kinds of land surface displacements are observable in the Alps. They occur with different surface velocities and are of different spatial extent (Table 12.1), and also their activity varies timely, ranging from sudden events such as rock fall to long-term surface movements such as from deep-seated landslides.

Besides the interest in understanding the mountain geomorphological processes in more detail over space and time, the mapping of land surface changes and displacements is an essential task in civil protection and engineering geology. When Alpine mass movements happen near inhabited mountain areas, they pose a risk to people, property and infrastructure. That is why there is a strong need for mapping and monitoring land surface displacements in mountain environments.

12.2.2 Use of Satellite SAR and Ground-based Radar Interferometry for Surface Displacement Mapping

Prominent measurement techniques for the observation of dynamics and relief changes in mountains are airborne stereo-photogrammetry, airborne and terrestrial laser scanning, differential GPS, classical geodetic surveying, as well as change detection from optical and radar satellite images [19]. In the last decade, satellite synthetic aperture radar interferometry (InSAR) [20] has been established as a remote sensing technique of use in exploring deformation and topographic mapping. More recently, measurements with ground-based radar/SAR interferometers have also become popular [6].

An imaging radar can be described as an active system sending out electromagnetic radiation to a target area of interest and collecting the received signal that is scattered back from the target area to form an image. With a SAR system, which is like a moving radar, the electromagnetic radiation backscatter is not collected at one time only, but pulses are sent out and received over the same target area for a few seconds while moving, forming a so-called synthetic aperture. The received signals are then filtered and processed to one image with the advantage of higher image resolution in moving direction (azimuth), which is desirable for remote targets observed from space.

In InSAR for surface displacement mapping, a so-called interferogram is formed from two images acquired at different times. From the phase information of the two complex images the phase difference is exploited which is sensitive to surface displacement, topography, acquisition geometry, atmospheric influences on the signal and noise sources. For the calculation of the surface displacement that happened over the time that passed between the first and the second acquisition, the basic data processing technique of (Two-Pass) Differential InSAR (DInSAR) [21] is applied. With more acquisitions available, displacement time-series can be calculated and information on non-linear movements obtained. The processing of displacement time-series with DInSAR methods is variably described as advanced DInSAR [22], as multi-temporal interferometry [23] or as time-series DInSAR [24]. An overview is given in [24]. The data processing techniques for multi-temporal analysis are divided in two main categories: the Small Baseline Subset (SBAS) approaches [25] and Persistent Scatterers Interferometry (PSI), for example [26–28]. A description and review of the techniques in the context of landslide displacement mapping is given by [23,29,30]. Also combinations of the time-series methods are used [24,31].

Using radar has the advantage of day and night monitoring possibilities, as well as signal penetration through clouds. Advantages of satellite InSAR are the wide spatial coverage and the high resolution of the satellite SAR data, which gives the possibility to map the Earth and produce precise deformation maps at regional scale with deformation measurement accuracies of millimetres to centimetres [4]. The technique of InSAR has limitations in mountains, because the satellite observations are affected by layover and shadow effects [29], restricting the spatial coverage. Another limitation is the temporal decorrelation over vegetated and snow covered areas because the predefined satellite revisit time is typically several days causing small changes in the repeated acquisitions. The most challenging aspect is to calculate the path delay of the signal caused by the troposphere. All these factors influence the quality of the interferometrically processed deformation maps [29]. Another important element is that the surface displacement is always measured in the line-of-sight (LOS) direction, defined as the direction from the target to the sensor and, it is a relative surface displacement measurement with respect to a defined reference point in the imaged area. This means that the acquisition geometry and the choice of the reference point play a large role for what is measured and mapped.

In order to calculate two- and three-dimensional surface displacement information, data fusion and integration approaches are applied [32], taking into account multi-sensor acquisitions, acquisitions from ascending and descending passes and certain side assumption or different data processing approaches such as along-track interferometry.

Table 12.2 A selection of European projects using radar interferometry techniques for landslide research.

Project	Topics of study	Reference
SAFELAND	Overview of techniques applied in landslide research.	[33]
GALAHAD	Development of advanced monitoring techniques and improvement of forecasting methods for landslides.	[34]
SLAM	Development and qualification of a landslide mapping and monitoring service based on Earth observation data.	[35]
TERRAFIRMA	PSI methodology applied for landslide research and subsidence.	[36]
DORIS	Detection, mapping, monitoring, and forecasting of landslides, including the use of satellite and ground-based radar interferometry.	[37]

There have been various research projects (Table 12.2) that analysed the different existing surface displacement measurement techniques in the context of landslide research. These have employed satellite InSAR and ground-based radar/SAR interferometry techniques, and their utilization advantages and disadvantages for landslide mapping and monitoring. Results from the projects listed in Table 12.2 confirm the suitability of satellite and ground-based InSAR for the detection and mapping of ground deformations in a number of different cases of landslide studies. The above-mentioned limited monitoring capabilities have also been discussed and the complementary use of radar interferometry with other measurement techniques has been stressed.

12.3 Study Area

Being characterized by many different Alpine geomorphological processes, a part of the periglacial environment at the eastern side of the Matter Valley in the municipalities Grächen and St. Niklaus is selected as the study site for this research work. Grächen and St. Niklaus are two neighbouring municipalities in the canton of Valais in Switzerland belonging to the Western Swiss Alps. The study area stretches from the Hannigalp (2114 m) in the north to the Platthorn (3246 m) in the south with slope orientations towards west/north-west above the forest line (Figure 12.1). It has several characteristics of a typical alpine region and is thus suitable for research on slope processes. With the majority of the mass movements located above the vegetation boundary at around 2200 m, this test site is well suited for InSAR studies of surface displacements with different velocities in Alpine terrain.

The village of Grächen is located on a deep-seated landslide with movements in the order of five millimetres per year [38]. This extremely slow movement will not be part of the current investigation. Here, the focus is on the surface processes in an area of 10 square kilometres above the forest line where rock glaciers, landslides, fluvial and erosional processes characterize the landscape. The Distelhorn and the Ritigraben rock glacier [39] are the two fastest moving rock glaciers in the study area (Figure 12.1).

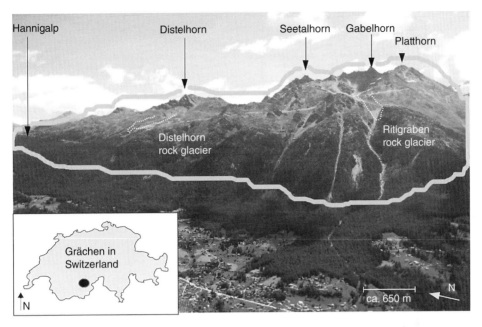

Figure 12.1 The study site is situated in Grächen and partly in St. Niklaus in the Western Swiss Alps stretching from the Hannigalp in the north to the Platthorn in the south. The Distelhorn and Ritigraben rock glaciers are the fastest moving rock glaciers in the study area.

12.4 Case Studies

12.4.1 Mapping Mass Movements in Grächen and St. Niklaus with Satellite SAR Interferometry

Mapping Alpine mass movements with satellite SAR interferometry using recent high-resolution satellite data has shown its advantages in several cases [40]. A higher spatial resolution can give more details on the form and extent of geomorphological elements and spatially smaller movements can also be recorded. With a higher temporal resolution, the amount of change between two acquisitions is reduced which can increase the coherence and thus the reliability of the calculated surface displacement in certain areas.

A stack of descending TerraSAR-X satellite data in Stripmap mode with a spatial resolution of 1.4 m in slant-range direction and of 2.0 m in azimuth direction and a continuous temporal resolution of 11 days is processed. The acquisitions are covering the whole study area in the snow free season from 8 June 2012 until 7 October 2012. Figure 12.2 shows the processing workflow. An interferogram is formed from two acquisitions with a multi looking factor of 3 pixels in range and 3 pixels in azimuth. Then, a digital elevation model (DEM) at 5 m resolution is used to remove the phase term related to topography and the interferometric phase is filtered with an adaptive spectral filter to reduce noise. In the next step, the DEM is used to calculate an atmospheric phase model based on a linear regression of the interferometric phase with respect to height. This phase term is subtracted and if necessary, also a 2D linear regression is applied to remove liner trends related to an inaccurate estimation of the satellite acquisition geometry. As a last filtering step, a low-pass filter is applied in order to reduce noise and large-scale

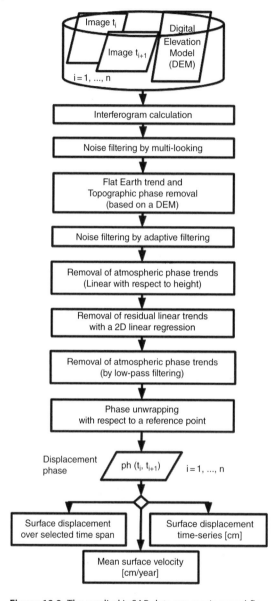

Figure 12.2 The applied InSAR data processing workflow.

atmospheric effects. The phase is then unwrapped to resolve the 2PI phase ambiguities and to retrieve a displacement phase, which is a relative measurement to a selected reference point in the imaged area. This is done for each consecutive 11-day interferogram. The displacement phase values are then converted to surface displacement in centimetres. Also the mean velocity is computed based on all calculated interferograms by applying a temporal filtering using a network inversion technique based on singular value decomposition [41]. For the calculation, also small noisy areas are put into the time-series in order to have a regularly spaced input and with the assumption that phase jumps that occur due to temporal decorrelation cancel out in the time-series.

Figure 12.3 gives an overview of the detected mass movements in the study site. The processing results were geocoded to the Swiss reference system CH1903 with a pixel size of 10 metres and overlaid with a topographic map. On the left side, an example of an unwrapped 11-day interferogram is shown. The interferometric phase represents a displacement measurement of about 1.55 cm for one fringe (one phase cycle). The map on the right side displays the annual mean velocity of the mass movements. It is calculated from eleven consecutive 11-day interferograms, which cover a time span of 121 days over the snow free season from June to October 2012. The surface displacement is mapped with a scale from 0 to 80 cm per year. The two fastest moving features are the Distelhorn and the Ritigraben rock glaciers. They reach velocities in the order of 80 cm per year and more in certain areas of their frontal parts. These measured velocities are in the same order of magnitude as is known from measurements of earlier years by [38,42,43]. Other visible movements can be addressed to slower moving rock glaciers, landslides and permafrost creep.

The above presented displacement mapping activities are part of an ongoing study to map the different geomorphological processes in Grächen and St. Niklaus in Switzerland

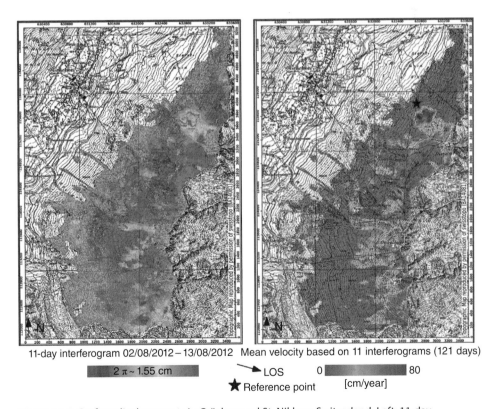

11-day interferogram 02/08/2012 – 13/08/2012   Mean velocity based on 11 interferograms (121 days)

2 π ~ 1.55 cm   LOS   0 ▬▬▬▬ 80
★ Reference point   [cm/year]

Figure 12.3 Surface displacements in Grächen and St. Niklaus, Switzerland. Left: 11-day TerraSAR-X interferogram between 2 August and 13 August 2012, right: annual mean velocity based on TerraSAR-X data available over the snow-free season from 08 June to 07 October 2012. (*See insert for color representation of the figure.*)

with satellite InSAR. The mass movements that are detected can be well aligned to geomorphological elements in the study area. The annual mean displacement map was calculated with the purpose to show the areas where surface displacement is present and to get a rough estimation in which order of magnitude the displacement takes place.

12.4.2 Mapping Rock Glacier Movements with a Ground-Based Radar Interferometer

In order to have a complementary surface displacement measurement for the Distelhorn rock glacier, a field campaign with a ground-based radar interferometer was performed in Grächen. The ground-based GAMMA Portable Radar Interferometer (GPRI) [44] (Figure 12.4) is an FM-CW radar working at 17.1–17.3 GHz (Ku-Band) and provides data with 0.75 m resolution in range and about 7 m resolution in azimuth at a distance of 1 km. The instrument is rotating around the vertical axis during acquisition. Consequently, there is a conical area covered with the measurements and the surface displacement is collected in different LOS directions (Figure 12.4). The processing of the data follows a similar workflow as for the satellite data described in Figure 12.2, except that two processing steps are left out. Because the instrument position remains the same, the topographic phase does not need to be removed and the linear detrending step is not necessary. The data that is acquired with the second receive antenna can function as a backup solution and can be employed for the calculation of the topography of the monitored scene [45]. The instrument has been successfully tested for the mapping of landslide displacements, rock glacier and glacier movement [37,44,45].

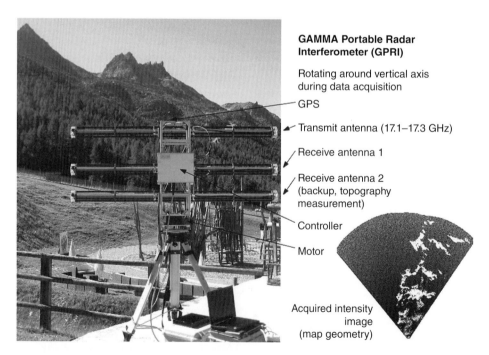

Figure 12.4 The ground-based GAMMA Portable Radar Interferometer (GPRI) is set up in the field for mapping the movement of the Distelhorn rock glacier.

In Grächen, the GPRI was set up at the Hannigalp with direct view from north to the Distelhorn rock glacier in 1.5–2 km distance (Figure 12.4) and data was collected over 24 hours in 10-minute intervals starting on 10 July 2012 12 pm until 11 July 2012 12 pm. The data was interferometrically processed to surface displacement over 24 hours and mapped to the Swiss reference system CH1903 with a pixel spacing of 10 metres (Figure 12.5, left image).

For the purpose of comparison and complementarity, a displacement map was also calculated from two satellite acquisitions spanning over the time of the fieldwork. This 11-day displacement map was scaled to displacement per day and is presented in Figure 12.5 (right image).

The movement of the Distelhorn rock glacier is clearly visible in both maps. The fastest velocity is measured with the ground-based radar interferometer at the fronts of the Distelhorn rock glacier with values in the order of 3–4 mm per day. The movement is clearly detectable, however the influence of the troposphere on the radar signal, which was only estimated, can have an effect on the accuracy of the result in the order of 1–2 mm per day. For future fieldwork at the Distelhorn rock glacier it is therefore recommended to take measurements over a longer time span in order to get a better estimate for the atmospheric phase term and with that a more accurate velocity measure.

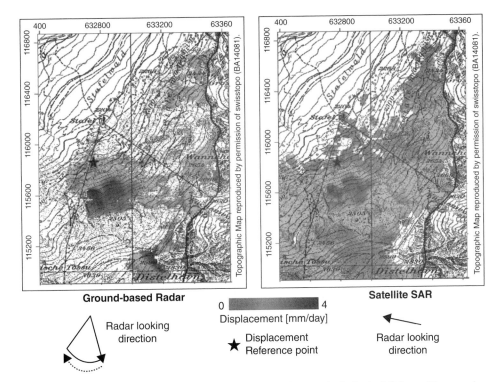

Figure 12.5 Surface displacement measured for the Distelhorn rock glacier in Grächen with ground-based radar and satellite SAR interferometry. Left: fieldwork results observing the Distelhorn rock glacier for 24 hours, right: 11-day satellite InSAR displacement map scaled to displacement per day. The results represent movement measured in different LOS directions as indicated below the figures. (*See insert for color representation of the figure.*)

With the different acquisition geometries of satellite SAR and ground-based radar interferometry a different and complementary coverage with surface displacement information is achieved. The upper most part of the Distelhorn rock glacier below the Distelhorn is for instance only visible in the results from the ground-based experiment, while it is masked by layover in the satellite InSAR results. On the contrary is the central part of the rock glacier masked out by shadow in the ground-based acquisitions but well visible from satellite. Because of the different acquisition geometries from ground and space, the surface displacement is also measured in different LOS directions providing the possibility of combining the data for a better understanding of the movement behaviour in three dimensions. The ground-based measurements can become also essential when there is movement in north–south direction, which can only minimally be captured with satellite InSAR due to the satellites' predefined orbits.

Finally, the much shorter acquisition time interval of ground-based radar interferometry makes the application of this technique useful when the velocity of a rock glacier or a landslide is faster than is possible to capture by satellite InSAR. This is for instance the case for parts of the rock glaciers with movements in the order of 3–4 mm per day, which scale up to 3.3–4.4 cm in 11 days with a partly decorrelated signal in TerraSAR-X interferograms (Figure 12.3).

12.5 Conclusions

The surface velocity of several Alpine mass movements in the study area in Grächen and St. Niklaus has been mapped by interferometrically processing satellite SAR and ground-based radar data. The satellite and ground-based surface displacement measurements are well suited to complement each other. With satellite InSAR surface displacement information can be provided over large spatial areas. The spatial coverage reached with a ground-based instrument is much smaller, but the advantages are that there is the possibility to choose the acquisition time interval, the viewing geometry and a different LOS direction to the target, other than from the satellite. With the ground-based radar data can be collected for time intervals from within a few minutes to hours, allowing also the measurement of movements up to several metres per year, whereas the data acquired by a satellite follows a certain temporal interval. On the other hand, there are already large archives of satellite data available which can help in studying the historical behaviour of geomorphological processes and to detect changes not only over several days, but also over seasons, years and decades as is useful when studying the kinematics of rock glaciers [13].

For the study site in Grächen and St. Niklaus, more TerraSAR-X data acquired between 2009 and 2013 will be processed with the goal to map the different existing processes and their movement velocity variations over the last years. Post-processing of the measured LOS displacements will then be performed by including information on slope and aspect angle of the terrain to calculate a downslope velocity.

Furthermore, it is planned to couple surface displacements from satellite (ascending and descending orbits) and ground-based measurements over the Distelhorn rock glacier area with the intention to derive its movement behaviour in three dimensions and to better understand the mechanism underlying the displacement (e.g. sliding, subsidence, creeping).

Another task will be to combine the measured displacements with results obtained by processing other satellite SAR data, or with displacement measurements from three GPS data loggers that were installed in the study area in July 2012 by the Swiss Federal Office of Environment and the Swiss Federal Institute of Technology (ETH Zurich). GPS data can be used for validation of the satellite or ground-based radar interferometric measurements and they can also be integrated for an enhanced 3D analysis.

The presented results can already serve as a complement to other geoscientific research activities in the study area [46,47] and can contribute to a better understanding of the kinematics of the ongoing geomorphological processes. Moreover, they can be considered for the compilation of an inventory of unstable slopes with indication of the class of activity [48,49] and serve as baseline information to document the ongoing mass movements in Grächen and St. Niklaus.

Acknowledgements

This work was developed under the framework of the Copernicus, formerly GMES, Initial Operations – Network on Earth Observation Research Training (GIONET) project between 2011 and 2014. GIONET was funded by the European Commission, Marie Curie Programme, Initial Training Networks, Grant Agreement number PITN-GA-2010-264509. TerraSAR-X data, © DLR, was kindly provided through the FP7 DORIS project [37]. The SwissAlti3D 5 m digital elevation model and the Swiss topographic map are copyright to the Swiss Federal Office of Topography (swisstopo). The topographic maps are reproduced by permission of swisstopo (BA14081).

References

1 J.F. Shroder Jr and M.P. Bishop, Mountain Geomorphic Systems, in *Geographic Information Science and Mountain Geomorphology*, M.P. Bishop, J.F. Shroder Jr (Eds.), Springer Verlag, Berlin, pp. 381–402, 2004.

2 M.J. Smith and C.F. Pain, Application of Remote Sensing in Geomorphology, *Progress in Physical Geography*, **33**(4), 568–582 (2009).

3 L. Schrott, J.-C. Otto, J. Götz and M. Geilhausen, Fundamental classic and modern field techniques in geomorphology: an overview, in *Treatise on Geomorphology*, J. Shroder (Editor in Chief), A.D. Switzer and D.M. Kennedy (Eds.), Academic Press, San Diego, CA, vol. **14**, Methods in Geomorphology, pp. 6–21, 2013.

4 P. Rosen, S. Hensley, I. Joughin, F. Li, S. Madsen, E. Rodriguez and R. Goldstein, Synthetic aperture radar interferometry, *Proceedings of the IEEE*, **88**, 333–382 (2000).

5 R. Bamler and P. Hartl, Synthetic Aperture Radar Interferometry, *Inverse Problems*, **14**, R1–R54 (1998).

6 O. Monserrat, M. Crosetto and G. Luzi, A review of ground-based SAR interferometry for deformation measurement, *ISPRS Journal of Photogrammetry and Remote Sensing*, **93**(C), 40–48 (2014).

7 J.R. Janke and L.W. Price, Mountain Landforms and Geomorphic Processes, in *Mountain Geography: Physical and Human Dimensions*, M.F. Price, A.C. Byers, D.A. Friend, T. Kohler and L.W. Price (Eds.), University of California Press, Berkeley and Los Angeles, California, 2013.

8 O. Hungr, S. Leroueil, and L. Picarelli, The Varnes classification of landslide types, an update, *Landslides*, **11**, 167–194 (2014).

9 W. Haeberli, J. Nötzli, L. Arenson, R. Delaloye, I. Gärtner-Roer, S. Gruber, K. Isaksen, C. Kneisel, M. Krautblatter and M. Phillips, Mountain permafrost – development and challenges of a young research field, *Journal of Glaciology*, **56** (200), 1043–1058 (2010).

10 W. Haeberli, Creep of Mountain Permafrost: Internal Structure and Flow of Alpine Rock Glaciers, *Mitteilungen der Versuchsanstalt fuer Wasserbau, Hydrologie und Glaziologie an der Eidgenoessischen Technischen Hochschule (ETH) Zuerich.*, **77**, 1–142 (1985).

11 D. Barsch, *Rockglaciers: Indicators for the present and former geoecology in high mountain environments*, Springer Series in Physical Environment 16, 1996.

12 T. Strozzi, A. Kääb and R. Frauenfelder, Detecting and quantifying mountain permafrost creep from in situ inventory, space-borne radar interferometry and airborne digital photogrammetry, *International Journal of Remote Sensing*, **25**(15), 2919–2931 (2004).

13 R. Delaloye, C. Lambiel and I. Gärtner-Roer, Overview of rock glacier kinematics research in the Swiss Alps: Seasonal rhythm, interannual variations and trends over several decades, *Geographica Helvetica*, **65** (2), 135–145 (2010).

14 W. Haeberli, M. Hoelzle, A. Kääb, F. Keller, D. Vonder Mühll and S. Wagner, Ten Years after drilling trough the permafrost of the active rock glacier Murtel, Eastern Swiss Alps: Answered questions and new perspectives, in *Proceedings of the Seventh International Permafrost Conference*, Yellowknife, Canada, Collection Nordicana 55, 1998.

15 A. Kääb, R. Frauenfelder and I. Roer, On the response of rock glacier creep to surface temperature increase, *Global and Planetary Change*, **56**, 172–187 (2007).

16 F. Paul, A. Kääb and W. Haeberli, Recent glacier changes in the Alps observed by satellite: Consequences for future monitoring strategies, *Global and Planetary Change*, **56**, 111–122 (2007).

17 W. Haeberli, M. Wegmann and D. Vonder Mühll, Slope stability problems related to glacier shrinkage and permafrost degradation in the Alps, *Ecologae Geologicae Helvetiae*, **90**, 407–414 (1997).

18 T. Strozzi, R. Delaloye, A. Kääb, C. Ambrosi, E. Perruchoud and U. Wegmüller, Combined observations of rock mass movements using satellite SAR interferometry, differential GPS, airborne digital photogrammetry and airborne photography interpretation, *Journal of Geophysical Research*, **115**, F01014 (2010).

19 M.J. Smith and C.F. Pain, Applications of remote sensing in geomorphology, *Progress of Physical Geography*, **33**(4), 568–582 (2009).

20 D. Massonnet and K.L. Feigl, Radar interferometry and its application to changes in the earth's surface, *Reviews of Geophysics*, **36**(4), 441–500 (1998).

21 A. Gabriel, R. Goldstein and H. Zebker, Mapping Small Elevation changes over large areas: Differential interferometry, *Journal of Geophysical Research*, **94**, 9183–9191 (1989).

22 E. Sansosti, F. Casu, M. Manzo and R. Lanari, Space-borne radar interferometry techniques for the generation of deformation time series: An advanced tool for Earth's surface displacement analysis, *Geophysical Research Letters*, **37**, L20305 (2010).

23 J. Wasowski and F. Bovenga, Investigating landslides and unstable slopes with satellite Multi Temporal Interferometry: Current issues and future perspectives, *Engineering Geology*, **174**, 103–138 (2014).

24 A. Hooper, D. Bekaert, K. Spaans and M. Arikan, Recent advances in SAR interferometry time series analysis for measuring crustal deformation, *Tectonophysics*, **514–517**, 1–13 (2012).

25 P. Berardino, G. Fornaro, R. Lanari and E. Sansosti, A new algorithm for surface deformation monitoring based on small baseline differential SAR interferograms, *IEEE Transactions on Geoscience and Remote Sensing*, **40**, 2375–2383 (2002).

26 A. Ferretti, C. Prati and F. Rocca, Permanent scatterers in SAR interferometry, *IEEE Transactions on Geoscience and Remote Sensing*, **39**, 8–20 (2001).

27 C. Werner, U. Wegmüller, T. Strozzi and A. Wiesmann, Interferometric point target analysis for deformation mapping, in *Proceedings of the IEEE International Geoscience and Remote Sensing Symposium (IGARSS), Toulouse, France*, 2003.

28 A. Hooper, H. Zebker, P. Segall and B. Kampes, A new method for measuring deformation on volcanoes and other natural terrains using InSAR persistent scatterers, *Geophysical Research Letters*, **31**, L23611 (2004).

29 C. Colesanti and J. Wasowski, Investigating landslides with space-borne Synthetic Aperture Radar (SAR) interferometry, *Engineering Geology*, **88**, 173–199 (2006).

30 A. Corsini, P. Farina, G. Antonello, M. Barbieri, N. Casagli, F. Coren, L. Guerri, F. Ronchetti, P. Sterzai and D. Tarchi, Space-borne and ground-based SAR interferometry as tools for landslide hazard management in civil protection, *International Journal of Remote Sensing*, **27**(12), 2351–2369 (2006).

31 A. Hooper, A multi-temporal InSAR method incorporating both persistent scatterer and small baseline approaches, *Geophysical Research Letters*, **35**, L16302 (2008).

32 J. Hu, Z.W. Li, X.L. Ding, J.J. Zhu, L. Zhang and Q. Sun, Resolving three-dimensional surface displacements from InSAR measurements: A review, *Earth-Science Reviews*, **133**, 1–17 (2014).

33 SafeLand project, deliverable 4.4, *Guidelines for the selection of appropriate remote sensing technologies for monitoring different types of landslides, 2011*, available at www.safeland-fp7.eu, accessed on 31.07.2014.

34 GALAHAD project, *Advanced Remote Monitoring Techniques for Glaciers, Avalanches and Landslides Hazard Mitigation*, deliverable 10-4, Final Technical Book, 2008.

35 P. Farina, D. Colombo, A. Fumagalli, F. Marks and S. Moretti, Permanent Scatterers for landslide investigations: outcomes from the ESA-SLAM project, *Engineering Geology*, **88**, 200–217 (2006).

36 TERRAFIRMA project, *A pan-European Ground Motion Hazard Service*, www.terrafirma.eu.com, accessed on 31.07.2014.

37 DORIS project, *An advanced downstream service for the detection, mapping, monitoring and forecasting of ground deformations*, www.doris-project.eu, accessed on 21.11.2016.

38 F. Noverraz, C. Bonnard, H. Dupraz and L. Huguenin, Grands glissements de terrain et climat, VERSINCLIM *Rapport final PNR31 (Programme National de Recherche)*, Zurich, vdf, 1998.

39 J. Papke, T. Strozzi, U. Wegmüller and N.J. Tate, Rock Glacier Monitoring with Spaceborne SAR in Graechen, Valais, Switzerland, in *Proceedings of the IEEE International Geoscience and Remote Sensing Symposium (IGARSS), Munich, Germany*, 2012.

40 T. Strozzi, U. Wegmueller, C. Werner, A. Wiesmann, R. Delaloye and H. Raetzo, Survey of landslide activity and rockglaciers movement in the Swiss Alps with TerraSAR-X, in *Proceedings of IEEE International Geoscience and Remote Sensing Symposium (IGARSS)*, Cape Town, South Africa, 2009.

41 D. Schmidt and R. Buergmann, Time-dependent land uplift and subsidence in the Santa Clara Valley, California, from a large interferometric synthetic aperture radar data set, *Journal of Geophysical Research*, **108**, B9, 2416 (2003).

42 R. Lugon and M. Stoffel, Rock glacier dynamics and magnitude-frequency relations of debris flows in a high-elevation watershed: Ritigraben, Swiss Alps, *Global and Planetary Change*, **73**, 202–210 (2010).

43 M. Rebetez, R. Lugon and P. Baeriswyl, Climatic change and debris flows in high mountain regions: the case of study of the Ritigraben torrent (Swiss Alps), *Climatic Change*, **36**, 371–389 (1997).

44 C. Werner, A. Wiesmann, T. Strozzi, R. Caduff, U. Wegmüller and A. Kos, The GPRI Multi-mode Differential Interferometric Radar for Ground-based Observations, *Proceedings of the 9th European Conference on Synthetic Aperture Radar (EUSAR)*, Nürnberg, Germany, 2012.

45 T. Strozzi, C. Werner, A. Wiesmann and U. Wegmüller, Topography Mapping With a Portable Real-Aperture Radar Interferometer, *IEEE Geoscience and Remote Sensing Letters*, **9** (2), 277–281 (2012).

46 M. Philipps, F. Ladner, M. Müller, U. Sambeth, J. Sorg and P. Teysseire, Monitoring and reconstruction of a chairlift midway station in creeping permafrost terrain, Graechen, Swiss Alps, *Cold Regions Science and Technology*, **47**, 32–42 (2007).

47 E. Zenklusen Mutter and M. Philipps, Thermal evidence of recent talik formation in Ritigraben rock glacier: Swiss Alps, in *Proceedings of the 10th International Conference on Permafrost*, Salekhard, Russia, 2012.

48 T. Strozzi, C. Ambrosi and H. Raetzo, Interpretation of aerial photographs and satellite SAR interferometry for the inventory of landslides, *Remote Sensing*, **5**(5), 2554–2570 (2013).

49 C. Barboux, R. Delaloye and C. Lambiel, Inventorying slope movements in an Alpine environment using DInSAR, *Earth Surface Processes and Landforms*, doi: 10.1002/esp.3603, 2014.

13

Sample Supervised Search-Centric Approaches in Geographic Object-Based Image Analysis: Concepts, State of the Art and a Future Outlook

C. Fourie and E. Schoepfer

German Aerospace Center (DLR), German Remote Sensing Data Center (DFD), Oberpfaffenhofen, Germany

13.1 Search-Centric Sample Supervised Image Analysis

13.1.1 Image Analysis as a Design Problem

Image analysis signifies the process of extracting required or useful information from a digital image, typically using various forms of image processing and classification. Research on image analysis spans several diverse and well established disciplines; some focused on specific problem domains and applied research [1] while others are concerned with more primary, problem insensitive inquiries [2,3]. Some example domains where image analysis research forms a prominent part of activities include digital image processing, satellite remote sensing, biomedical image analysis, astronomy, mathematical imaging, image information mining and computer vision. The intrinsic characteristics of the images in these domains and thus the developed approaches to image analysis are as diverse as the fields that are concerned with it.

Image analysis processes can range from simple expert system classifier based approaches observing the spectral values of basic elements in an image (e.g. pixels), to complex processes considering spatial [4] and hierarchical context relationships, using numerous feature descriptors (attributes of image elements) [5], combining various primary analysis paradigms such as semantic segmentation, template matching, discriminative models and parts based models [6] and also involving various approaches taken from machine learning, e.g. as applied in context based image retrieval [7] and optimization [8].

Remote sensing image analysis poses as an interesting domain where the nature of the used analysis techniques evolved due to the progression of the fidelity of the spatial resolution of the captured imagery [9]. Classically image analysis within this domain was conducted with simple supervised, unsupervised or expert system's classification approaches functioning on the spectral values of pixels in multi-band images. Typically on modern very high resolution (VHR) optical imagery, approaches attempting to conduct semantic segmentation along with using context information are more common [10].

Earth Observation for Land and Emergency Monitoring, First Edition. Edited by Heiko Balzter.
© 2017 John Wiley & Sons Ltd. Published 2017 by John Wiley & Sons Ltd.

Irrespective of the domain in which the image analysis is performed, the method, consisting of image processing and/or classification techniques, used to conduct the information extraction needs to be designed.

13.1.2   Metaheuristics as Tools for Design

Metaheuristics, which signify a family of efficient problem independent stochastic search algorithms are well suited and extensively studied in various fields as efficient tools to aid in design processes [11,12], despite the non-deterministic nature of the methods and lack of theoretical models to many algorithms [13]. Within the fields concerned with image analysis and especially computer vision, metaheuristics have been studied and applied extensively in extremely varied ways to design solutions [14,15].

Figure 13.1 illustrates a simple general architecture common to most population based metaheuristics. An initial population is defined, typically randomly initiated, consisting of a group of candidate solution vectors or agents to a design problem. An individual agent within the population contains all the building blocks for a method, encoded as real, integer, discrete or binary values, depending on the specific details of the given metaheuristic. These values can control, amongst others, parameters for pre-existing image processing components. Although problem specific, values within agents are typically interdependent.

A metaheuristic search method proceeds iteratively. These iterations are referred to as generations if the metaheuristic is nature inspired (e.g. as with evolutionary inspired algorithms such as genetic algorithms and differential evolution). Each agent of the population undergoes a fitness evaluation during iterations of the algorithm, which denotes the process of calculating how well the agent performs on the given problem. This evaluation process is defined by a fitness function, also called an objective function. After the fitness evaluations of all agents in the population the method termination condition is queried. If any agent in the population has a fitness score (quality of the solution) greater than a specified threshold, if a certain number of method iterations have passed or if very little change is observed among subsequent method iterations, the method terminates. The agent possessing the best solution vector represents the result of the search process. If the method does not terminate, a new population is created. The agents of the new population may be defined via combining algorithms

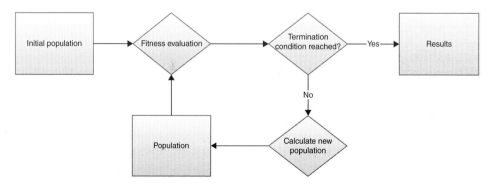

Figure 13.1 Generalized architecture of simple population based metaheuristics.

utilising different agents in the population, some randomization and knowledge of how an agent performed compared to previous iterations.

The domain or space explored by metaheuristics is often referred to as the search landscape. Figure 13.2 illustrates a two dimensional search landscape, defined by a common benchmark function (Ackley's function). The red dots represent positions visited by a metaheuristic during the search process, in this instance an adaptation of modified cuckoo search [16]. Lower z-axis values imply better solutions.

Figure 13.2 also illustrates the concept of local optima, where the central global optimal position in terms of solution quality is surrounded by eight solution positions of lesser quality, although performing better than their immediate surroundings. Note the agglomeration of agents around the central global optimal position. The use of metaheuristics are routinely justified on such problems, where simpler single agent strategies, either using search landscape gradient information or not, struggles to escape from such local optima [17]. Metaheuristics use the concepts of exploitation and exploration to efficiently traverse search landscapes in search of the global optimum: agents typically use information from their own best performances in addition to information from the performances of other good performing agents to modify their own values [11,17]. From a design perspective, the ways in which one could encode solution vectors as control processes to generate search landscapes, such as in Figure 13.2, are boundless.

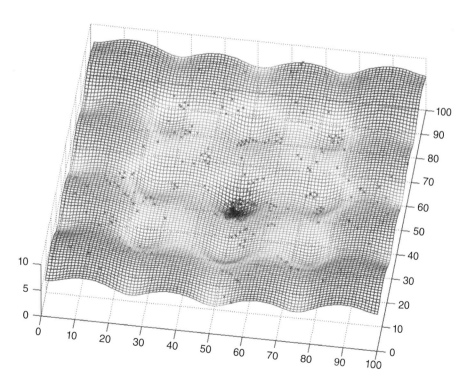

Figure 13.2 An arbitrary search landscape, with positions visited by a metaheuristic during the search process denoted with red dots.

Specific solution vector encodings related to image processing and image analysis could be extremely complex and require extensive fitness evaluations and computing time. Metaheuristics may assist in addressing such problems by attempting to solve them faster and obtaining more robust results compared to classical search methods [11]. Figure 13.3 illustrates two fitness traces of two search algorithms. A fitness trace signifies the quality of the best agent/solution vector as the iterative search process proceeds (e.g. the solution vector quality of the best performing red dots in Figure 13.2 as a function of iterations). The fitness traces in Figure 13.3 are generated by the Differential Evolution (DE) metaheuristic and Random Search (RND) for an arbitrary seven parameter design problem in image analysis. For this design problem, the use of the DE metaheuristic is justified over such a simpler search strategy. A lower solution quality signifies better results.

In a search process, each fitness evaluation is assigned with a unit of computational cost. Commonly for many problems, including problems in the image analysis domains, such potentially expensive fitness evaluation costs are the main drivers for attempting to search as efficiently as possible. Metaheuristics typically only explore a fraction of the search landscape compared to brute force search or simpler search strategies [18] to obtain optimal or near optimal solutions. In many practical instances this allows for design to be performed with metaheuristics. The time taken to search for a solution is also influenced by the problem difficulty or the nature of the search landscape [13] as defined by the fitness function. Metaheuristics are typically evaluated with black box optimization benchmarks [19] using experimental research methodologies [20].

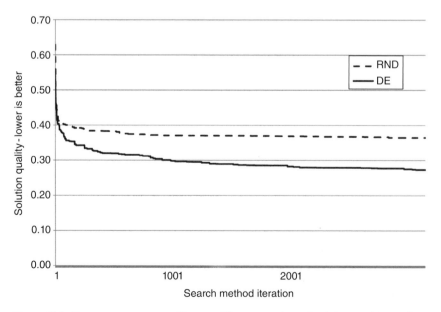

Figure 13.3 Fitness traces generated by two different search methods for an arbitrary design problem in an image analysis context.

13.1.3   Granularity and Fitness Functions in Sample Supervised Search-Centric Image Analysis Method Design

It can be argued, and has been alluded to [12,14,18], that the way in which a metaheuristic is used in image analysis method design may be described by the granularity of method construction. Granularity defines the fineness or coarseness of the basic building blocks of the method design process. At a coarse level, agent values could simply define free parameters of an image processing method that needs to be tuned, for example the parameters of a segmentation algorithm [18,21–23] or the parameters (e.g. a threshold value to identify water in a satellite image) of an expert system's classification approach. At a finer granularity, the building blocks could be more basic units or definition processes, such as defining structuring elements in classical mathematical morphology image processing [24], defining cellular automata rules [25], finding a sequence of good image processing using genetic programming [8] or even for defining the nature of attributes used in classification processes [26].

More complex information extraction frameworks are also possible, for example approaches that interleave object recognition and machine learning in the search process [27,28], or even approaches that search for coarse grained methods within the search process [29]. It can be debated that the finer the granularity of the building blocks, the higher the dimensionality of the search landscape, and the more difficult the search problem becomes. Alternatively stated, the finer the granularity, the more search time would be needed. On the other hand, finer grained processes might execute faster (per fitness evaluation) than coarse grained processes and would probably be more flexible to various problem instances.

Keeping in mind the granularity of method building blocks that defines the search landscape complexity, another major consideration to using metaheuristics in design is how to define the fitness function. In image analysis method design, solution quality can be defined by various quality measures describing the quality of the final information product or only intermediary steps in the image analysis process. The output could thus be new intermediary data (data in, data out) or information.

Two common fitness function groupings are identified, measures observing low- and mid-level processes, specifically segmentation results, and classification accuracy measures. Empirical goodness methods [30,31] define unsupervised notions of image simplification or segmentation quality, typically by evaluating statistical measures from generated image areas or segments. Empirical discrepancy measures [30] are supervised quality measures, where generated results are measured against provided ground truth, reference or gold standard examples. With classification, user's and producer's accuracy, kappa, geometric means, f-score and various other quality criteria could measure the generated results against the provided reference results.

Thus, search-centric sample supervised image analysis and search-centric sample supervised segment generation denotes method design processes where a sample or selection of reference elements is provided. The result of a fitness evaluation is how well certain generated elements match with provided reference elements. An interesting, but potentially computationally prohibitively expensive alternative to searching for a solution is to calculate all possible solutions in an ordered manner (e.g. tree hierarchies as used with interactive image information mining) and extract the set that matches most closely with the provided reference elements [32].

13.2 Approaches in Geographic Object-Based Image Analysis

GEOBIA defines a general image analysis philosophy centred on the concept of semantic segmentation [9,33,34]. As an important objective GEOBIA addresses the development of methods and accompanying theory to replicate a human's ability to interpret remote sensing imagery in automated and semi-automated ways [35]. The proliferation of VHR data, increased computing power and closer integration between Geographic Information Systems (GIS) and remote sensing processes are often cited as driving forces to the development of GEOBIA [9,33]. When concerned with VHR data, basic thematic units may be much larger than individual pixel sizes. Attempting to classify such imagery with pixel-based approaches often leads to the so-called speckle or salt and pepper effect in classification results [33].

Central to GEOBIA approaches is the generation of image objects [36], typically using various image segmentation algorithms. Other common constituents in GEOBIA include classification, attribution and information presentation considerations [37]. A distinction is made between image objects, which are arbitrary image segments and geographic or geo-objects, which denote thematic elements of interest [35,36]. Progressing from image objects to thematic objects typically proceeds in two ways, either via attempting to segment elements of interest correctly from the start, or by using various post segmentation processes to progress to semantic objects. In remote sensing image analysis problems, elements of interest may be numerous, and numerous types of elements might also be of interest.

In some problem instances, especially if the spatial and spectral characteristics of elements of interest are similar, a thematic segmentation or geographic object genera-tion process may be attempted with a segmentation algorithm by tailoring its parameters to the problem. Figure 13.4 illustrates such a scenario and depicts white

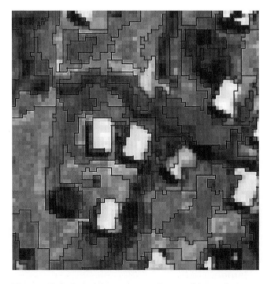

Figure 13.4 A simple segmentation problem where spectrally and geometrically similar elements of interest are segmented with a single segment layer.

tents of similar geometric and spectral characteristics, segmented with an arbitrary segmentation algorithm. In other scenarios, elements of interest that are thematically identical might be too different to allow for simple segmentation.

For more complex image analysis problems where semantic segmentation cannot be achieved easily, various approaches within GEOBIA have been proposed. A multi-scale image analysis philosophy might be advocated, where a scale-scape constraint segmentation algorithm is run multiple times to create a hierarchy of image objects [32,38]. Information through the hierarchy might be used to identify proper thematic objects (classification) in addition to using context and attribute information within a specific layer, e.g. [39].

A prominent strategy for addressing both simple and complex problems within GEOBIA is an expert system's or rule set/rule based approach [9,10,33]. The various image processing and classification, and sequences of processes needed to identify elements are defined by an expert user. The set of methods or procedures designed are encapsulated in a rule set. Such a rule set is typically created on a specific input image, but it may be used to extract information on other imagery.

Common procedures in rule set development include performing general image processing, defining image segmentation operations with various parameters to tune, merging and splitting segments, assigning or defining attributes attached to segments and classifying segments using expert knowledge or supervised/unsupervised classifiers. Using segment relations as information in hierarchical segmentation layers and in segments of a single layer is also common. Rule sets may contain additional processes and may also display complexity in orderings or sequences of processes [10]. Rule set development is commonly facilitated with commercial or free image analysis software [33,40,41].

Although this general strategy in GEOBIA is highly advocated, especially in problem instances with highly complex image details [9], two general concerns could be raised. Firstly, the transferability (generalizability) of rule sets to other geographic areas or by using data captured by different sensors may be a concern and are not routinely, explicitly incorporated into method designs. Secondly the process of designing a rule set is time consuming and highly dependent on the skill of the designer. One of the primary problems identified in this regard is selecting and tuning a segmentation algorithm for a specific scenario. Some approaches address this problem by using various empirical goodness measures to automatically define good segment layers [38,42]. Another general approach utilizes search methods to find good segment layers, which is presented next.

13.3   Search-Centric Sample Supervision within GEOBIA

13.3.1   Automatic Segmentation Algorithm Parameter Tuning

Within GEOBIA a general search-centric automatic segmentation algorithm parameter tuning approach has been proposed due to the need for more streamlined processes [21,23,43,44], especially when considering rule set approaches. Due to the computationally expensive nature of image segmentation a thorough search of the search space is commonly not feasible, requiring the use of efficient search methods. This general approach draws inspiration from earlier works from outside the domain of remote

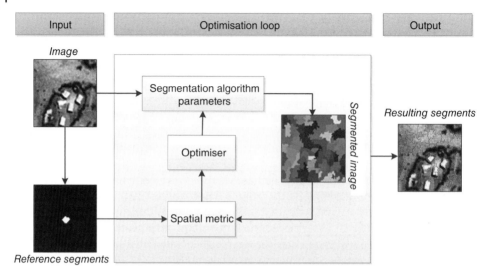

Figure 13.5 Generalized architecture of a search centric sample supervised segment generation approach.

sensing [18,22] and is also actively pursued in other imaging disciplines, e.g. [28]. Figure 13.5 illustrates a generalized architecture of such a method. The method attempts to find good segments based on a limited number of user provided reference segments or objects. A user provides reference segments either via digitizing or with other available input methods [45]. Other uses of metaheuristics have also been demonstrated in this domain, e.g. for feature selection [46].

During the optimization loop iterations, the search algorithm gives as output a parameter set for the given segmentation algorithm. Next the image, or subsets of the image for efficiency, is segmented with the tuned segmentation algorithm. The segmented image is compared with the provided reference segments to generate a quantitative measure of quality. This constitutes the fitness evaluation in the search process. The result of the fitness evaluation is passed on to the optimizer, which uses the information to direct the search process in subsequent iterations of the search method. Various stopping criteria could be specified for this general method, as is common for metaheuristic approaches.

Figure 13.6 illustrates an example scenario of how such a general approach may be used. A user might select or digitize elements of interest (Figure 13.6a), in this example tents in a refugee camp. The user may view the results of the automatic segmentation algorithm parameter tuning process (Figure 13.6b), followed by a classification procedure to identify similar elements, for example by using a novelty detector and the reference objects as training data (Figure 13.6c).

Research surrounding this general approach strives to show its practical feasibility and performance in a remote sensing and GEOBIA context. Research focussed tools with graphical user interfaces have been developed [47–49] for such an approach due to its interactive and visual nature. The freeware Segmentation Parameter Tuning (SPT) tool [49] allows a user to automatically tune a range of segmentation algorithms under different metric and search method conditions, along with the ability to export

(a)

(b)

(c)

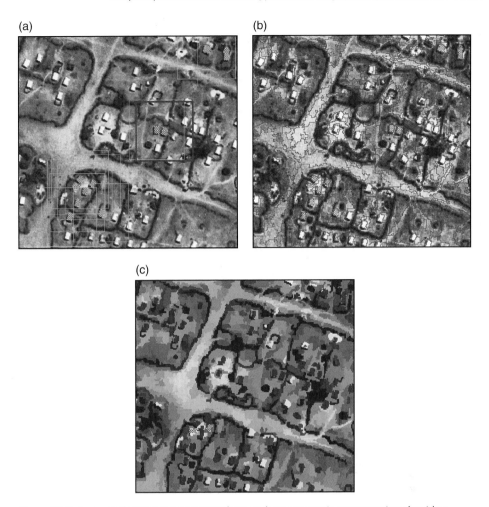

Figure 13.6 An example mapping scenario where such an automatic segmentation algorithm parameter tuning process may form part of the image analysis workflow.

results for further use. Geographic Object Novelty Detector (Geo-ND) [47] is a simple proof-of-concept tool illustrating a complete workflow incorporating target detection using automatic segmentation algorithm parameter tuning in addition to classification processes.

13.3.2 Measuring Segment Quality

An important aspect of this general approach described is how to measure the amount of similarity between the generated segments and the reference segments. Quality is typically measured with empirical discrepancy methods [30] with various imaging disciplines making use of such techniques. In GEOBIA and more generally in remote sensing numerous quality measures have also been proposed, although not always used as fitness functions in search processes [21,50–58]. Measures typically observe notions of area overlap (e.g. Figure 13.7) where a reference segment (R) is matched for

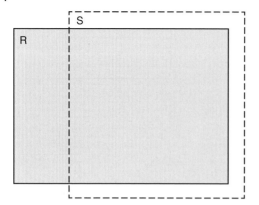

Figure 13.7 A representation of area overlap. Empirical discrepancy measures or spatial metrics could measure the amount of spatial overlap between the reference (R) and a generated (S) segment.

spatial coherence with a generated segment (S). Notions of over-segmentation, under-segmentation, geometry, edge-offsets and spectral properties may also be encoded into such quality metrics.

Figure 13.8 illustrates three image subsets with an example reference segmented delineated by a bold red polyline. The objective could be to accurately segment buildings. During the search iterations of the method depicted in Figure 13.5, such evaluations are performed. Quality scores of the Reference Weighted Jaccard (RWJ) measure is also shown [48], with lower values signifying better segmentation results, with a range of [0,1]. Figure 13.8(a) illustrates a segmentation algorithm parameter set and accompanying image segments (green lines) resulting in an over-segmentation. Figure 13.8(b) represents a better or adequate segmentation, while Figure 13.8(c) represents an under-segmentation scenario.

13.3.3 Advances in Search-centric Sample Supervised Segment Generation

Various aspects of such a sample supervised search-centric method have been subjected to further investigations within GEOBIA and remote sensing. The behaviour of using spatial metrics have been shown to be highly correlated [55] on a metric test bed consisting of general area-overlap metric variants (e.g. Figure 13.7). Multi-objective optimization, where various metrics are used that typically observe different quality aspects have also been demonstrated [53].

Another major consideration is the choice of search method or optimizer. The selection of an appropriate method is typically conducted using experimentalism [20]. The utilized optimizers progressed from classic genetic algorithms [21] to proven faster methods such as differential evolution [59] and direct search derivative free methods [60]. For more complex optimization problems the use of metaheuristics are warranted over simpler search strategies [48]. Future implementations might select appropriate methods based on both experimentalism [20] and method performances on black box optimization benchmarks [19]. Parallel computing implementations for fitness evaluation processes such as segmentation have also been demonstrated that could reduce the required computing times [61].

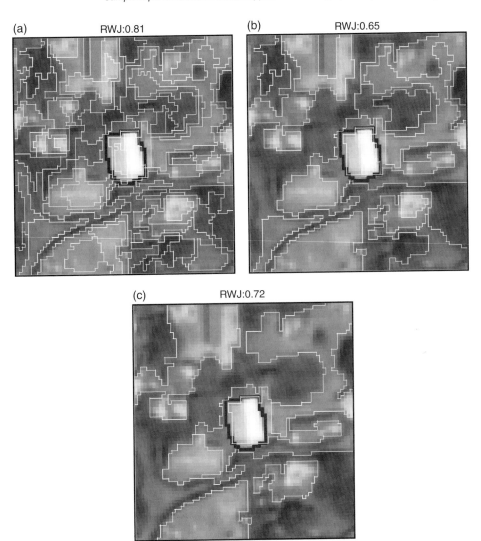

Figure 13.8 Three segmentation evaluation examples using the RWJ metric evaluating the mismatch between reference (Black line) and generated (White lines) segments.

A general objective in search-centric sample supervised method research is to create methods that are either faster or lead to better quality solutions than current methods. The generalizability and adaptability performances of methods to problems have received little research attention. Some methods address problem complexity by incorporating algorithms to merge segments that are unable to be segmented correctly – irrespective of the tuning of the segmentation algorithm [62,63]. Segmentation algorithms may also be defined that are more modular and thus better able to fit a range of problem instances [64,65].

Another line of thinking that addresses the general drive to generate improved results suggests the creation of enlarged or higher dimensional search spaces. These enlarged search spaces not only consider or contain segmentation algorithm parameters, but also other parameterizable processes surrounding segmentation that might influence the results [48,66]. Data transformation functions or low-level image processing processes could be appended to the search spaces [43,48], potentially resulting in better segmentation results. This notion has been extended to mathematical morphology inspired segmentation approaches, where attribute tuning, segmentation algorithm tuning and data transformation function tuning are combined in a single search space [65]. Metaheuristics could also be used to select the appropriate attributes in such approaches [67]. Another approach propose the selection, or searching of, mathematical morphology connected filters for a given problem [68].

Research methods surrounding sample supervised segment generation typically focus on comparing a new formulation against previous approaches in terms of achieved segment quality, required processing time and the robustness of results [23,48,69]. Generalizability performances and sampling size requirements may also be considered as performance indicators. Initial contributions focussed on proof-of-concept, commonly as conference papers (e.g. [21]), with more recent work (e.g. [48]) borrowing heavily from comparative research methods [20] from the optimization community.

13.4 Limitations of Search-centric Sample Supervised Segment Generation Approaches in GEOBIA

Although the general technique of search-centric sample supervised segment generation within GEOBIA shows promise, some limitations should be highlighted. A common problem or concern with a search-centric sample supervised method is its ability to adapt to difficult or unseen problems [21]. A method might just be unable to address the given problem. Having expert knowledge of the capabilities of the used segmentation algorithm(s) could allow a user to judge the feasibility of the method beforehand. Another concern is the definition of a class of interest or the similarity of elements within a class. Thematically similar elements of interest might have vastly different spectral and geometric characteristics, suggesting that a segmentation algorithm may not be effectively tuned to solve such a problem [70]. For example, "buildings" in Figure 13.9 range from large rectangular bright objects to small grass roofed huts, suggesting multiple runs (sub-classes) of a method for a single class.

The number of samples needed for such an approach to generalize well to the elements of interest has not been the topic of an investigation. Typically a small selection of reference samples are used in experimentation, with generalizability mechanisms such as cross-validation optional [48]. Methods from supervised classification could potentially be transferred to address such concerns [69,71]. Another concern related to sampling is the processing time required and the extent of the areas that are segmented. More samples imply more processing time. Subsets of the image, centred on the reference segments can be extracted and used as input to a method, provided that segmentation results are identical compared to subsets (segmentation algorithm dependent).

Although various metrics have been used in such a general approach, little is known how metric values relate to final classification accuracies. It can be assumed that better

Figure 13.9 A segmentation scenario where the object of interest could be buildings of vastly different spectral and geometric characteristics.

quality segments leads to better quality classification results and information products. The utilized classification methods and attributes (features) should also be considered. Profiling the relations or correlations among used metrics and classifiers considering various (attribute) conditions might be fruitful. The value of different metrics or observed aspects or even of multi-objective optimization strategies might be better understood, specifically how the notions of under- and over-segmentation relate to classification results.

There are substantial uncertainties surrounding the usage and evaluation of metaheuristics. Although the general methods described here are considered applications of metaheuristics, the concerns raised in basic research should be noted. Metrics define the characteristics of the search landscape (e.g. Figure 13.2) and encoding or creating additional processes in the search landscape also inherently influences its characteristics [48]. Theory on how metaheuristics behave on various search landscapes are in its infancy [13,72] with models able to predict problem difficulty a topic of active research [73]. In the context of search-centric sample supervised segment generation approaches a handful of search methods have been experimentally profiled in terms of fitness traces over different search landscapes [48]. Simpler search methods such as a hill climber and random search displayed varied performances on a range of typical search landscapes [48], suggesting further investigations into the appropriateness of using specific search methods in these problem contexts.

Based on the uncertainties regarding performance theories and models, empirical research or experimentalism is generally advocated when conducting research on metaheuristic applications [13,20,74]. Such approaches typically require extensive experimentation to calculate measures of quality, robustness and generalizability, with numerous experimental runs required on various problem instances and under various conditions or parameter sets. Such considerations should be noted when conducting experimentation within image analysis and especially when considering segmentation, which is typically a computationally expensive process.

13.5 An Outlook on Search-Centric Sample Supervision in GEOBIA Approaches – Epistatic Links and Search Granularity

In the context of GEOBIA and considering advances in image processing and metaheuristics, the question could be raised how one could define components of GEOBIA workflows as optimization problems. A current philosophy in rule set development suggest an expert user explores or investigates ways to obtain good results or build methods [10]. Efficient search methods could assist in such method construction or tuning processes. A very specific application of metaheuristics in GEOBIA is to tune the parameters of segmentation algorithms (as described above), but more encompassing or integrative implementations could be considered.

Classical GEOBIA processes such as attribution, segmentation, low-level image processing and classification [37] typically display high degrees of complex interactions [10] or high degrees of epistasis (dependencies) [75]. Such processes are parameterizable and their interactions could also be encoded in some manner. A simple example from the domain of machine learning suggests simultaneous classifier free parameter tuning and attribute subset selection [76]. Classification and attributes are not only dependent on each other (unless attribute selection is implicit in the classifier), but in all likelihood on the nature of the generated segments (segment parameters and metric scores). In addition, attributes themselves could be highly modular and designable [26] or tuneable [77]. The domain of mathematical morphology inspired image processing and analysis encompasses numerous highly modular, parameterizable or designable components related to segmentation (e.g. [78]).

Ways in which one could encode combined search problems from such varied components for efficient methods are unclear, although having different data quantizations in a single optimization problem are feasible [79]. Figure 13.10 illustrates an artificial or toy problem (Figure 13.10(a)) where the encoding of an arbitrary search-centric sample supervised method consists of more elaborate parameter domains [66]. In this instance two classification processes (a one-class and a two class support vector machine, Figure 13.10(b, c)) interacts with a data layering and segmentation process (Figure 13.10(d)) to aim to correctly segment and find the elements of interest.

Another point of contention is how such processes could elegantly integrate with existing GEOBIA workflows or philosophies (rule set development), and how good synergies between user experience/interaction and automatic search could be advocated. This is an important consideration in the context of emergency response or rapid mapping. In such scenarios information products need to be generated quickly. It is unclear if even the most advanced method designs could successfully address moderately difficult remote sensing image analysis problems on their own. A basic approach as depicted in Figure 13.6 would probably not deliver results of sufficient quality in most problem instances. In addition, there are numerous ways or intermediate steps to measure quality at the image processing and classification levels. These measures (or combinations of them) should correlate with user judgment. Another line of thinking suggests the creation of agent based modelling for rule-set adaptation [80].

The general processes described above and specifically the image processing could be considered as coarse grained processes. It is known that integrative approaches

(a) (b)

(c) (d)

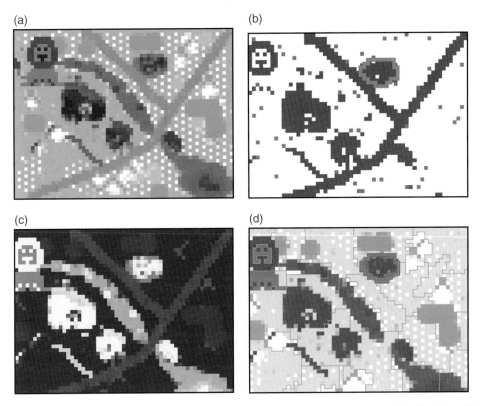

Figure 13.10 An arbitrary search-centric image analysis method executed on an artificial problem (a) considering pixel-based classification processes (b, c) and their interaction with data transformation and segmentation processes (d).

combining various aspects of image processing are needed to address complex image analysis problems [6,81]. It would be interesting to see how fine grained (search-centric) processes such as cellular automata [24] and genetic programming based image processing [8] would compare with segmentation algorithm parameter tuning considering the required search times, quality of results achieved and generalizability performances. Fine grained search-centric methods might have interesting influences on, or synergies with traditional rule set based approaches, which currently only consider coarse grained parameter tuning processes with high generalizability performances.

13.6 Conclusion

Search-centric sample supervised image analysis approaches might find use alongside traditional image analysis approaches within software environments, even if initially only to remove the guesswork from parameter tuning processes. Ultimately the involvement of a metaheuristic could range from simple parameter tuning to designing the entire image analysis method [14]. Although any mapping scenario might benefit from such strategies (e.g. land cover mapping), time critical mapping tasks (e.g. Copernicus Emergency Management Services) where user interaction is actively encouraged might benefit substantially. In such instances image analysis solutions (supervised classifiers, rule sets) to various problems cannot readily be designed beforehand.

An overview of metaheuristics as tools for design was presented, with a focus on how such approaches have been applied within the domain of GEOBIA. Usage considerations and limitations of such a general approach were also highlighted. Based on developments from outside the domain of GEOBIA and acknowledging the unique processes within it, an outlook was given on potential avenues for future research, focusing on the concept of epistatic links among processes. How search-centric sample supervised methods could integrate with current GEOBIA approaches to create efficient workflows should be a topic for future investigation.

Currently search-centric sample supervised approaches remain steadfastly within the domain of research, as performances are not sufficiently adequate (as of yet) and method nuisances are plentiful. Systematic investigations on the various aspects of search-centric sample supervised methods will probably continue. Such approaches touch upon concepts from various academic disciplines or domains. In the case of GEOBIA and considering the current and potential designs of search-centric sample supervised methods, concepts from machine learning, optimization, computer vision, image processing and remote sensing are involved (among others). This may also be evident based on the diversity of citations in this manuscript.

References

1 S.M. De Jong and F.D. Van Der Meer, *Remote Sensing Image Analysis: Including the Spatial Domain*, Springer, Dordrecht, (2004).

2 B. Jähne, *Digital Image Processing*, Springer, Berlin, (2005).

3 M.H.F. Wilkinson and J.B.T.M. Roerdink, *Mathematical Morphology and Its Application to Signal and Image Processing*, Springer, Berlin, (2009).

4 M. Pedergnana, P.R. Marpu, M.D. Mura, J.A. Benediktsson and L. Bruzzone, Classification of remote sensing optical and lidar data using extended attribute profiles, *IEEE Journal of Selected Topics in Signal Processing*, **6**, 856–865, (2012).

5 K. Mikolajczyk and C. Schmid, A performance evaluation of local descriptors, *Pattern Analysis and Machine Intelligence, IEEE Transactions on*, **27**, 1615–1630, (2005).

6 S. Zheng, A. Yuille and Z. Tu, Detecting object boundaries using low-, mid-, and high-level information, *Computer Vision and Image Understanding*, **114**, 1055–1067, (2010).

7 R. Datta, D. Joshi, J. Li and J.Z. Wang, Image retrieval: Ideas, influences, and trends of the new age, *ACM Computing Surveys*, **40**, Article 5, (2012).

8 S. Harding, J. Leitner and J. Schmidhuber, Cartesian genetic programming for image processing, in *Genetic Programming Theory and Practice X*, R. Riolo, E. Vladislavleva, M.D. Ritchie and J.H. Moore (Eds), Springer, New York, (2013).

9 T. Blaschke, G.J. Hay, M. Kelly, S. Lang, P. Hofmann, E. Addink, R. Queiroz Feitosa, F. Van Der Meer, H. Van Der Werff, F. Van Coillie and D. Tiede, Geographic Object-Based Image Analysis – Towards a new paradigm, *ISPRS Journal of Photogrammetry and Remote Sensing*, **87**, 180–191, (2014).

10 M. Baatz, C. Hoffman and G. Willhauck, Progressing from object-based to object-oriented image analysis, in *Object-Based Image Analysis: Spatial Concepts for Knowledge-Driven Remote Sensing Applications*, T. Blaschke, S. Lang and G.J. Hay (Eds), Springer, Berlin, (2008).

11 E. Talbi, *Metaheuristics: From Design to Implementation*, Wiley, Hoboken, (2009).

12 S.J. Louis, *Genetic Algorithms as a Computational Tool for Design*, Doctor of Philosophy, Indiana University, (1993).

13 L. Vanneschi, L. Mussi and S. Cagnoni, Hot topics in evolutionary computation, *Intelligenza Artificiale*, **5**, 5–17, (2011).

14 S. Cagnoni, Evolutionary computer vision: a taxonomic tutorial, *Hybrid Intelligent Systems (HIS 2008)*, 10–12 September, Barcelona, Spain, (2008).

15 S. Cagnoni, E. Lutton and G. Olague, *Genetic and Evolutionary Computation for Image Processing and Analysis*, Hindawi Publishing Corporation, New York, (2007).

16 S. Walton, O. Hassan, K. Morgan and M. Brown, Modified cuckoo search: a new gradient free optimisation algorithm, *Chaos, Solitons & Fractals*, **44**, 710–718, (2011).

17 K.V. Price, R.M. Storn and J.A. Lampinen, *Differential Evolution: A Practical Approach to Global Optimization*, Springer, Berlin, (2005).

18 B. Bhanu, S. Lee and J. Ming, Adaptive image segmentation using a genetic algorithm, *Systems, Man and Cybernetics, IEEE Transactions on*, **25**, 1543–1567, (1995).

19 N. Hansen, A. Auger, S. Finck and R. Ros, Real-Parameter Black-Box Optimization Benchmarking 2010: Experimental Setup, *INRIA Research Report RR-7215*, (2010).

20 T. Bartz-Beielstein, *Experimental Research in Evolutionary Computation*, Springer, Berlin, Germany, (2006).

21 R.Q. Feitosa, G.A.O.P. Costa, T.B. Cazes and B. Feijo, A genetic approach for the automatic adaptation of segmentation parameters, in *Geographic Object-Based Image Analysis (GEOBIA 2006)*, S. Lang, T. Blaschke and E. Schoepfer (Eds), ISPRS, Salzburg, (2006).

22 G. Pignalberi, R. Cucchiara, L. Cinque and S. Levialdi, Tuning range image segmentation by genetic algorthm, *EURASIP Journal on Applied Signal Processing*, **2003**, 780–790, (2003).

23 S. Derivaux, G. Forestier, C. Wemmert and S. Lefevre, Supervised image segmentation using watershed transform, fuzzy classification and evolutionary computation, *Pattern Recognition Letters*, **31**, 2364–2374, (2010).

24 I. Yoda, K. Yamamoto and H. Yamada, Automatic acquisition of hierarchical mathematical morphology procedures by genetic algorithms, *Image and Vision Computing*, **17**, 749–760, (1999).

25 P.L. Rosin, Image processing using 3-state cellular automata, *Computer Vision and Image Understanding*, **114**, 790–802, (2010).

26 T. Kowaliw, W. Banzhaf, N. Kharma and S. Harding, Evolving novel image features using genetic programming-based image transforms, *in Evolutionary Computation (CEC'09), IEEE Congress on*, IEEE, 2502–2507, Trondheim, Norway, 2009.

27 B. Bhanu and J. Peng, Adaptive integrated image segmentation and object recognition, *IEEE Transactions on Systems, Man, and Cybernetics*, **30**, 427–441, (2000).

28 V. Martin and M. Thonnat, A cognitive vision approach to image segmentation, *Tools in Artificial Intelligence*, **1**, 265–294, (2008).

29 M. Ebner, A real-time evolutionary object recognition system, in *Genetic Programming*, L. Vanneschi, S. Gustafson, A. Moraglio, I. De Falco and M. Ebner (Eds), Springer, Berlin, 2009.

30 Y.J. Zhang, A survey on evaluation methods for image segmentation, *Pattern Recognition*, **29**, 1335–1346, (1996).

31 S. Chabrier, B. Emile, C. Rosenberger and H. Laurent, Unsupervised performance evaluation of image segmentation, *EURASIP Journal on Applied Signal Processing*, **2006**, 1–12, (2006).

32 C. Kurtz, N. Passat, P. Gancarski and A. Puissant, Extraction of complex patterns from multiresolution remote sensing images: A hierarchical top-down methodology, *Pattern Recognition*, **45**, 685–706, (2012).

33 T. Blaschke, Object based image analysis for remote sensing, *ISPRS Journal of Photogrammetry and Remote Sensing*, **65**, 2–16, (2010).

34 E.A. Addink, F.M.B. Van Coillie and S.M. De Jong, Introduction to the GEOBIA 2010 special issue: From pixels to geographic objects in remote sensing image analysis, *International Journal of Applied Earth Observation and Geoinformation*, **15**, 1–6, (2012).

35 G.J. Hay and G. Castilla, Geographic object-based image analysis (GEOBIA): A new name for a new discipline, in *Object-Based Image Analysis: Spatial Concepts for Knowledge-Driven Remote Sensing Applications*, T. Blaschke, S. Lang and G.J. Hay (Eds), Springer, Berlin, Germany, (2008).

36 G. Castilla and G.J. Hay, Image objects and geographic objects, *in Object-Based Image Analysis: Spatial Concepts for Knowledge-Driven Remote Sensing Applications*, T. Blaschke, S. Lang and G.J. Hay (Eds), Springer, Berlin, Germany, (2008).

37 T. Lübker and G. Schaab, A work-flow design for large-area multilevel GEOBIA: Integrating statistical measures and expert knowledge, in *Geographic Object-Based Image Analysis (GEOBIA 2010)*, E.A. Addink and F.M.B. Van Coillie (Eds), ISPRS, Ghent, Belgium, (2010).

38 L. Drăguţ, D. Tiede and S.R. Levick, ESP: a tool to estimate scale parameter for multiresolution image segmentation of remotely sensed data, *International Journal of Geographical Information Science*, **24**, 859–871, (2010).

39 J. Kersten, M. Gähler and S. Voigt, A general framework for fast and interactive classification of optical VHR satellite imagery using hierarchical and planar Markov random fields, *Photogrammetrie, Fernerkundung, Geoinformation*, **2010**, 439–449, (2010).

40 C. Doulaverakis, A. Tzotsos, I. Tsampoulatidis, N. Gerontidis, A. Argyridis, K. Karantzalos, I. Kompatsiaris and D. Argialas, Gnorasi: A modular knowledge-based platform for object-based image analysis, *South-Eastern European Journal of Earth Observation and Geomatics*, **3**, 473–476, (2014).

41 D. Passo, E. Bias, R. Brites and G. Costa, Comparison of the accuracy of classifications generated by interimage and by interimage integrated with data mining, *South-Eastern European Journal of Earth Observation and Geomatics*, **3**, 93–97, (2014).

42 G.J. Hay, G. Castilla, M.A. Wulder and J.R. Ruiz, An automated object-based approach for the multiscale image segmentation of forest scenes, *International Journal of Applied Earth Observation and Geoinformation*, **7**, 339–359, (2005).

43 C. Fourie and E. Schoepfer, Combining the heuristic and spectral domains in semi-automated segment generation, Geographic Object-based Image Analysis (*GEOBIA 2012*), 7–9 May 2012, Brazil, (2012).

44 H. Herold and G. Meinel, Adaptive segmentation of heterogeneous images using metaheuristic evolutionary optimization, *South-Eastern European Journal of Earth Observation and Geomatics*, **3**, 461–464, (2014).

45 J. Osman, J. Inglada and E. Christophe, Interactive object segmentation in high resolution satellite images, *IEEE International Geoscience and Remote Sensing Symposium*, 12–17 July 2009, Cape Town, (2009).

46 F. Van Coillie, L.P. Verbeke and R.R. De Wulf, Feature selection by genetic algorithms in object-based classification of IKONOS imagery for forest mapping in Flanders, Belgium, *Remote Sensing of Environment*, **110**, 476–487, (2007).

47 C. Fourie, *A One Class Object Based System for Sparse Geographic Feature Identification*, M.Sc., Stellenbosch University, (2011).

48 C. Fourie and E. Schoepfer, Data transformation functions for expanded search spaces in geographic sample supervised segment generation, *Remote Sensing*, **5**, 3791–3821, (2014).

49 P. Achanccaray, V. Ayma, L. Jimenez, S. Garcia, P. Happ, R. Feitosa and A. Plaza, A free software tool for automatic tuning of segmentation parameters, *South-Eastern European Journal of Earth Observation and Geomatics*, **3**, 707–712, (2014).

50 A. Carleer, O. Debeir and E. Wolff, Assessment of very high spatial resolution satellite image segmentations, *Photogrammetric Engineering and Remote Sensing*, **71**, 1285–1294, (2005).

51 Q. Zhan, M. Molenaar, K. Tempfli and W. Shi, Quality assessment for geo-spatial objects derived from remotely sensed data, *International Journal of Remote Sensing*, **26**, 2953–2974, (2005).

52 U. Weidner, Contribution to the assessment of segmentation quality for remote sensing applications, *International Archives of Photogrammetry, Remote Sensing and Spatial Information Sciences*, **37**, 479–484, (2008).

53 C. Persello and L. Bruzzone, A novel protocol for accuracy assessment in classification of very high resolution images, *Geoscience and Remote Sensing, IEEE Transactions on*, **48**, 1232–1244, (2010).

54 B. Özdemir, S. Aksoy, S. Eckert, M. Pesaresi and D. Ehrlich, Performance measures for object detection evaluation, *Pattern Recognition Letters*, **31**, 1128–1137, (2010).

55 R.Q. Feitosa, R.S. Ferreira, C.M. Almeida, F.F. Camargo and G.A.O.P. Costa, Similarity metrics for genetic adaptation of segmentation parameters, in *Geographic Object-Based Image Analysis (GEOBIA 2010)*, E. Addink and F. Van Coillie (Eds), ISPRS, Ghent, Belgium, (2010).

56 P. Corcoran, A. Winstanley and P. Mooney, Segmentation performance evaluation for object-based remotely sensed image analysis, *International Journal of Remote Sensing*, **31**, (2010).

57 M. Neubert, H. Herold and G. Meinel, Evaluation of remote sensing image segmentation quality – further results and concepts, *International Archives of Photogrammetry, Remote Sensing and Spatial Information Sciences*, **36**, (2006).

58 R. Trias-Sanz, G. Stamon and J. Louchet, Using colour, texture, and hierarchical segmentation for high-resolution remote sensing, *ISPRS Journal of Photogrammetry and Remote Sensing*, **63**, 156–168, (2008).

59 L.M. Melo, G.A.O.P. Costa, R.Q. Feitosa and A.V. Da Cruz, Quantum-inspired evolutionary algorithm and differential evolution used in the adaptation of segmentation parameters, in *Geographic Object-Based Image Analysis (GEOBIA 2008)*, G.J. Hay, T. Blaschke and D. Marceau (Eds), ISPRS, 6–7, Calgary, Canada, (2008).

60 P. Happ, R.Q. Feitosa and A. Street, Assessment of optimization methods for automatic tuning of segmentation parameters, *Geographic Object-based Image Analysis (GEOBIA 2012)*, 7–9 May 2012, Rio De Janeiro, Brazil, (2012).

61 P. Happ, R. Feitosa, C. Bentes and R. Farias, A Region-Growing Segmentation Algorithm for GPUs, *IEEE Geoscience and Remote Sensing Letters*, **10**, 1612–1616, (2013).

62 C.M.B. Freddrich and R.Q. Feitosa, Automatic adaptation of segmentation parameters applied to non-homogeneous object detection, in *Geographic Object-based Image Analysis (GEOBIA 2008)*, G.J. Hay, T. Blaschke and D. Marceau (Eds), ISPRS, Calgary, Canada, (2008).

63 J. Michel, M. Grizonnet and O. Canevet, Supervised re-segmentation for very high-resolution satellite images, Geoscience and Remote Sensing Symposium (IGARSS 2012), *IEEE International*, 22–27 July 2012, Munich, Germany, 2012).

64 R.S. Ferreira, R.Q. Feitosa and G.A.O.P. Costa, A multiscalar, multicriteria approach for image segmentation, *Geographic Object-Based Image Analysis (GEOBIA 2012)*, 7–9 May 2012, Rio de Janeiro, Brazil, (2012).

65 C. Fourie and E. Schoepfer, Connectivity Thresholds and Data Transformations in Sample Supervised Segment Generation, Geoscience and Remote Sensing Symposium (IGARSS 2012), *IEEE International*, 21–26 July, Melbourne, (2013).

66 C. Fourie and E. Schoepfer, Classifier Directed Data Transformations in Sample Supervised Segment Generation, *Geographic Object-based Image Analysis (GEOBIA 2014)*, 21–24 May, Thessaloniki, Greece, (2014).

67 M. Pedergnana, P.R. Marpu, M. Dalla Mura, J.A. Benediktsson and L. Bruzzone, A novel technique for optimal feature selection in attribute profiles based on genetic algorithms, *IEEE Transactions on Geoscience and Remote Sensing*, **51**, 3514–3528, (2013).

68 M. Dalla Mura, J.A. Benediktsson and L. Bruzzone, A general approach to the spatial simplification of remote sensing images based on morphological connected filters, Geoscience and Remote Sensing Symposium (IGARSS 2012), *IEEE International*, 24–29 July, Vancouver, Canada, (2011).

69 G.M. Foody and A. Mathur, A relative evaluation of multiclass image classification by support vector machines, *IEEE Transactions on Geoscience and Remote Sensing*, **42**, 1335–1343, (2004).

70 M. Musci, R.Q. Feitosa and G.a.O.P. Costa, An object-based image analysis approach based on independent segmentations, *Joint Urban Remote Sensing Event*, 21–23 April, Sao Paulo, Brazil, (2013).

71 M. Birattari, M. Zlochin and M. Dorigo, Towards a theory of practice in metaheuristics design: A machine learning perspective, *RAIRO-Theoretical Informatics and Applications*, **40**, 353–369, (2006).

72 J. Watson, An introduction to fitness landscape analysis and cost models for local search, in *Handbook of Metaheuristics*, M. Gendreau and J. Potvin (Eds), Springer US, (2010).

73 M. Graff and R. Poli, Practical performance models of algorithms in evolutionary program induction and other domains, *Artificial Intelligence*, **174**, 1254–1276, (2010).

74 R.L. Rardin and R. Uzsoy, Experimental Evaluation of Heuristic Optimization Algorithms: A Tutorial, *Journal of Heuristics*, 7, 261–304, (2001).

75 K. Weicker and N. Weicker, On the improvement of co-evolutionary optimizers by learning variable interdependencies, *Evolutionary Computation (CEC 1999)*, 6–9 July 1999, Washington DC, United States of America, (1999).

76 S.-W. Lin, K.-C. Ying, S.-C. Chen and Z.-J. Lee, Particle swarm optimization for parameter determination and feature selection of support vector machines, *Expert Systems with Applications*, **35**, 1817–1824, (2008).

77 T. Ojala, M. Pietikainen and T. Maenpaa, Multiresolution gray-scale and rotation invariant texture classification with local binary patterns, *Pattern Analysis and Machine Intelligence, IEEE Transactions on*, **24**, 971–987, (2002).

78 G.K. Ouzounis and M.H. Wilkinson, Hyperconnected attribute filters based on k-flat zones, *Pattern Analysis and Machine Intelligence, IEEE Transactions on*, **33**, 224–239, (2011).

79 J. Lampinen and I. Zelinka, Mixed integer-discrete-continuous optimization by differential evolution, *in 5th International Conference on Soft Computing*, Citeseer, 77–81, (1999).

80 P. Hofmann, P. Lettmayer, T. Blaschke, M. Belgiu, S. Wegenkittl, R. Graf, T. Lampoltshammer and V. Andrejchenko, ABIA – A Conceptual Framework for Agent Based Image Analysis, *South-Eastern European Journal of Earth Observation and Geomatics*, **3**, 125–130, (2014).

81 C. Li, A. Kowdle, A. Saxena and T. Chen, Toward holistic scene understanding: Feedback enabled cascaded classification models, *Pattern Analysis and Machine Intelligence, IEEE Transactions on*, **34**, 1394–1408, (2012).

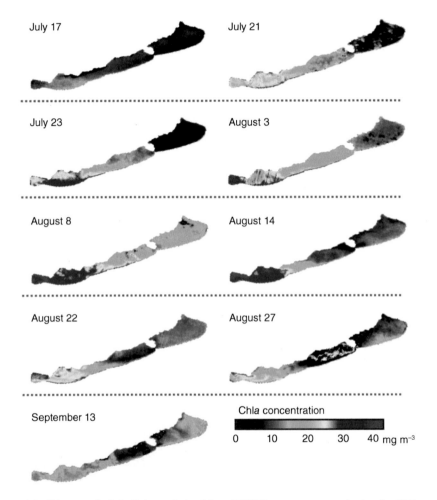

Figure 10.5 Chl*a* maps for Lake Balaton derived from MERIS imagery processed using the FLH algorithm, capturing a summer bloom event beginning in late July 2010.

Earth Observation for Land and Emergency Monitoring, First Edition. Edited by Heiko Balzter.
© 2017 John Wiley & Sons Ltd. Published 2017 by John Wiley & Sons Ltd.

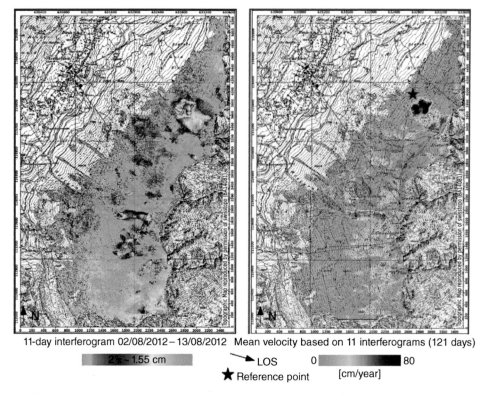

11-day interferogram 02/08/2012 – 13/08/2012   Mean velocity based on 11 interferograms (121 days)

2.5 – 1.55 cm   ➜ LOS   0 ▬▬▬▬▬ 80
                     ★ Reference point   [cm/year]

Figure 12.3 Surface displacements in Grächen and St. Niklaus, Switzerland. Left: 11-day TerraSAR-X interferogram between 2 August and 13 August 2012, right: annual mean velocity based on TerraSAR-X data available over the snow-free season from 08 June to 07 October 2012.

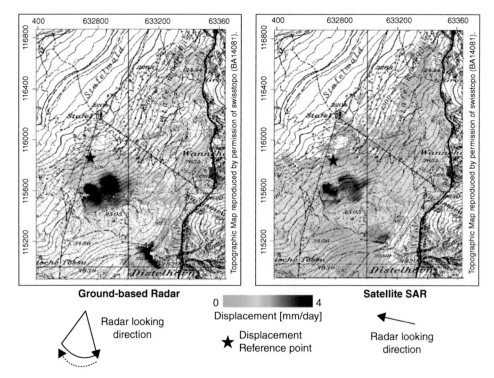

Ground-based Radar   0 ▬▬▬▬▬ 4   **Satellite SAR**
                         Displacement [mm/day]

⟨ Radar looking   ★ Displacement   ← Radar looking
   direction   Reference point   direction

Figure 12.5 Surface displacement measured for the Distelhorn rock glacier in Grächen with ground-based radar and satellite SAR interferometry. Left: fieldwork results observing the Distelhorn rock glacier for 24 hours, right: 11-day satellite InSAR displacement map scaled to displacement per day. The results represent movement measured in different LOS directions as indicated below the figures.

(a)

(b)

(g)

☐ Lac Bam

km
0   200

(c)

(d)

(h)

(e)

(f)

km
0   2.5   5

Figure 14.2 Lac Bam in Burkina Faso provides examples of the different land cover types of sub-Saharan wetlands. (a): Sediment-rich water causes brown-coloured water; (b): Bare soil following retreating water at the shoreline, flooded grasses in the background; (c): Standing and floating vegetation (water depth 80 cm) approx. 100 m from shoreline; (d): Standing vegetation and trees next to the small dam in the south; (e): Motor pump; (f): Irrigated gardening close to shore; (g): Map of site relative to African continent; (h): The geographical locations of Figures 14.2 (a)–(f) are indicated with arrows on a RapidEye false colour image (band 5–3–2) from 19 October 2013 (Source: BlackBridge). All photos are by L. Moser, Oct 2013.

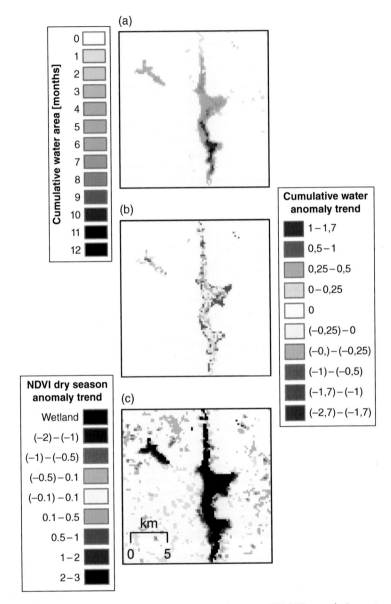

Figure 14.6 (a): Moderate Resolution Imaging Spectroradiometer (MODIS) cumulative water-covered surface area; (b): anomaly trend of cumulative water-covered surface area; (c): NDVI trend of surrounding areas, showing a strong increase in irrigated cultivation (Source: USGS). Adapted from Moser *et al.*, 2014 [84].

(a)   (b)   (c)   (d)   (e)

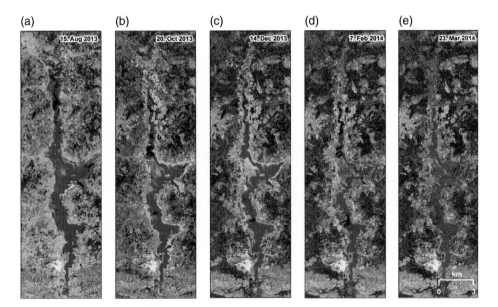

Figure 14.7 An example of a time series of five selected TerraSAR-X (TSX) acquisitions from (a): 15 Aug., (b): 20 Oct. and (c): 14 Dec. 2013, as well as (d): 07 Feb. and (e): 23 Mar. 2014 (Source: DLR 2013, 2014). A false colour composite has been created (K4–K0–K3). Open water is displayed in blue, with blue colours indicating a dominance of HH over VV backscatter. Green colours appear where the sum of the surface intensity of HH and VV is particularly strong, and pink colours are dominant where double-bounce scattering dominates over surface scattering, such as in the areas of flooded vegetation.

(a)

(b)

(c)

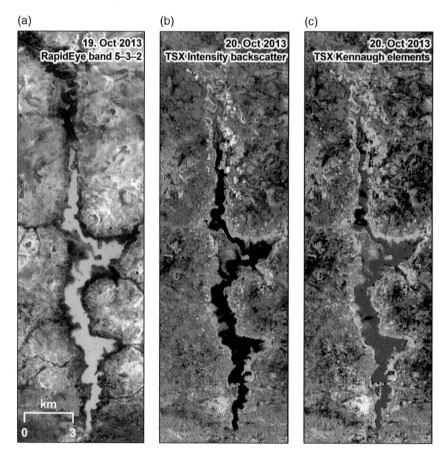

Figure 14.8 Comparison of RapidEye and TerraSAR-X imagery; (a): a RapidEye image false colour composite (5–3–2) taken on 19 Oct 2013 (Source: BlackBridge); (b): a TerraSAR-X Stripmap intensity image (K0) taken on 20 Oct. 2013 2013 (Source: DLR 2013); (c): the Kennaugh elements from HH and VV data from a TerraSAR-X image taken on 20 Oct. 2013 (Source: DLR 2013).

Part V

Earth Observation for Climate Adaptation

14

Remote Sensing of Wetland Dynamics as Indicators of Water Availability in Semi-Arid Africa

L. Moser, A. Schmitt, S. Voigt and E. Schoepfer

German Aerospace Center (DLR), German Remote Sensing Data Center (DFD), Oberpfaffenhofen, Germany

14.1 Wetland Monitoring in West Africa Using Remote Sensing

14.1.1 Introduction

This chapter focuses on remote sensing techniques and applications for monitoring wetlands in sub-Saharan Africa, and provides three examples using different sensors. The role wetlands play in water availability, as well as their impact on people's livelihoods, is described based on the example of Burkina Faso, a country in West Africa. The rationale for using earth observation for wetland detection, mapping, inventorying, classification, assessment, and monitoring is presented, and advantages, limitations, and gaps are discussed. Moreover, the parameters to be observed in wetlands are described based on pictures from a field study at Lac Bam, Burkina Faso. Three groups of remote sensing sensor types and the applicability of their data and existing methods for wetland mapping and monitoring are reviewed: optical high-resolution imagery, optical medium-resolution time series, and synthetic aperture radar data. The Lac Bam case study is used to show applications of these three remote sensing sensors types. Conclusions and future perspectives for wetland remote sensing, as well as possibilities and limitations, are assessed. Remaining gaps and possible future solutions are discussed, with a focus on the European Space Agency's new Sentinel missions. The role of wetland monitoring for both, estimating water availability in semi-arid areas and indicating drought is also noted.

14.1.2 Water Scarcity, Wetlands, and Drought

Semi-arid African regions are often prone to water scarcity and drought. The Sahel region experienced droughts in the mid-1970s and mid-1980s, as well as, most recently, in 2012; these droughts lead to food crises. The total renewable water resources relative to the number of inhabitants of a country are a measure of physical water scarcity. This is calculated using the Falkenmark water stress index [1,2], for which a quantity of water below $1000\,\mathrm{m}^3$ inhabitant^{-1} year^{-1} serves as the international water scarcity threshold.

Earth Observation for Land and Emergency Monitoring, First Edition. Edited by Heiko Balzter.
© 2017 John Wiley & Sons Ltd. Published 2017 by John Wiley & Sons Ltd.

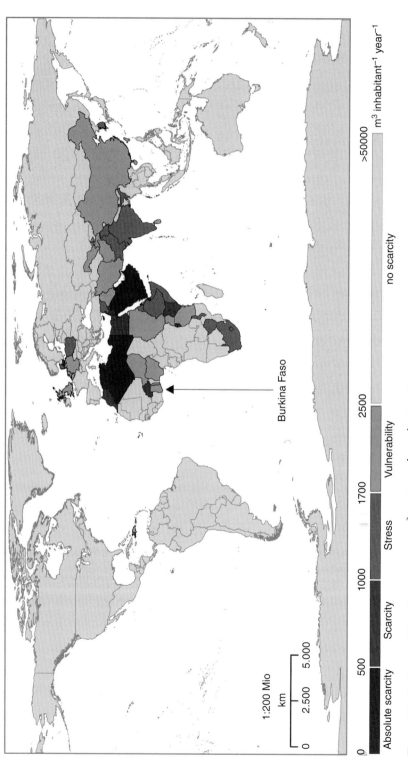

Figure 14.1 Total renewable water resources in m^3 inhabitant^{-1} year^{-1} defined based on United Nations (UN) water thresholds (data based on Food and Agriculture Organization (FAO) AQUASTAT data for the year 2011).

Figure 14.1 shows a map of countries suffering from physical water scarcity. Sub-Saharan African countries are particularly faced with not only physical, but also economic water scarcity, with limited water access due to limited financial and human resources, along with insufficient infrastructure for water access. This economic water scarcity is possible even when water resources might not actually be physically scarce [1]. Burkina Faso is a landlocked country in West Africa with only 820.5 m^3 inhabitants^{-1} year^{-1} of total renewable water, and which is experiencing economic water stress in conjunction with drought and strong population growth.

The definition of *wetlands* varies widely among authors. One internationally established definition is that given by the Ramsar Convention on Wetlands in 1971 [3]. Ramsar is an intergovernmental treaty, which maintains a global list of Wetlands of International Importance (the "Ramsar List"), which, as of January 2017, comprises 2251 wetlands corresponding to a surface of approx. 215 million ha – including 18 Ramsar wetlands in Burkina Faso. These wetlands are selected based on nine criteria: their status as representative, rare, or unique wetland types, as well as any of the eight criteria of biological diversity [4]. Wetlands fulfil different ecosystem functions. Their function as water reservoirs is of great importance for local populations, and they also serve as buffers for floods. Further functions include providing habitats for various animal and plant species; protecting shorelines; serving as carbon sinks; recharging and discharging groundwater; retaining nutrients, sediments, and pollutants; and stabilizing local climate conditions [4–6]. The categories predominant in sub-Saharan semi-arid West Africa are natural freshwater lakes that can be permanent or seasonal, human-made reservoirs and ponds, irrigated agriculture around natural or artificial lakes, and riparian wetlands located along rivers. Almost all rivers are only seasonal rivers and carry water during and at the end of the rainy season. Only 8% of the water is transported to wetlands by surface runoff in the Sahelian climate, and evapotranspiration rates are as high as 92% [7]. The maximum water extent is reached at the end of the rainy season in September or October. Larger artificial water bodies exist throughout the whole year, but vary greatly in size, while smaller water bodies may dry out during the dry season. One exception of a large permanent inland water body and wetland is the *Delta Intérieur du Niger* (Inland Niger Delta) in Mali, which is connected to the Niger River during the whole year, and therefore does not follow a water regime consistent with local climate conditions.

In semi-arid African regions like Burkina Faso, wetlands are very important in providing water availability for people's livelihoods [6,8], and each year new dams are built to create new artificial water bodies where rainwater and water from runoff can be collected and stored. The dry season starts around October and lasts until May or June in the sub-Saharan Sahel and Savannah of West Africa. There is a strong precipitation gradient by latitude, ranging from less than 300 mm per annum in the northern Sahel to more than 1000 mm in the southern Guinea Savannah; this wide precipitation gradient results in different climatic zones and amounts of available water. The most important livelihoods are farming, herding, and fishing, all of which critically depend on the availability of water. Water is also extracted from wetlands for generating electricity, supplying cities, or, in some cases, for mining or industrial use. Irrigation from water bodies and wetlands is the only way to farm during the long dry season. Climate and anthropogenic change are threatening sub-Saharan wetlands. In addition to water extraction, increasing cultivation and irrigation is leading to the removal of natural vegetation, which then causes more sediment input into wetlands. Siltation on the bottom of lakes is a major concern; this causes water bodies to become shallower, which allows water to spread out further and evaporate faster. Water availability is critical for a range of livelihoods, and a critical lack of water can occur during seasons of drought. Population pressure and exploitation of resources can lead to wetland degradation [8]. African wetlands can also be subjected to conflicts between different livelihood groups, as people compete for access to limited land and water resources [1,6]. The spatio-temporal dynamics of water surface areas in Burkina Faso's wetlands have been found to coincide with the occurrence of drought seasons [9].

14.2   Remote Sensing of Wetlands

14.2.1   Why Perform Satellite Monitoring of Wetlands?

Wetlands are highly dynamic in terms of both episodic and periodic variations; this results in a need for frequent observations to monitor and quantify the variations. Frequently, wetlands and their surrounding areas cover large areas, ranging from less than a square kilometre to hundreds of square kilometres in size, and they can be situated in remote locations or be difficult to access. For these reasons, wetlands are often unsuitable for systematic ground surveying, which cannot cover larger areas at short, regular intervals. In sub-Saharan Africa, ground data are very scarce. A few wetlands in Burkina Faso are surveyed in terms of the water level, as measured at their dam [10]; however, their surface water area or changing lake floor topography due to sedimentation are not assessed. Despite their ecological significance and importance for human livelihoods, wetlands in Africa are scarcely monitored at all. For these reasons, remote sensing provides a promising tool for wetland inventorying, classification, and monitoring.

Satellite monitoring can capture remote or large areas and repeat observations frequently, two key advantages for inventorying or monitoring the spatio-temporal dynamics of wetlands. Improved access to increasingly large satellite data archives have made it possible to derive changes and trends using time series analysis. Remote sensing analysis may also be less costly and time consuming than field campaigns. However, remote sensing also has limitations, such as reduced data availability during cloudy situations missing data continuity after the lifetime of satellite systems, possible sensor failure, high data costs for commercial applications, and the limited availability of computing power and technical knowledge. Local knowledge is required for interpretation and validation of remote sensing studies, and, depending on the application, ground surveys cannot always be replaced by remote sensing.

Beyond wetland monitoring, wetland classification is very relevant for land use/land cover (LULC) mapping. Wetland classification has the highest rates of inaccuracy for global land cover products such as GlobCover and MODIS Land Cover [11,12], as wetland mapping needs to take into account input imagery from different seasons. GlobCover (spatial resolution: 300 m) contains one class for water bodies, one class for post-flooding or irrigated croplands, two classes for regularly flooded forests (one for forests flooded with fresh or brackish water and one for forests flooded with saline water), and one class for grassland or shrub land that is regularly flooded or that has waterlogged soil [13]. The recent Land Cover (LC) project of the Climate Change Initiative (CCI) follows the same legend as GlobCover, and contains an additional water body product (CCI-LC WB) [14]. MODIS Land Cover (MCD12Q1) (spatial resolution: 500 m) has only two classes associated with wetlands: water, and permanent wetlands [15]. The widely used Land Cover Classification System (LCCS) of the Food and Agricultural Organization (FAO) distinguishes between cultivated aquatic or regularly flooded areas, natural and semi-natural aquatic or regularly flooded vegetation, artificial water bodies, natural water bodies, and snow and ice. Moreover, water seasonality can be divided into the sub-categories of permanent, semi-permanent, temporary or seasonal, or waterlogged [16]. In the recently completed project GlobWetland-II a new legend for a hybrid wetland classification was developed, based on three hierarchical levels of Corine Land Cover (CLC) classes, complemented by two levels of Ramsar wetland types [17]. The Mapping and Assessment of Ecosystems and their Services (MAES) as part of the European Union (EU) Biodiversity Strategy to 2020 foresees a classification scheme oriented towards the description of wetlands from an ecosystem service point of view. Wetland classification requirements differ according to the user perspective, covering a range from water/no water or wetland/no wetland classification, traditional land cover classification (e.g., LCCS), the classification of wetland types (e.g., Ramsar) or the characterization by ecosystem services (e.g., MAES). In general, not much research has been published on the dynamics of African wetlands: wetlands and water bodies in the Sahel have been mapped and monitored [9,11,18–22], the Inland Niger Delta has been specifically investigated [11,19,23], and water bodies over the African continent were recently monitored [23].

14.2.2 Detection, Mapping, Inventorying, Classification, Assessment or Monitoring?

Davidson and Finlayson 2007 [24] describe and define wetland inventorying, assessment, monitoring, and management as follows: *inventorying* is making note of the location and ecological characteristics; *assessment* is assessing the status, trends, and threats; *monitoring* is observing the status and trends, including the decrease and increase of threats; while *management* is taking actions against potential damages. In terms of remote sensing analysis, however, *monitoring* often implies making repeated observations and analysing time series data for a certain parameter; while these activities may result in identifying trends, the results do not need to include information about threats. *Detection and mapping* often refer to deriving results without any temporal character, and in the case of *classification*, different classes and attributes are assigned to wetlands. At present, there is no consistent African or even global wetland inventory. Available wetland inventories are often incomplete, inconsistent, or inaccessible, and remote sensing analyses are based on different classification schemes and methods [8,11]. Global inventory databases use different approaches, which result in different indications of total wetland area [24,25]. For several regions in Africa, South America, and Russia the extent of major wetlands is not even known, as a result of varying definitions and the lack of detailed inventories. This may lead to incorrect estimations in the size of wetlands; for example, riparian wetlands along rivers that may sum up to a larger surface area are reflected in some studies, but are not considered in others [6].

14.2.3 Earth Observation Projects to Support Wetland Initiatives

Global initiatives for water and wetland inventorying as well as assessment are summarized in this section. An extensive overview of initiatives and organizations that conduct global wetland inventorying has been presented [8], including possibilities for remote sensing [6,24,26]. Recent advances to combine wetland research and EO methods by supporting the Ramsar Convention have been proposed by a number of researchers [24,27] and were implemented via the GlobWetland I project by the European Space Agency (ESA) [28,29]. A follow-up project, GlobWetland II, aims to map Mediterranean wetlands on a high-resolution scale, using Landsat imagery from the archives for the years 1975, 1990, and 2000 [12,17,30,31]. The recently initiated GlobWetland-Africa (GW-A) project will concentrate on pilot sites on the African continent [32]. Another initiative of ESA is the TIGER Initiative to support the African earth observation capacity for water resource monitoring, in close collaboration with African water authorities and experts. TIGER-NET is a major part of developing an information system for monitoring, assessing, and inventorying water resources [33,34]. Wetlands and water bodies are two out of five themes within the Pan-European High Resolution Layers (HRL) 2012 of Europe's Copernicus project. Using high-resolution optical data a wetland/no wetland product (20 m) and percentage wetland occurrence product (100 m) was generated, as well as a permanent water bodies and water body occurrence product [35,36]. The new Copernicus HRLs 2015 feature a water/wetness product, consisting of water bodies and areas of wetness (in terms of soil moisture), both of permanent and temporary character. The Japanese Space Agency (JAXA) has implemented the Kyoto & Carbon (K&C) Initiative, which includes a wetland theme and emphasizes the use of ALOS PALSAR [37]. The NASA JPL wetland project aims to build up a global-scale Earth System Data Record (ESDR) of inundated wetlands based on SAR imagery-derived wetland extent, vegetation type and seasonal inundation dynamics (100 m), as well as global monthly mapping of inundation extent at approx. 25 km resolution [38]. Among the goals of the recently initiated project (June 2015), Satellite-based Wetland Observation Service (SWOS), are the development of an operational, standardized monitoring service for wetlands, and to further develop existing approaches, in particular the GlobWetland II approach, based on Sentinel data [39]. The GloboLakes project (2012–17) uses time series data from MERIS, MODIS, AATSR, and SeaWifs (going back to 1997), with planned continuation using Sentinel-2 and Sentinel-3 using near-real time data processing to observe the state of lakes and the response to climate drivers [40]. Other earth observation related projects that are relevant to wetlands and inland waters include the Mediterranean Wetlands Initiative (MedWet), the Asian Wetland Inventory, the pan-European wetland inventory project, the recent EartH2Observe and Diversity II projects, and local Ramsar initiatives in different countries.

14.3 Remote Sensing Principles for Wetland Monitoring

14.3.1 Parameters of Wetland Remote Sensing

Wetlands consist of a mix of different land cover types. In sub-Saharan West Africa wetlands typically comprise open water surfaces, vegetation floating on water, vegetation standing in water or flooded vegetation (e.g. macrophytes such as grasses, water lilies, or trees in the vicinity of the coast), natural coastal vegetation (e.g. shrubs and trees), irrigated cultivation (e.g. vegetable gardening or rice) close to the shoreline, bare soil (e.g. exposed soil after water retreat, ploughed agricultural fields), or urban areas (e.g. dwellings, settlements) or infrastructures (e.g. dams, water pumps, bridges). Particularly in seasonally dynamic environments like the West African Savannah and the Sahel, these parameters can change rapidly within the course of a single season. The Lac Bam case study (section 14.4) exemplifies this effect. Wetland land cover types are demonstrated by pictures taken during a field study at Lac Bam in October 2013 (Figure 14.2(a–f)).

Water detection, delineation, and monitoring are of vital importance for functions as wide-ranging as human use, maintaining biodiversity and animal habitats, conducting drought assessment, and as input for hydrological models. Detection of open water can be performed using a variety of sensors. However, the spectral

Figure 14.2 Lac Bam in Burkina Faso provides examples of the different land cover types of sub-Saharan wetlands. (a): Sediment-rich water causes brown-coloured water; (b): Bare soil following retreating water at the shoreline, flooded grasses in the background; (c): Standing and floating vegetation (water depth 80 cm) approx. 100 m from shoreline; (d): Standing vegetation and trees next to the small dam in the south; (e): Motor pump; (f): Irrigated gardening close to shore; (g): Map of site relative to African continent; (h): The geographical locations of Figures 14.2 (a)–(f) are indicated with arrows on a RapidEye false colour image (band 5–3–2) from 19 October 2013 (Source: BlackBridge). All photos are by L. Moser, Oct 2013. (*See insert for color representation of the figure.*)

properties and visual colour of water change according to the absorption and scattering of light energy. Optically active materials like sediment, organic molecules, and plankton affect scattering and absorption. The spectral properties of water are characterized by high absorption (i.e. low reflectance values) at near infrared (NIR) wavelength range and beyond in the mid-infrared (MIR), and by higher reflectance values in the visible range, especially in the blue band. Particularly in the visible region, water has higher reflectance when there is high turbidity, chlorophyll concentration, or sediment content. The sediment-rich, turbid water of Lac Bam is clearly demonstrated in Figure 14.2.

The existence of a NIR or MIR band is vital for monitoring water and NIR is available for most remote sensing devices. Water detection through the use of NIR thresholding is possible due to both the strong absorption of water and the contrast between water and land surfaces in the NIR range. If a red band exists, one can make use of the different characteristics of the visible and NIR bands to create a normalized index. This index, the *Normalized Difference Vegetation Index* (NDVI) [41], is calculated using the equation (NIR − red)/(NIR + red), which yields low values over water; however, this equation yields very high values when, for example, there is vegetation floating on the water, making this index inaccurate in the case of vegetated wetlands. If there is a green band, the *Normalized Difference Water Index* (NDWI) [42] (NIR − green)/(NIR + green) can be used to detect water. Another method for calculating the NDWI, (NIR − MIR)/(NIR + MIR), has also been proposed [43]. This proposed index has very promising results for detecting water by adding a band in the mid-infrared (MIR, also referred to as shortwave-infrared (SWIR)) region, which also allows the *modified Normalized Difference Water Index* (mNDWI) to be calculated by replacing the NIR band with the MIR band: (MIR − green)/(MIR + green) [44]. The *Normalized Difference Pond Index* (NDPI) and the *Normalized Difference Turbidity Index* (NDTI) were applied for Sahelian areas [20,45]. Recently, the hue-saturation-value (HSV) colour transformation was applied by transferring the red–green–blue (RGB) colour space into the HSV colour space using the MIR, NIR and red bands from MODIS [23], PROBA-V [46], and Landsat [47,48]. Other recent advances useful for water and wetland detection are the use of the *Automated Water Extraction Index* (AWEI) that has been applied on Landsat imagery [49,50], and the Landsat 8 tasselled cap (TC) [51].

To monitor wetland vegetation, which can be floating on water, standing in water, growing on the land, or persisting as irrigated or rain-fed cultivation, the NIR range is again of high importance since chlorophyll activity and healthy green vegetation lead to high NIR reflectance values. The NDVI index presented in the preceding paragraph is a common way to detect high vegetation chlorophyll activity content, and can be applied using almost every optical remote sensing sensor. Additional vegetation indices, such as the *Enhanced Vegetation Index* (EVI), *Soil Adjusted Vegetation Index* (SAVI), *Global Environment Monitoring Index* (GEMI), or *Leaf Area Index* (LAI), etc., have also been developed [52–54]. The application of these vegetation indices for identifying wetland vegetation has been discussed extensively in the literature [55]. Similar to vegetation on land, leaf morphology and water content are the deciding parameters in determining the infrared reflection of aquatic vegetation, with leaf pigments accounting for the reflectance. In the green region greater penetration of energy into the water occurs, submerged plants can be detected, and emergent species show higher reflectance [56].

Mixed pixels of aquatic emergent vegetation and water, vegetation on land, or soil persist, all of which can lead to poor automatic classification results, particularly as the NIR and MIR bands are strongly influenced by the absorption of water [57]. To address uncertainties caused by mixed pixels, spectral mixture analysis (SMA) has become popular in wetland research [12,58]. Downscaling approaches are applied to improve the spatial resolution of water body maps [59–61]. Figure 14.3 shows a selection of commonly used remote sensing sensors that have been used or that show potential for wetland mapping, classification, or monitoring. Based on the three groups of sensors, data and methods for wetland monitoring are described in the preceding three sub-sections.

14.3.2 Optical High Resolution Imagery

Optical high resolution (HR) imagery is characterized by its spatial resolution of between about 5 m and 30 m, and spectral resolution between around 3 to 15 spectral bands in the visual (VIS), near infrared (NIR), mid-infrared (MIR; also called short-wave infrared (SWIR)), or thermal infrared (TIR) range of the electromagnetic spectrum. High-resolution data do not yet exist for regular temporal intervals in data archives, and are therefore primarily used for wetland change detection or classification applications, making use of data from two or more points in time. Such data can be acquired by multispectral Earth observation satellites, such as Landsat, SPOT, the Advanced Spaceborne Thermal Emission and Reflection Radiometer (ASTER), the Indian Remote Sensing Satellite (IRS) series, satellites of the Disaster Monitoring Constellation (DMC), or the RapidEye satellites [62]. A well suited HR data source for wetland mapping and monitoring is Landsat imagery, which has been frequently used due to its long-term record [8,12,20,24,31,58,63–66] and the free availability of the Landsat archive. Even in data-scarce areas like West Africa, multi-temporal coverage for different seasons is available for several years. Also other HR sensors like ASTER [8,58], Beijing-1 [67,68], RapidEye [22,69], and SPOT-5 [69–71] have been used for wetland mapping, change detection, or classification. Very high-resolution (VHR) data from satellites, as well as aerial imagery, both of which have a resolution that is typically between 0.5 m and 1 m, have been used for classification and validation [58,72–75].

High-resolution data have been used for mapping and classifying wetlands. Wetland classification can use either a mono-temporal analysis using images acquired at a single point in time [8], or a multi-temporal classification based on two (bi-seasonal) [12,64] or multiple images per year [66,68]. Multitemporal approaches aim at characterizing dynamic classes by capturing their seasonal variations. Change detection between classification results of two or more points in time that should have preferably been acquired in the same season of different years has been applied by various authors [8,20,58,69]. Multitemporal classification based on high-resolution time series data is still a new area of research in the domain of seasonal wetland monitoring. The full temporal range of the Landsat archive was exploited only in one study for wetland time series analysis [69]; this study had a target of using one image per year since 1984, which resulted in a time series of about 20 Landsat images, with some gaps of up to two or three years between available acquisitions. Recently, dynamic water body mapping using the Landsat archive on a global scale has been introduced [48]. Data which capture seasonal temporal frequency over many years and different seasons are, however, not yet

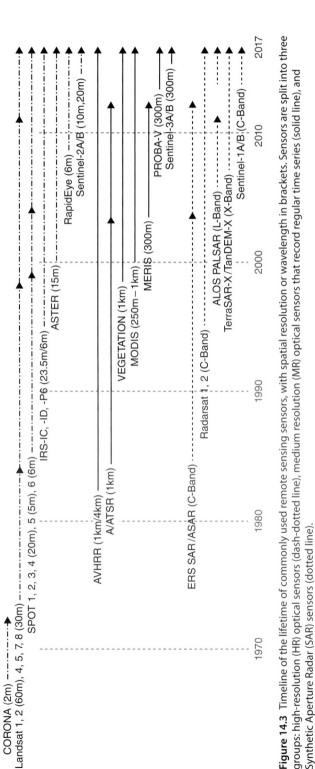

Figure 14.3 Timeline of the lifetime of commonly used remote sensing sensors, with spatial resolution or wavelength in brackets. Sensors are split into three groups: high-resolution (HR) optical sensors (dash-dotted line), medium resolution (MR) optical sensors that record regular time series (solid line), and Synthetic Aperture Radar (SAR) sensors (dotted line).

available from HR sensors, and data for West Africa are particularly scarce. Therefore, HR based monitoring activities cannot yet be carried in similar frequent and dynamic ways as it is possible with medium resolution EO data.

Some studies have focused on open water classification [63,69], such as estimating the size of water bodies [20]. In some cases temporal classes are built distinguishing permanent and seasonal open water [8]. Land cover, vegetation types, and crop types in wetland areas have been classified [8,70,71,74], and some studies have focused on the plant functional types in wetland vegetation [65,68]. Very few studies have focused on HR wetland mapping in the Sahel. One of these studies, Gardelle *et al.* 2010 [20], used a multi-sensor approach for detecting surface water area between 1954 and 2007, using data from SPOT, Formosat, Landsat, Corona, and MODIS. Recently, efforts in temporal mapping of water bodies in the Sahel using HR imagery have been introduced [76]. Classification techniques for wetland delineation and mapping have been reviewed [5,55,57,65]. Supervised classification using the Maximum Likelihood Algorithm has been applied to HR data [8,63]; decision tree approaches such as random forest [12,77] have also been used to study land cover of wetland areas or [8,70]. Change vector analysis (CVA) was used in one study to derive wetland change and no-change classes between two points in time [72]. Diverse machine learning algorithms were compared in one study [65], while support vector machines have been used in other research to map and monitor wetlands [68]. More recently, research in remote sensing of wetlands is moving towards geographic object-based image analysis (GEOBIA) [67,71,73–75,78], as well as spectral mixture analysis [12,58].

14.3.3 Optical Medium Resolution Imagery

Medium resolution (MR) earth observation data have the capability to both track land surface processes over large areas and to extract temporal information at short time intervals (i.e. days). Existing satellite data archives contain both, raw data and higher level processed products recorded at regular time intervals; this began in the early 1980s with the launch of the Advanced Very High Resolution Radiometer (AVHRR) by the National Oceanic and Atmospheric Administration (NOAA), a device mounted on board of different NOAA satellites and which samples to a pixel size of 1.1 km and 4.4 km respectively. Over the past 15 years other multispectral sensors like SPOT-VEGETATION, which is operated on board SPOT 4 and SPOT 5 (1 km pixel size), the Medium Resolution Imaging Spectrometer (MERIS) (on board Envisat; 300 m), the Advanced Along Track Scanning Radiometer (AATSR) (on board Envisat; 1 km), and the Moderate Resolution Imaging Spectroradiometer (MODIS) (on board Terra and Aqua) have been put into orbit. MODIS is a sensor that is currently mounted on two platforms – Terra since 1999 and Aqua since 2002 – and can therefore provide coverage twice per day for regions at the higher latitudes. MODIS land surface products are available beginning in February 2000, and are provided for 250 m, 500 m, and 1 km spatial resolution. Sensors mounted on geostationary meteorological satellites (e.g. SEVIRI on Meteosat) provide very frequent datasets (i.e. every 15 minutes); however, spatial resolution is lower: SEVIRI has 1 km resolution at NADIR for one visible band, and 3 km for the VNIR and IR bands [62].

Monitoring over large areas can be challenging; temporal dynamics, regional and latitudinal differences, and mixed pixels are all constant challenges in analysing MR data. Several studies have focused on the spatio-temporal development of wetlands or water bodies, taking into account dynamic aspects [9,19,23] and seasonality [19,79]. The availability of MODIS surface reflectance products for daily, eight-daily, or monthly composites allows dynamic processes to be captured for larger wetlands [9,21]. Time series analysis has been conducted on wetlands or open water bodies over Africa using MODIS [9,23,80] and SPOT VEGETATION [19,21,81]. Multi-temporal classification based on MODIS time series data from one season was also performed for an area in Burkina Faso and Ghana [11].

Medium resolution sensors usually offer between two and 36 bands in the visible, NIR, SWIR, and TIR ranges: MODIS 250 m (1 VIS and 1 NIR band), MODIS 500 m (3 VIS, 2 NIR, and 2 TIR bands), MODIS 1 km (36 bands covering all ranges), MERIS 300 m (15 VNIR bands, only operational from 2003–12), VEGETATION 1 km (2 VIS, 1 NIR, and 1 SWIR band), and AVHRR 1.1 km/4.4 km (1 VIS, 1 NIR, 1 SWIR, and 2 TIR bands). Time series analysis is often based on indices that can then be applied over the whole time series for over many years and seasons. Water is well detectable in the NIR band due to the strong absorption of the NIR energy of water. The water/land contrast is higher in NIR then in visual wavelengths, as only topographic shadows from steep slopes, cloud shadows or burn scars can lead to similar low NIR values. While some studies have tried to capture water in time series on a 250 m scale using the NIR reflectance band of MODIS [9,11,23,82,83], others have focused on vegetation and have combined NDVI analysis with water detection [11,80,84,85]. The *Land Surface Water Index* (LSWI) [86,87], *Open Water Index* (OWI) and *Open Water Likelihood Algorithm* (OWL) [88,89], and *Floating Algae Index* (FAI) [90,91] have been used to analyse MODIS data. Recent studies have exploited a dynamically varying threshold, based on the spectral statistics of each image using AATSR imagery [92], the MODIS 250-m NIR band [82,83], as well as AVHRR data [82]. A transformation from the RGB (red–green–blue) colour space into HSV (hue–saturation–value) using the MIR, NIR, and red bands was recently developed for MODIS water monitoring in Africa [23] and applied onto SPOT VEGETATION data [46].

Medium resolution offers potential for global or continental applications. Available global water body products are the Shuttle Radar Topography Mission (SRTM) Water Body data (SWBD) [93] and the Moderate Resolution Imaging Spectroradiometer (MODIS) 250 m land-water mask (MOD44W) [94]. Within the Geoland-2 project a water body product has been created based on SPOT-VEGETATION (1 km resolution) and is available every 10 days; this product also provides information about seasonal aspects [79]. Currently, the 250 m MODIS Global WaterPack is under development [83]. The first global water body product based on SAR data is the Water Bodies Product developed in the framework of the CCI (CCI-LC WB) which is based on Envisat Advanced SAR (ASAR) Wide Swath Mode and the SRTM SWBD [14]. Low-resolution sensors like the passive microwave sensors Advanced Microwave Scanning Radiometer – Earth Observing System (AMSR-E) and Special Sensor Microwave Imager (SSMI), the Advanced Scatterometer (ASCAT), and multispectral sensors (AVHRR, MODIS) are combined into the Global Inundation Extent from Multi-Satellites (GIEMS) [60]. While global initiatives for systematic monitoring of temporary and permanent water bodies exist, this is not the case in global wetland monitoring. In this field to date, only a few case studies are addressing this topic at global scale.

14.3.4   Synthetic Aperture Radar Imagery

Synthetic aperture radar (SAR) sensors are active sensors; i.e., they use their antenna both to transmit and receive polarized pulses in the microwave range of the electro-magnetic spectrum. Thus, data can be recorded during the day as well as during the night, and almost independently of the weather conditions; this allows SAR sensors to take projectable acquisitions at regular time intervals, and under specified and very stable imaging conditions. Alternating polarization during the measurement leads to multi-polarized data sets – currently, mostly composed of horizontally- (H) and verti-cally- (V) oriented linear polarization – which enables the identification of physical scattering mechanisms, typical of specific land cover classes, such as surface or single-bounce scattering for grasslands, diplane or double-bounce for buildings, and volume or diffuse scattering for forests.

Multiple studies have shown that the mapping and monitoring of open, calm surface water or flooded areas can be done using single-polarized acquisitions. Due to its high reflectance, smooth water acts like a mirror, reflecting almost all energy away from the sensor; this causes water-covered areas to appear very dark in the intensity image. Studies using C-band SAR sensors such as Envisat Advanced SAR (ASAR) [14,95–100] or RADARSAT-1 [101,102], and X-band SAR sensors such as TerraSAR-X [97,103–106] have been carried out. The possibility of using L-band SAR sensors for wetland moni-toring in support of the Ramsar Convention has been explored using the ALOS PALSAR sensor [37]. Global monitoring of wetlands using Envisat ASAR Global Mode (GM) has also been investigated [95], and a method for deriving global open standing water from multi-temporal Envisat ASAR Wide Swath Mode (WSM) data was recently introduced [14,99]. Dense time series of SAR imagery have been applied a few times in the context of wetland and water monitoring [98,107,108].

Despite the suitability of SAR sensors for detecting open, calm surface water, rough water surfaces – which can be caused by factors such as wind or currents, as well as vegetation that is flooded, standing in water, or floating – often appear very bright because the small-scale structures form very strong corner-like reflectors, making them indistinguishable from land. These small-scale structures present a significant chal-lenge when mapping and monitoring wetlands [104]. Multi-polarized SAR data have shown strong potential in solving this problem [109–117], and different methods for monitoring wetlands with polarimetric SAR imagery have been recently summarized [118]. Flooded vegetation, for example, is characterized by a significant double-bounce component, while bare soil or grassland generally show dominant surface scattering [114,115]. Schmitt and Brisco 2013 [115] have demonstrated that partial polarimetric data (i.e. two channels instead of four) is sufficient for the monitoring of flooded vegeta-tion, which is a great advantage of SAR missions like TerraSAR-X, Sentinel-1, and Envisat ASAR that provide, at most, linearly dual-polarized data. An alternative SAR remote sensing technique is interferometric synthetic aperture radar (InSAR), which exploits the phase shift when imaging the same landscape, using a shifting viewing angle; The InSAR technique has been demonstrated to be sensitive to water level change in wetlands [119–122]. Little research has been done on the combination of optical and SAR imagery for wetland monitoring [77,123,124], but it has been concluded that opti-cal imagery might not be suitable for detecting flooded vegetation when the water is below the vegetation canopy, or when there is floating vegetation on the water [115].

To date, there are only few methods applied for analysing dual-polarimetric SAR data [115,125], and a lack of studies using dense polarimetric SAR time series for water and wetland monitoring persists. An approach for wetland monitoring based on dual-polarimetric time series is introduced in Chapter 7.

14.4 The Case Study of Lac Bam: A Natural Ramsar Wetland Site in Burkina Faso

Lac Bam is the largest natural lake in Burkina Faso, with an area of about 25 km^2 [4,9]. It is a permanent freshwater lake and is one of the Ramsar wetland sites of international importance, as discussed in Chapter 1. Irrigated cultivation around the lake has increased and, particularly in the past couple of years, more and more motor pumps have been installed to irrigate fields for vegetable gardening up to 2 km from the lake (see Figures 14.2 and 14.4). Several different livelihoods, such as fishing and herding, depend on Lac Bam, since livestock corridors pass through the area. Lac Bam is an important habitat for many species of birds and fish. The lake is strongly prone to siltation, as increasing cultivation around the lake has led to deforestation and increasing sediment input into the lake. If water extraction and siltation continue, in about 25 years the largest natural freshwater lake of Burkina Faso might have turned into a river [126]. Recent RapidEye imagery from 2013–14; Landsat archive imagery from 1986, 2000, and 2010; a MODIS time series from 2000–13; and a TerraSAR-X time series from 2013–14 are used for this case study.

14.4.1 Change Detection Using High Resolution Imagery

RapidEye data are multispectral data with five bands (blue, green, red, redEdge, and NIR), with a resolution of around 6.5 m. Landsat 5 has a resolution of 30 m and seven multispectral bands (blue, green, red, two NIR bands, two SWIR bands, and one TIR band), and Landsat 7 adds to that a panchromatic band with 15 m resolution. In order to coregister Landsat and RapidEye images, the RapidEye Level 1b data were atmospherically and geometrically corrected using the DLR in-house processing chain Catena [127], which utilizes ATCOR-2/-3 for atmospheric correction [128,129] and performs orthorectification and coregistration [130]. Change detection from HR time series data can be applied in two ways: (1) inter-annually, using images of the same season from multiple years (Figure 14.4); or (2) intra-annually, capturing the seasonal changes during one season (Figure 14.5). Figure 14.4 shows a false colour composite (NIR–red–green) of four images of Lac Bam during the dry season in February: Landsat 5 in 1986, Landsat 7 in 2000, Landsat 5 in 2010, and RapidEye in 2014. The irrigated areas clearly did not change much between 1986 and 2000, but increased between 2000 and 2010. Then, from 2010 to 2014, the area of irrigated cultivation increased markedly. This same effect is also visible for the smaller Koumbango reservoir in the West of Lac Bam between 2010 and 2014. However, conclusions regarding changing water surface areas have to be treated with caution if they are based solely on one satellite image, since seasonal effects could have caused these apparent changes (as is clearly demonstrated in Figure 14.5). A typical weather (i.e. particularly wet or dry year) could also be responsible for such changes.

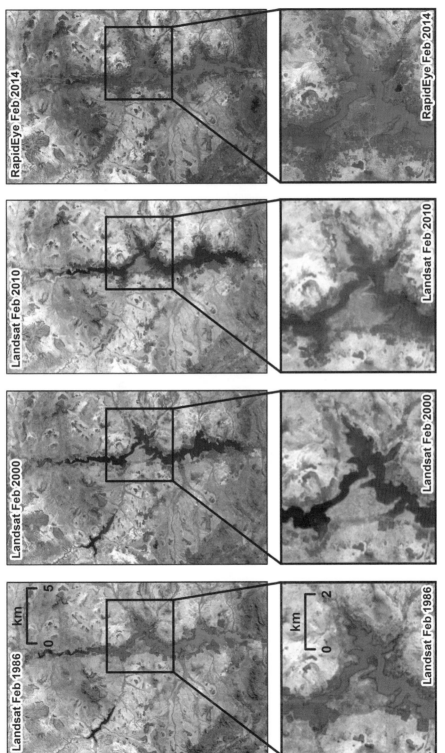

Figure 14.4 Landsat 4–3–2 (Source: USGS) and RapidEye 5–3–2 (Source: BlackBridge) from the dry season: Landsat 5; 03 Feb. 1986; Landsat 7; 18 Feb. 2000; Landsat 5; 21 Feb. 2010; and RapidEye; 07 Feb 2014. (*See insert for color representation of the figure.*)

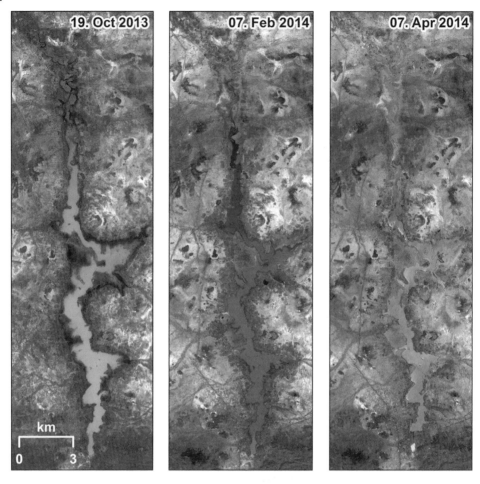

Figure 14.5 RapidEye images (band 5–3–2) from 19 Oct. 2013, 07 Feb. 2014, and 07 Apr. 2014 showing seasonal variations in water-covered area, irrigated cultivation, and surrounding natural vegetation (Source: BlackBridge). (*See insert for color representation of the figure.*)

14.4.2 Time Series Analysis Using Medium Resolution Imagery

Spatio-temporal changes of wetlands located in different climate zones over a north–south gradient in Burkina Faso were investigated using MODIS 250 m data. Yearly calculated cumulative water coverage, as well as the water-coverage trend and time series for water-covered areas, were investigated over a period of 13 years [9]. Based on work completed by Moser *et al.* 2014 [9], MODIS 250 m red and NIR surface reflectance time series with an 8-day interval (MOD09Q1 product) [131] from 2000–12 were used to conduct surface-water and NDVI time series analyses for Lac Bam. Figure 14.6 shows the mean cumulative water-covered surface area of Lac Bam for 2000–12, with the number of flooded months per year, with a year being defined as May–April of the consecutive year, starting with the rainy season in May. The area that is near-permanently flooded (i.e. 9–12 months per year) is colour-coded in blue, and the area that is dynamically flooded (i.e. 2–8 months per year) is represented by shades of green. Based

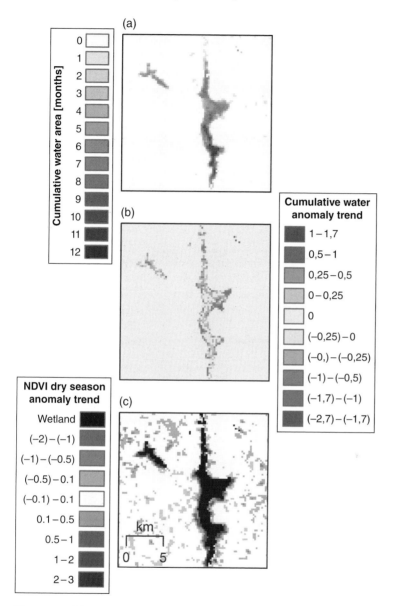

Figure 14.6 (a): Moderate Resolution Imaging Spectroradiometer (MODIS) cumulative water-covered surface area; (b): anomaly trend of cumulative water-covered surface area; (c): NDVI trend of surrounding areas, showing a strong increase in irrigated cultivation (Source: USGS). Adapted from Moser *et al.*, 2014 [84]. (*See insert for color representation of the figure.*)

on yearly derivations of the cumulative water-covered surface area, a standardized anomaly trend was computed, which shows the anomalies in the monthly duration of water-covered areas for Lac Bam over 12 years. Negative anomalies are displayed in shades of red in the centre graphic of Figure 14.6, and a few pixels with positive anomalies are displayed in shades of blue, concluding to shorter but wider water coverage of

Lac Bam in the east and the Koumbango reservoir. The increase in irrigated cultivation is visible in the strongly positive NDVI trend (i.e. blue colours) around the lake, for which all pixels water-covered for more than two months per year are masked out (Figure 14.6). This matches the findings from Landsat/RapidEye time series (Figure 14.4), confirming that MODIS is able to capture trends of irrigated agriculture around wetlands, as well as hot spots of positive or negative vegetation trends in the surrounding areas [84]. Changing flooding regimes in terms of shorter but wider water coverage, which are likely related to sedimentation, pose a threat to the lake and the water availability of the region, which has been enhanced by increasing water extraction for irrigation [9,84] with motor pumps as seen in Figure 14.2e.

14.4.3 Time Series Analysis Using Polarimetric Synthetic Aperture Radar Data

In our case study, a time series of TerraSAR-X HH-VV polarized Stripmap data was acquired for the RAMSAR site Lac Bam in Burkina Faso. Stripmap data covers a swatch with a 15 km width and up to a 50 km length. Acquisitions under the same geometric conditions and with a regular time interval of 11 days, starting in the rainy season and running until the end of the dry season, were used. In total, 22 SAR images of Lac Bam were acquired between August 2013 and April 2014; RapidEye imagery (see Figure 14.5) acquired on the same or approximately the same day was available for three dates. A decomposition technique based on the Kennaugh matrix [115,125] was adapted for the TerraSAR-X imagery and was applied on the time series. This results in four Normalized Kennaugh elements, which split the backscatter signal into (a) the total intensity of HH and VV (K0); (b) the intensity ratio between double bounce and surface intensity (K3); (c) the ratio between HH and VV intensity (K4); and (d) the phase shift between double-bounce and surface scattering (K7). This method is demonstrated for five selected images along a time series with 11-day intervals. A false colour composite of Kennaugh elements K3, K0, and K4 was used to visualize the data (Figure 14.7). Throughout the entire rainy season (i.e. June to October) the blue colour (i.e. HH backscatter dominates over VV backscatter) appears dominant over open water, and a significant reduction of the water surface can be observed over the course of the rainy season. Pink (i.e. double-bounce scattering dominates over surface scattering) indicates flooded vegetation, which can also be separated from the environment. Green colour (i.e. where the sum of the surface intensity of HH and VV is particularly strong) indicates vegetated surfaces on the land during the rainy season; this can be observed in the image from August. Some irrigated fields appear as light green pixels towards the dry season.

Figure 14.8a shows the false-colour RapidEye image from 19 October 2013 (as used in the previous chapter and as seen in Figure 14.5a) next to a SAR intensity image (displayed as K0) from 20 October in Figure 14.8b; Figure 14.8c also gives the corresponding colour composite using the three above-mentioned Kennaugh elements (i.e. K4–K0–K3). While, in the optical image, water could also be detected over shallow flooded areas using the NDWI [42], the SAR intensity image (K0) was only able to detect open water when there was no interference from vegetation of any kind; the high backscatter in the surrounding areas with flooded vegetation made it impossible to detect water in the SAR intensity image (K0). Water containing flooded or standing vegetation, however, can be captured using the polarized content of the SAR image, as seen in Figure 14.8c.

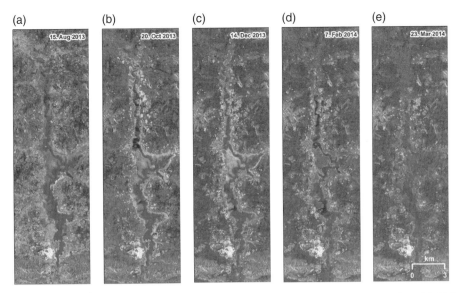

Figure 14.7 An example of a time series of five selected TerraSAR-X (TSX) acquisitions from (a): 15 Aug., (b): 20 Oct. and (c): 14 Dec. 2013, as well as (d): 07 Feb. and (e): 23 Mar. 2014 (Source: DLR 2013, 2014). A false colour composite has been created (K4–K0–K3). Open water is displayed in blue, with blue colours indicating a dominance of HH over VV backscatter. Green colours appear where the sum of the surface intensity of HH and VV is particularly strong, and pink colours are dominant where double-bounce scattering dominates over surface scattering, such as in the areas of flooded vegetation. (*See insert for color representation of the figure.*)

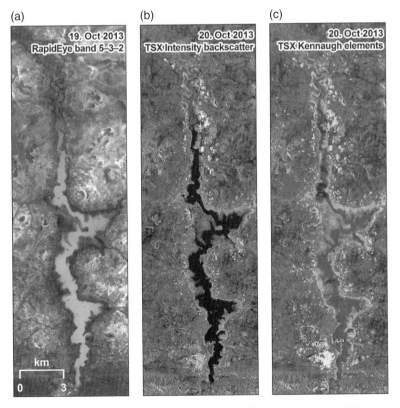

Figure 14.8 Comparison of RapidEye and TerraSAR-X imagery; (a): a RapidEye image false colour composite (5–3–2) taken on 19 Oct 2013 (Source: BlackBridge); (b): a TerraSAR-X Stripmap intensity image (K0) taken on 20 Oct. 2013 2013 (Source: DLR 2013); (c): the Kennaugh elements from HH and VV data from a TerraSAR-X image taken on 20 Oct. 2013 (Source: DLR 2013). (*See insert for color representation of the figure.*)

Water-covered areas that appear almost black in the Kennaugh image might indicate the presence of wind, which affects X-Band SAR but not optical data. Strongly vegetated areas of standing vegetation in water show a strong infrared response (displayed in red colour) in the optical image, and water can often not be detected below dense vegetation. Depending on the season and environmental conditions it can be difficult to distinguish between vegetation on land and floating or standing vegetation in water; however, polarimetric SAR imagery shows potential in its ability to capture standing and floating vegetation, which is visible in shades of pink in Figure 14.7 and Figure 14.8c. The synergetic use of optical and SAR imagery at similar spatial resolutions to exploit their complementary potential is still an open field of research.

14.5 Perspectives on Wetland Remote Sensing for Water Availability

14.5.1 Possibilities and Limitations, Gaps and Opportunities

Wetland mapping, inventorying, classifying, and monitoring applications are carried out using a variety of remote sensing sensors, for a variety of different applications. These applications include the monitoring of open water, aquatic vegetation, plant ecology, habitats, water availability, water quality, infrastructure, and irrigated agriculture; monitoring can also assess the full LULC of larger wetland areas and their surroundings, including agricultural, forest, and urban areas. The task of assessing wetlands is made more difficult by the fact that the definition of wetlands varies from source to source, both among different remote sensing applications and among the different branches of natural science. Size matters in terms of the scale of a wetland or case study area, affecting the required spatial resolution for monitoring and impacting what remote sensing sensor is most suitable for the desired analysis. The size of the wetland under study can vary with the definition of *wetland* that is used, and may, for example, mean that a case study comprises a single lake and its surrounding areas, or large areas such as deltaic or coastal areas. The study of wetlands has a strong temporal component, as seasonal changes can be significant, depending on geographic areas; for example, sub-Saharan areas, which have marked rainy and dry seasons; areas with strong summer–winter differences; areas strongly influenced by snow melt or flash floods; and areas affected by meteorological phenomena like monsoon all show great seasonal variation. The recent availability of time series data from satellite archives has opened up opportunities to compute trends (see Figure 14.6) and to drive models of temporal behaviour for future long-term drought mitigation. Depending on the desired application, the relative importance of sensors characteristics such as spatial resolution, aerial coverage, spectral resolution, and temporal resolution have to be considered in order to determine the right trade-off for a sensor or sensor combination for each specific application.

The importance of the relationship between the temporal and spatial components is illustrated in Figure 14.9, which depicts different applications related to wetland monitoring. Medium resolution time series (i.e. daily or near-daily observations with medium spatial resolution) allow surface water dynamics and vegetation dynamics to be derived (as seen in Figure 14.6), as well as parameters such as water quality, soil moisture, or

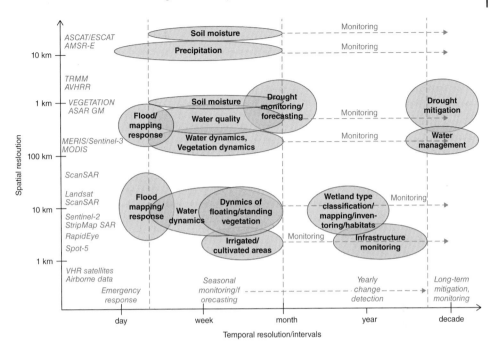

Figure 14.9 Relationship between different spatial and temporal resolutions required for different wetland monitoring parameters. The spatial axis (*y*-axis) displays geometric resolution on a logarithmic scale, including some example remote sensing sensors; the temporal axis (*x*-axis) displays required temporal scales in terms of data collection frequency for a given application within 1 year. Arrows indicate the timescale for useful long-term monitoring based on seasonal observations. Applications with current limitations are located in the lower left-hand corner of the diagram.

precipitation (upper left-hand corner). Data need to be collected at intervals between daily and monthly for seasonal monitoring. Carried out over many years and analysed as time series (indicated by the arrows in Figure 14.9), such monitoring can allow trends for long-term drought mitigation to be derived. Applications like wetland type classification, inventorying, and infrastructure monitoring (lower right-hand corner) do not necessarily require data collected at short time intervals, and can therefore be observed with HR data. Applications for which there are still gaps (lower left-hand corner) cannot be satisfactorily completed with the spatial or temporal scales of existing satellite sensors. Such applications include the dynamics of surface water, floating and flooded vegetation, and irrigation on a high-resolution scale, either for a single season (as exemplified in Figures 14.5 and 14.7) or as time series over a series of years. Figure 14.10 illustrates a selected area of Lac Bam displayed using different remote sensing data with different spatial resolutions in April 2014: MODIS 500 m (a); MODIS 250 m (b); Landsat 30 m (c); and RapidEye 5 m (d). Below, zooming into the area covered by a 500 × 500 m MODIS pixel shows again the same Landsat 30 m (e); RapidEye 5 m (f); and WorldView-2 0.5 m (g) resolved image. Figure 14.11 shows the water covered area of Lac Bam derived from the 8-day MODIS time series from 2000–12 (bold line) paired with available dates of high resolution Landsat TM and ETM+ images with less than 30% cloud cover from the Landsat archive (marked with vertical lines).

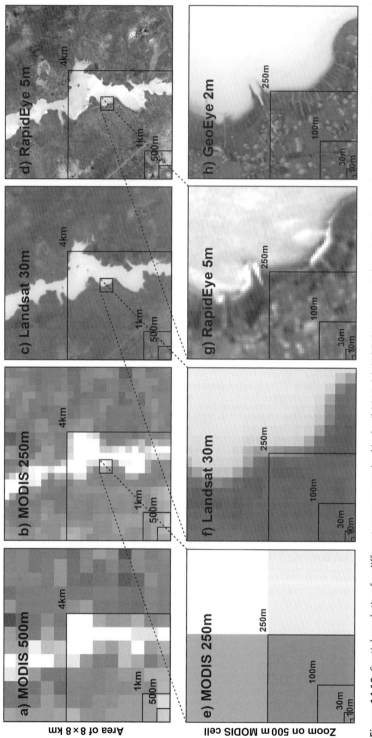

Figure 14.10 Spatial resolution for different sensors acquired in April 2014: (a): MODIS 500 m spatial resolution for Lac Bam, source: USGS; (b): MODIS (250 m), source: USGS; (c) High resolution Landsat (30 m), source: USGS; (d): High resolution RapidEye (5 m), source: BlackBridge; Spatial resolutions for AATSR and VEGETATION (1 km) and AVHRR (4 km) are indicated as rectangles; (e): Zoom into the extent of a 500 m MODIS pixel showing again the MODIS (250 m) image; (f) high resolution Landsat image (30 m); (g): RapidEye image (5 m); (h): GeoEye-1 image (2 m), source: EUSI/Digital Globe; Spatial resolutions for Sentinel-2 (10 m), Landsat (30 m), 100 m, and MODIS (250 m) are indicated as rectangles.

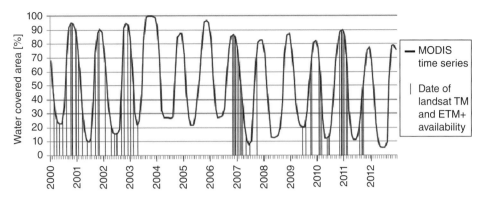

Figure 14.11 MODIS-derived water covered area time series of Lac Bam from 2000–12 (bold line) paired with available dates of high resolution Landsat TM and ETM+ images with less than 30% cloud cover from the Landsat archive (marked with vertical lines).

Table 14.1 provides a synthesis of the advantages and limitations of different types of sensors for wetland monitoring:

Future opportunities for wetland remote sensing should focus on newly launched or planned satellite missions. With a focus on a European perspective these are ESA's Sentinel-1, -2, and -3 satellites [132]. The Sentinel missions will ensure the continuity of existing data archives or open up opportunities for new analysis techniques. Two of each satellite will initially be launched, followed by two more satellites towards the end of satellite lifetime, to assure data continuity.

Up to 15 years of archives of suitable data exist for MODIS, MERIS, and VEGETATION, and up to around 30 years from (A)ATSR and AVHRR. Data continuity is planned or guaranteed in most cases: MODIS is already used in conjunction with AVHRR, and its NDVI products continue the legacy of long-term (since 1981) vegetation products from AVHRR [133], such as the long-term data record (LTDR), which uses AVHRR and MODIS in synergy [134]. AVHRR and MODIS have been recently used in synergy for delineating inland water bodies [82]. The Visible/Infrared Imager Radiometer Suite (VIIRS) of NOAA's Joint Polar Satellite System (JPSS) will serve as a continuation of both the MODIS and AVHRR sensors for the next 20 to 30 years, providing data with a 375 m resolution. The first satellite carrying VIIRS was launched on 30 October 2011 (Suomi NPP), and further planned launches are set for 2017 (JPSS-1) and 2022 (JPSS-2); both of these satellites will carry the VIIRS instrument [62,135–137]. To date, no specific programme using VIIRS for wetlands or water bodies has been announced, but studies to evaluate surface water detection were carried out [138]. The continuation of MERIS and (A)ATSR will be provided through the ESA Sentinel-3 satellites. The Ocean and Land Colour Instrument (OLCI) has a 300 m resolution will be used with MERIS; and the Sea and Land Surface Temperature Radiometer (SLSTR), with a 500 m resolution for VNIR and SWIR bands, and 1 km resolution for TIR bands, will be used with (A)ATSR. Together, these two instruments left a gap of several years between the loss of contact with the Envisat satellite in April 2012 and the launch of Sentinel-3 in February 2016 [62,139,140]. The GloboLakes project intends to supplement MERIS data with OLCI data to continue near real-time lake monitoring, and will

Table 14.1 Summary of the advantages and limitations of the three different sensor groups: optical medium resolution (MR), optical high resolution (HR), and synthetic aperture radar (SAR).

Remote sensing sensor type	Advantages for wetland monitoring	Limitations for wetland monitoring
Medium resolution optical	• Existence of NIR and (at times) SWIR bands • Temporal availability: regular, short intervals; repeatable; time series ~15y backwards; data continuity • Temporal information: Seasonal cycle derivation, long-terms trends • Straightforward methodology, transferable to other regions and sensors • Large area coverage, for national/continental/global applications • Potential for operational applications	• Low geometric resolution, important small wetlands might not be detected or monitored • No interpretation of details (only water/non water and vegetation indices) • Mixed pixel issues • Cloud-cover dependent, product flags needed • Large amount of data and computing power needed • Different viewing angles due to large area coverage with large field of view
High resolution optical	• Good opportunities for water detection (classification or indices) due to existence of NIR and (at times) SWIR bands • Detection and dynamics of small wetlands of high significance • Validation of MR data, interpretation of trends • Possibilities for detailed LULC classification • Possibilities for multitemporal classification	• Smaller areas only, allowing case studies rather than large coverage • Cloud-cover dependent • Irregular and less frequent temporal coverage • Gap in data archives, especially over data-scarce regions in Africa, due to, e.g., lack of ground stations • Sun glint effects • Strongly varying reflectance values in the case of sediment-rich waters, differences between rainy dry seasons
Synthetic aperture radar	• Good for water detection based on low backscatter intensity • All weather, including rainy season • Operates day and night, yielding higher repetition and more image acquisition opportunities • Good response to open surface water • ScanSAR can cover larger areas • Polarimetric SAR has potential for standing vegetation, flooded vegetation, and water below vegetation	• No regular time series archive, no guarantee for regular acquisitions or data continuity • Wind influence may lead to water not being detected (especially for short wavelength sensors) • Dry sands may lead to incorrect water classification • Interruption in data archives for L-Band and C-band SAR data

also exploit SLSTR and data from ESA's HR mission (Sentinel-2) launched in June 2015. The SPOT/VEGETATION instrument on board the SPOT-4 and SPOT-5 satellites, which has been used so far for dynamic water-body mapping [19,21,79], will be extended by the PROBA-V mission, which was launched in May 2013 by ESA and the Belgian Science Policy Office (BELSPO). The PROBA-V mission provides data continuity until 2016, to be then continued with Sentinel-3 data [62,141,142].

Entering a new era of high-resolution time series, these data show great potential for monitoring selected wetland areas using time series with a regular data collection of 1–2 images per month. This is of particular interest for the GlobWetland projects that intend to use ESA's Sentinel-2 satellite that has been recently launched on 23 June [140,143] as the source of their main dataset for wetland monitoring on a HR scale [32], as well as for the Copernicus Land Monitoring Service, which aims to produce *High Resolution layers* (HRLs) of water and wetness areas. Advantages of Sentinel-2 are its large swath (290 km), high resolution (bands with 10 m, 20 m, and 60 m resolution), and 13 spectral bands (including three SWIR bands and three red-edge bands). Data archives from other HR data show significant gaps over remote areas that lack the infrastructure for downlink stations. For example, in Burkina Faso, seasonal Landsat observations exist for the period 1986–87, almost no Landsat data are available between 1990 and 1999, and, after a period of good coverage from 2000–02 from Landsat 7 ETM+, there is again a large gap between 2003 and 2007, and 2007 and 2009 – the only data available for this time period are Landsat 7 images corrupted by striping following a scan line corrector failure in 2003. The period from 2000–12 is displayed in Figure 14.11 in terms of Landsat archive imagery availability with less than 30% cloud cover. From April 2014 onwards, data from Landsat 8 are available in 16-day intervals. This dataset shows great potential for wetland mapping: it has 11 bands, including a shortwave band for coastal and aerosol studies, which could be used to map shallow waters and detect fine dust particles, and two SWIR and two thermal bands [144]. ESA and NASA launched initiatives to use Landsat 8 and Sentinel-2 together to build up more dense time series, or to provide continuation for Landsat and SPOT satellite data. Both Landsat 8 and Sentinel-2 data are available free of charge for scientific applications, and are therefore very suitable for wetland monitoring in developing countries and operational applications.

There are some gaps in the availability of SAR data. Although there are large archives for ERS-1 SAR, ERS-2 SAR, and Envisat ASAR C-band SAR data, ESA SAR data has been interrupted since the loss of Envisat in April 2012. Sentinel-1A [140,145] was launched in April 2014 and data are now provided free of charge. Likewise, the L-Band SAR satellite ALOS PALSAR stopped providing data in April 2011, and ALOS-2 was launched in May 2014. L-Band SAR data have shown to be very promising for wetland mapping, not only due to the wavelength, but also due to the fact that this satellite is operational in quad-pol mode (i.e. recording HH, HV, VH, and VV data at the same time). TerraSAR-X and TanDEM-X have been proven to be successfully applied for wetland studies profiting from their dual-co-pol (HH-VV) acquisition capabilities. [146,147]. TerraSAR-X, TanDEM-X, and COSMO-SkyMed are expected to operate for a few more years as X-band SAR satellites, and a future X-band follow-on satellite mission of TerraSAR-X is planned, as well as the Spanish X-band PAZ satellite is to fly in constellation with TerraSAR-X and TanDEM-X. RADARSAT-2 is an operational C-band SAR satellite with dual- and quad-pol capability and will continue in the RADARSAT Constellation Mission (RCM), which will consist of three satellites [62]. Upcoming SAR missions carrying instruments operating in longer SAR wavelengths with greater ability to penetrate vegetation offer potential for water and wetland applications: The L-Band SAR constellation SAOCOM (Argentina), TanDEM-L (Germany) and the P-Band SAR mission Biomass (ESA). Along with other existing processing techniques for polarimetric SAR, the methodology proposed in Chapter 7 for polarimetric SAR processing [125] could be applied to a variety of SAR sensors covering all wavelengths and polarization combinations for wetland monitoring. The Synthetic Aperture Radar Altimeter (SRAL) on board the Sentinel-3 satellite has been designed to measure the surface water levels of lakes [140]. NASA's upcoming Surface Water Ocean Topography (SWOT) Mission carrying wide-swath radar altimetry technology is currently targeted to launch in 2020. Among the SWOT mission objectives are the study of the dynamics of floodplains and wetlands, as well as deriving water surface elevation, lake and reservoir storage, lake inventories, river discharge, water cycle and water resource assessment on a global scale [148].

14.5.2 Wetland Remote Sensing as Indicator of Water Availability and Drought

Despite the importance of water for so many people's livelihoods, wetlands are hardly monitored at all in Africa. In Burkina Faso, few selected strategic water bodies are surveyed in terms of the water level as measured at their dam, and surface water area is not monitored at all. The water distribution is strongly influenced by human activities such as the ongoing construction of dams to create artificial reservoirs, which tremendously influences downstream water availability. Wetland studies often focus on LULC classification or biodiversity applications; applications for the dynamic monitoring of water availability are much less common. Wetland monitoring using MR optical or SAR time series, using weekly to bi-weekly observations, would enable dynamic water monitoring and would allow researchers to estimate the water surface at the end of the rainy season. This would allow conclusions to be made regarding the amount of available water for the dry season. Such monitoring would be possible for large areas by employing sensors such as MODIS, VIIRS, Sentinel-3 OLCI, or Sentinel-1 wide-swath mode. The presented case study of the Lac Bam wetland site in semi-arid Burkina shows how different remote sensing data could be used together for more effective wetland monitoring.

A synthesis of remote sensing studies to assess four decades of land degradation in the Sahel includes water resource loss and water erosion as factors feasible to be monitored by remote sensing into their scheme of land degradation linkages [149]. Other than the work presented in Moser *et al.* 2014 [9], studies connecting remotely sensed water body or wetland dynamics with drought or land degradation are still scarce. Wetland monitoring has not yet been incorporated into drought index models, but the correlation between drought occurrence and negative anomalies for both water coverage duration and vegetation indices shows potential for further research [9,84]. Drought indices like the *Standard Precipitation Index* (SPI) mostly contain meteorological data such as precipitation and temperature; soil type or soil water content may also be included, like when using the commonly applied *Palmer Drought Severity Index* (PDSI) [150]. In terms of remote sensing, the *Vegetation Condition Index* (VCI) makes use of the NDVI to build a new index, and the *Temperature Condition Index* (TCI) requires brightness temperature retrieval; both of these indices can be applied to sensors like MODIS or AVHRR [151,152] to estimate drought. Combined approaches using meteorological and remote sensing data from vegetation indices are applied in the *Combined Drought Index* (CDI) [153]. Other remotely sensed parameters that could also contribute to drought assessment include soil moisture, precipitation, or water level height. Soil moisture has been estimated for the past 30 years using data from scatterometer and passive microwave instruments on a coarse resolution scale (around 25–50 km spatial resolution), as well as using SAR data on a higher resolution scale. Some satellites provide rainfall estimates, such as the Tropical Rainfall Measuring Mission (TRMM). The SWOT mission as well as the SRAL instrument on board the Sentinel-3 satellite has been designed to measure the surface water levels of lakes [140].

To sum up, remote sensing data analysis has the potential to greatly contribute to water management in Africa. Synergistic combinations of multiple sensor systems are encouraged. Despite the fact that remote sensing might not replace field data for certain applications, remote sensing can certainly complement ground observations, and can be used alone for applications such as water body detection, flood detection, large area monitoring, inventorying, and detection of the time and location of newly built dams and artificial reservoirs. The application of time series analysis to data archives opens up a range of opportunities, such as deriving spatio-temporal trends of water coverage, seasonal cycles of wetlands, large-scale vegetation trends, and hot spots, as well as irrigation development monitoring. Spatio-temporal monitoring of surface water dynamics can aid researchers in assessing the long-term impact of water extraction, sedimentation, and climate change on wetlands. Monitoring the Sahelian wetlands remains a challenge due to the spatial and temporal requirements of observing small wetlands with strong seasonal differences. However, such effort is well worth it, as such small wetlands are of great importance for biodiversity and human use, and might strongly reflect global climatic and anthropogenic changes.

Acknowledgements

The authors acknowledge the financial support of GIONET, which is funded by the European Commission, Marie Curie Programme Initial Training Network, Grant Agreement PITN-GA-2010-264509. Many thanks go to Christian Hüttich for providing helpful input and discussions, and to Raymond Ouedraogo and Francesc Betorz Martínez for assistance during field work in Burkina Faso. The authors acknowledge the use of RapidEye data which has been provided on behalf of the German Aerospace Center through funding of the German Federal Ministry of Economy and Energy with the proposal "Wetland Monitoring of Lac Bam"; TerraSAR-X data, which have been made available for scientific use via the TerraSAR Science Service proposal "LAN2000: Wetland monitoring and water stress in sub-Saharan West Africa"; a GeoEye-1 image, copyright DigitalGlobe provided by European Space Imaging; and Landsat data and MODIS data, which have been made available for scientific use free of charge via the USGS.

References

1 UNEP, *Africa Water Atlas*, Division of Early Warning and Assessment (DEWA), United Nations Environment Programme (UNEP), Nairobi, Kenya, 2010.

2 M. Falkenmark, The Massive Water Scarcity Now Threatening Africa: Why Isn't It Being Addressed? *Ambio*, **18**, 112–118, (1989).

3 Ramsar, *Convention on Wetlands of International Importance Especially as Waterfowl Habitat*. Ramsar (Iran), 2 February 1971. UN Treaty Series No. 14583. As amended by the Paris Protocol, 3 December 1982, and Regina Amendments, 28 May 1987, 1971.

4 Ramsar, *Ramsar Convention on Wetlands*, Available: www.ramsar.org/; Accessed: January 2017.

5 J.G. Lyon, R.D. Lopez, L.K. Lyon and D.K. Lopez, *Wetland landscape characterization: GIS, remote sensing and image analysis*, CRC Press, 2001.

6 W.J. Junk, S. An, C. Finlayson, B. Gopal, J. Květ, S.A. Mitchell, W.J. Mitsch and R.D. Robarts, Current state of knowledge regarding the world's wetlands and their future under global climate change: a synthesis, *Aquatic Sciences*, **75**, 151–167, (2013).

7 C. Peugeot, B. Cappelaere, B.E. Vieux, L. Séguis and A. Maia, Hydrologic process simulation of a semiarid, endoreic catchment in Sahelian West Niger. 1. Model-aided data analysis and screening, *Journal of Hydrology*, **279**, 224–243, (2003).

8 L.M. Rebelo, C.M. Finlayson and N. Nagabhatla, Remote sensing and GIS for wetland inventory, mapping and change analysis, *Journal of Environmental Management*, **90**, 2144–2153, (2009).

9 L. Moser, S. Voigt, E. Schoepfer and S. Palmer, Multitemporal Wetland Monitoring in Sub-Saharan West-Africa Using Medium Resolution Optical Satellite Data, *IEEE Journal of Selected Topics in Applied Earth Observations and Remote Sensing (J-STARS)*, **7**, 3402–3415 (2014).

10 Direction Générale des Ressources en Eau (DGRE), *Data about Water Levels and Volumes*, Ouagadougou, Burkina Faso, 2013.

11 T. Landmann, M. Schramm, R.R. Colditz, A. Dietz and S. Dech, Wide area wetland mapping in semi-arid Africa using 250-meter MODIS metrics and topographic variables, *Remote Sensing*, **2**, 1751–1766, (2010).

12 J. Reschke and C. Hüttich, Continuous field mapping of Mediterranean wetlands using sub-pixel spectral signatures and multi-temporal Landsat data, *International Journal of Applied Earth Observation and Geoinformation*, **28**, 220–229, (2014).

13 S. Bontemps, P. Defourny, E. Van Bogaert, O. Arino, V. Kalogirou and J. Ramos Perez, *GLOBCOVER 2009 Products Description and Validation Report*, European Space Agency, Paris, France, 2011.

14 M. Santoro, C. Lamarche, S. Bontemps, U. Wegmüller, V. Kalogirou, O. Arino and P. Defourny, Introducing a global dataset of open permanent water bodies, in *ESA Living Planet Symposium*, Edinburgh, United Kingdom, (2013).

15 A. Strahler, D. Muchoney, J. Borak, M. Friedl, S. Gopal, E. Lambin and A. Moody, *MODIS Land Cover Product Algorithm Theoretical Basis Document (ATBD) Version 5.0*, 1999.

16 A. Di Gregorio and L.J.M. Jansen, *Land Cover Classification System (LCCS): Classification concepts and user manual*, Available: http://www.fao.org/docrep/008/y7220e/y7220e00.htm; Accessed: January 2017.

17 K. Weise, B. Wolf, M. Schwarz, M. Paganini, E. Fitoka, E. Van Valkengoed, H. Hansen and M. Keil, DUE GlobWetland II project – Monitoring Wetlands for Sustainable Water Management. *Living Planet Symposium 2013* Edinburgh, United Kingdom, 2013.

18 P. Cecchi, A. Meunier-Nikiema, N. Moiroux and B. Sanou, *Towards an Atlas of Lakes and Reservoirs in Burkina Faso*, Montpellier Cedex, France, 2008.

19 E. Haas, E. Bartholomé and B. Combal, Time series analysis of optical remote sensing data for the mapping of temporary surface water bodies in sub-Saharan western Africa, *Journal of Hydrology*, **370**, 52–63, (2009).

20 J. Gardelle, P. Hiernaux, L. Kergoat and M. Grippa, Less rain, more water in ponds: a remote sensing study of the dynamics of surface water from 1950 to present in pastoral Sahel (Gourma region, Mali), *Hydrology and Earth System Sciences*, **14**, 309–324, (2010).

21 E.M. Haas, E. Bartholomé, E.F. Lambin and V. Vanacker, Remotely sensed surface water extent as an indicator of short-term changes in ecohydrological processes in sub-Saharan Western Africa, *Remote Sensing of Environment*, **115**, 3436–3445, (2011).

22 Y. Walz, M. Wegmann and S. Dech, Schistosomiasis risk assessment from space using high resolution Rapid Eye data, *in International Geoscience and Remote Sensing Symposium (IGARSS)*, IEEE, 7224–7227, (2012).

23 J.-F. Pekel, C. Vancutsem, L. Bastin, M. Clerici, E. Vanbogaert, E. Bartholomé and P. Defourny, A near real-time water surface detection method based on HSV transformation of MODIS multi-spectral time series data, *Remote Sensing of Environment*, **140**, 704–716, (2014).

24 N. Davidson and C. Finlayson, Earth Observation for wetland inventory, assessment and monitoring, *Aquatic Conservation: Marine and Freshwater Ecosystems*, **17**, 219–228, (2007).

25 B. Lehner and P. Döll, Development and validation of a global database of lakes, reservoirs and wetlands, *Journal of Hydrology*, **296**, 1–22, (2004).

26 E. Fitoka and I. Keramitsoglou, *Inventory, assessment and monitoring of Mediterranean Wetlands: Mapping wetlands using Earth Observation techniques*, EBKY & NOA. MedWet publication, 2008.

27 D. Fernandez-Prieto, O. Arino, T. Borges, N. Davidson, M. Finlayson, H. Grassl, H. Mackay, C. Prigent, D. Pritchard and G. Zalidis, The Glob Wetland Symposium: Summary and way forward, in *GlobWetland Symposium, Looking at Wetlands from Space, October 2006*, European Space Agency, Frascati, Italy, (2006).

28 K. Jones, Y. Lanthier, P. Van Der Voet, E. Van Valkengoed, D. Taylor and D. Fernandez-Prieto, Monitoring and assessment of wetlands using Earth Observation: the GlobWetland project, *Journal of Environmental Management*, **90**, 2154–2169, (2009).

29 H. Mackay, C.M. Finlayson, D. Fernandez-Prieto, N. Davidson, D. Pritchard and L.M. Rebelo, The role of Earth Observation (EO) technologies in supporting implementation of the Ramsar Convention on Wetlands, *Journal of Environmental Management*, **90**, 2234–2242, (2009).

30 M. Paganini, K. Weise, E. Fitoka, H. Hansen, D. Fernandez Prieto and O. Arino, The DUE Globwetland-2 Project, 2010 *Living Planet Symposium 2010*, Bergen, Norway, (2010).

31 K. Weise, B. Wolf and M. Schwarz, DUE Globwetland II project results – Monitoring wetlands for sustainable water management, *International Geoscience and Remote Sensing Symposium (IGARSS 2014)*, IEEE, Québec, Canada, (2014).

32 M. Paganini, GlobWetland III, towards a Global Wetland Observing System in Africa, in *Globwetland III User Consultation Workshop*, Frascati, Italy, (2013).

33 A. Walli, C. Tottrup, V. Naeimi, P. Bauer-Gottwein, M. Bila, P. Mufeti, J. Tumbulto, C. Rajah, L. Moloele and B. Koetz, TIGER-NET – enabling an Earth Observation capacity for Integrated Water *Resource Management in Africa, Living Planet Symposium 2010*, Bergen, Norway, (2010).

34 TIGER-NET, *TIGER-NET Satellite Observations Supporting Integrated Water Resources Management in Africa*, Available: http://www.tiger-net.org; Accessed: January 2017.

35 T. Langanke, *GIO land (GMES/Copernicus initial operations land) High Resolution Layers (HRLs) – summary of product specifications*, GIO land team at the EEA, (2013).

36 T. Langanke, E. Schuren, H. Dufourmont and G. Vaitkus, Mapping Water for Europe: Operational Mapping and Data Products in the Context of Copernicus: EU Hydro and

High Resolution Layer on Water Bodies. *Status, Outline, Developments, Lessons Learned, Mapping Water Bodies from Space (MWBS)*, 18–19 March 2015, Frascati, Italy, (2015).

37 A. Rosenqvist, C.M. Finlayson, J. Lowry and D. Taylor, The potential of long-wavelength satellite-borne radar to support implementation of the Ramsar Wetlands Convention, *Aquatic Conservation: Marine and Freshwater Ecosystems*, **17**, 229–244, (2007).

38 NASA JPL, Wetlands – *Global monitoring of wetland extend and dynamics*, Available: http://wetlands.jpl.nasa.gov/science/index.html; Accessed: January 2017.

39 Satellite-based Wetland Observation Service (SWOS), Available: http://swos-service.eu; Accessed: January 2017.

40 Globolakes, *Globolakes – Global Observatory of Lake Responses to Environmental Change*, Available: www.globolakes.ac.uk/; Accessed: January 2017.

41 C.J. Tucker, Red and photographic infrared linear combinations for monitoring vegetation, *Remote Sensing of Environment*, **8**, 127–150, (1979).

42 S. Mcfeeters, The use of the Normalized Difference Water Index (NDWI) in the delineation of open water features, *International Journal of Remote Sensing*, **17**, 1425–1432, (1996).

43 B.-C. Gao, NDWI – a normalized difference water index for remote sensing of vegetation liquid water from space, *Remote Sensing of Environment*, **58**, 257–266, (1996).

44 H. Xu, Modification of normalised difference water index (NDWI) to enhance open water features in remotely sensed imagery, *International Journal of Remote Sensing*, **27**, 3025–3033, (2006).

45 J. Lacaux, Y. Tourre, C. Vignolles, J. Ndione and M. Lafaye, Classification of ponds from high-spatial resolution remote sensing: Application to Rift Valley Fever epidemics in Senegal, *Remote Sensing of Environment*, **106**, 66–74, (2007).

46 L. Bertels, B. Smets and D. Wolfs, Dynamic Water Surface Detection Algorithm Applied on PROBA-V Multispectral Data, Remote Sensing, 8.12, 1010, (2016).

47 J.-F. Pekel, A. Cottam, M. Clerici, A. Belward, G. Dubois, E. Bartholome and N. Gorelick, A Global Scale 30 m Water Surface Detection Optimized and Validated for Landsat 8, in *AGU Fall Meeting Abstracts*, 01P, (2014).

48 J.-F. Pekel, A. Cottam, N. Gorelick and A. Belward, 30 Years' Global Scale Mapping of surface water dynamics at 30 m resolution, *Mapping Water Bodies from Space (MWBS)*, 18–19 March 2015, Frascati, Italy, (2015).

49 G.L. Feyisa, H. Meilby, R. Fensholt and S.R. Proud, Automated Water Extraction Index: A new technique for surface water mapping using Landsat imagery, *Remote Sensing of Environment*, **140**, 23–35, (2014).

50 L. Ji, X. Geng, K. Sun, Y. Zhao and P. Gong, Target Detection Method for Water Mapping Using Landsat 8 OLI/TIRS Imagery, *Water*, **7**, 794–817, (2015).

51 Q. Liu, G. Liu, C. Huang, S. Liu and J. Zhao, A tasseled cap transformation for Landsat 8 OLI TOA reflectance images, in *Geoscience and Remote Sensing Symposium (IGARSS), 2014 IEEE International*, IEEE, 541–544, (2014).

52 R.D. Jackson and A.R. Huete, Interpreting vegetation indices, *Preventive Veterinary Medicine*, **11**, 185–200, (1991).

53 R.B. Myneni, F.G. Hall, P.J. Sellers and A.L. Marshak, The interpretation of spectral vegetation indexes, *IEEE Transactions on Geoscience and Remote Sensing*, **33**, 481–486, (1995).

54 E.P. Glenn, A.R. Huete, P.L. Nagler and S.G. Nelson, Relationship between remotely-sensed vegetation indices, canopy attributes and plant physiological processes: what vegetation indices can and cannot tell us about the landscape, *Sensors*, **8**, 2136–2160, (2008).

55 E. Adam, O. Mutanga and D. Rugege, Multispectral and hyperspectral remote sensing for identification and mapping of wetland vegetation: a review, *Wetlands Ecology and Management*, **18**, 281–296, (2010).

56 T.S. Silva, M.P. Costa, J.M. Melack and E.M. Novo, Remote sensing of aquatic vegetation: theory and applications, *Environmental Monitoring and Assessment*, **140**, 131–145, (2008).

57 S. Ozesmi and M. Bauer, Satellite remote sensing of wetlands, *Wetlands Ecology and Management*, **10**, 381–402, (2002).

58 T. Schmid, M. Koch and J. Gumuzzio, Multisensor approach to determine changes of wetland characteristics in semiarid environments (Central Spain), *IEEE Transactions on Geoscience and Remote Sensing*, **43**, 2516–2525, (2005).

59 S. Li, D. Sun, M. Goldberg and A. Stefanidis, Derivation of 30 m resolution water maps from TERRA/MODIS and SRTM, *Remote Sensing of Environment*, **134**, 417–430, (2013).

60 F. Aires, C. Prigent, F. Papa, E. Fluet-Chouinard and B. Lehner, A global and long-term inundation extent dataset from multiple-satellite observations: towards high spatial resolution, *Mapping Water Bodies from Space (MWBS)*, 18–19 March 2015, Frascacti, Italy, (2015).

61 X. Che, M. Feng, H. Jiang, J. Song and B. Jia, Downscaling MODIS Surface Reflectance to Improve Water Body Extraction, *Advances in Meteorology*, (2015).

62 CEOS, *CEOS EO Handbook – CEOS mission, instruments and measurements database online*, Available: http://database.eohandbook.com/; Accessed: January 2017.

63 E.R. De Roeck, N.E. Verhoest, M.H. Miya, H. Lievens, O. Batelaan, A. Thomas and L. Brendonck, Remote sensing and wetland ecology: a South African case study, *Sensors*, **8**, 3542–3556, (2008).

64 C. Hüttich, J. Reschke, M. Keil, S. Dech, K. Weise, C. Beltrame, E. Fitoka and M. Paganini, *Using the Landsat Archive for the Monitoring of Mediterranean Coastal Wetlands: Examples from the Glob-Wetland-II Project*, Available: https://earthzine.org/2011/12/20/using-the-landsat-archive-for-the-monitoring-of-mediterranean-coastal-wetlands-examples-from-the-globwetland-ii-project/; Accessed: January 2017.

65 I. Dronova, P. Gong, N.E. Clinton, L. Wang, W. Fu, S. Qi and Y. Liu, Landscape analysis of wetland plant functional types: The effects of image segmentation scale, vegetation classes and classification methods, *Remote Sensing of Environment*, **127**, 357–369, (2012).

66 A. Sanchez, D.A. Malak, A. Guelmami and C. Perennou, Development of an Indicator to Monitor Mediterranean Wetlands, *PLOS ONE*, **10**, (2015).

67 I. Dronova, P. Gong and L. Wang, Object-based analysis and change detection of major wetland cover types and their classification uncertainty during the low water period at Poyang Lake, China, *Remote Sensing of Environment*, **115**, 3220–3236, (2011).

68 L. Wang, I. Dronova, P. Gong, W. Yang, Y. Li and Q. Liu, A new time series vegetation – water index of phenological – hydrological trait across species and functional types for Poyang Lake wetland ecosystem, *Remote Sensing of Environment*, **125**, 49–63, (2012).

69 P. Maillard, M.O. Pivari and C.H.P. Luis, Remote Sensing for Mapping and Monitoring Wetlands and Small Lakes in Southeast Brazil, *Remote Sensing of Planet Earth. Croatia, Intechweb. org*, 23–46, (2011).

70 A. Davranche, G. Lefebvre and B. Poulin, Wetland monitoring using classification trees and SPOT-5 seasonal time series, *Remote Sensing of Environment*, **114**, 552–562, (2010).

71 R.P. Powers, G.J. Hay and G. Chen, How wetland type and area differ through scale: A GEOBIA case study in Alberta's Boreal Plains, *Remote Sensing of Environment*, **117**, 135–145, (2012).

72 C. Baker, R.L. Lawrence, C. Montagne and D. Patten, Change detection of wetland ecosystems using Landsat imagery and change vector analysis, *Wetlands*, **27**, 610–619, (2007).

73 M. Halabisky, L.M. Moskal and S.A. Hall, Object-based classification of semi-arid wetlands, *Journal of Applied Remote Sensing*, **5**, (05351), p.13, (2011).

74 T. Whiteside, R. Bartolo and G. Staben, A rule-based approach to segment and classify floodplain vegetation from WorldView-2 imagery, in *16th Australasian Remote Sensing and Photogrammetry Conference*, 2012.

75 K.B. Moffett and S.M. Gorelick, Distinguishing wetland vegetation and channel features with object-based image segmentation, *International Journal of Remote Sensing*, **34**, 1332–1354, (2013).

76 M. Grippa, L. Gal, E. Robert, L. Kergoat, P. Hiernaux, É. Mougin and J.-M. Martinez, The temporal mapping of water bodies from the perspective of climate modelling, *Mapping Water Bodies from Space (MWBS)*, Frascati, Italy, (2015).

77 X. Na, S. Zang, Y. Zhang and L. Liu, Wetland mapping and flood extent monitoring using optical and radar remotely sensed data and ancillary topographical data in the Zhalong National Natural Reserve, China, in *SPIE Remote Sensing*, International Society for Optics and Photonics, 88931M–88931M-7, (2013).

78 I. Dronova, P. Gong, L. Wang and L. Zhong, Mapping dynamic cover types in a large seasonally flooded wetland using extended principal component analysis and object-based classification, *Remote Sensing of Environment*, **158**, 193–206, (2015).

79 B. Smets, R. D'Andrimont and P. Claes, *GIO Global Land Component – Lot I "Operation of the Global Land Component". Product User Manual, Area of Water Bodies – Version 1*, geoland2, (2011).

80 T. Landmann, M. Schramm, C. Huettich and S. Dech, MODIS-based change vector analysis for assessing wetland dynamics in Southern Africa, *Remote Sensing Letters*, **4**, 104–113, (2013).

81 B. Fichtelmann, K.P. Guenther and E. Borg, Adaption of a self-learning algorithm for dynamic classification of water bodies to spot vegetation Data, in *Computational Science and Its Applications – ICCSA 2015*, O. Gervasi *et al.* (Eds), Springer, (2015).

82 I. Klein, A.J. Dietz, U. Gessner, A. Galayeva, A. Myrzakhmetov and C. Kuenzer, Evaluation of seasonal water body extents in Central Asia over the past 27 years derived from medium-resolution remote sensing data, *International Journal of Applied Earth Observation and Geoinformation*, **26**, 335–349, (2014).

83 I. Klein, A. Dietz, U. Gessner, S. Dech and C. Kuenzer, Results of the Global WaterPack: a novel product to assess inland water body dynamics on a daily basis, *Remote Sensing Letters*, **6**, 78–87, (2015).

84 L. Moser, S. Voigt and E. Schoepfer, Monitoring of critical water and vegetation anomalies of sub-Saharan west-African wetlands, in *International Geoscience and Remote Sensing Symposium (IGARSS 2014)*, Québec, Canada, (2014).

85 D.O. Fuller and Y. Wang, Recent trends in satellite vegetation index observations indicate decreasing vegetation biomass in the southeastern saline everglades wetlands, *Wetlands*, **34**, 67–77, (2014).

86 T. Sakamoto, N. Van Nguyen, A. Kotera, H. Ohno, N. Ishitsuka and M. Yokozawa, Detecting temporal changes in the extent of annual flooding within the Cambodia and the Vietnamese Mekong Delta from MODIS time-series imagery, *Remote Sensing of Environment*, **109**, 295–313, (2007).

87 Y.-E. Yan, Z.-T. Ouyang, H.-Q. Guo, S.-S. Jin and B. Zhao, Detecting the spatiotemporal changes of tidal flood in the estuarine wetland by using MODIS time series data, *Journal of Hydrology*, **384**, 156–163, (2010).

88 B. Gouweleeuw, C. Ticehurst, P. Dyce, J. Guerschmana, A.I. Van Dijk and P. Thew, An experimental satellite-based flood monitoring system for Southern Queensland, Australia, in *IRSE 2011 Symposium, Sydney*, 10–15, Sydney, Australia, (2011).

89 C. Ticehurst, J.P. Guerschman and Y. Chen, The strengths and limitations in using the daily MODIS open water likelihood algorithm for identifying flood events, *Remote Sensing*, **6**, 11791–11809, (2014).

90 C. Hu, A novel ocean color index to detect floating algae in the global oceans, *Remote Sensing of Environment*, **113**, 2118–2129, (2009).

91 L. Feng, C. Hu, X. Chen, X. Cai, L. Tian and W. Gan, Assessment of inundation changes of Poyang Lake using MODIS observations between 2000 and 2010, *Remote Sensing of Environment*, **121**, 80–92, (2012).

92 B. Fichtelmann and E. Borg, A New Self-Learning Algorithm for Dynamic Classification of Water Bodies, in *Computational Science and Its Applications – ICCSA 2012, Part III, LNCS 7335*, B. Murgante *et al.* (Eds), Springer-Verlag Berlin Heidelberg, Salvador de Bahia, Brazil, (2012).

93 NASA JPL, *Shuttle radar topography mission water body data set (SWBD)*, Available: www2.jpl.nasa.gov/srtm/; November 2016.

94 M. Carroll, J.R. Townshend, C.M. Dimiceli, P. Noojipady and R. Sohlberg, A new global raster water mask at 250 m resolution, *International Journal of Digital Earth*, **2**, 291–308, (2009).

95 A. Bartsch, W. Wagner, K. Scipal, C. Pathe, D. Sabel and P. Wolski, Global monitoring of wetlands – the value of ENVISAT ASAR Global mode, *Journal of Environmental Management*, **90**, 2226–2233, (2009).

96 J. Reschke, A. Bartsch, S. Schlaffer and D. Schepaschenko, Capability of C-Band SAR for operational wetland monitoring at high latitudes, *Remote Sensing*, **4**, 2923–2943, (2012).

97 V. Gstaiger, J. Huth, S. Gebhardt, T. Wehrmann and C. Kuenzer, Multi-sensoral and automated derivation of inundated areas using TerraSAR-X and ENVISAT ASAR data, *International Journal of Remote Sensing*, **33**, 7291–7304, (2012).

98 C. Kuenzer, H. Guo, J. Huth, P. Leinenkugel, X. Li and S. Dech, Flood mapping and flood dynamics of the Mekong delta: Envisat-ASAR-WSM based time series analyses, *Remote Sensing*, **5**, 687–715, (2013).

99 M. Santoro and U. Wegmuller, Multi-temporal synthetic aperture radar metrics applied to map open water bodies, *IEEE Journal of Selected Topics in Applied Earth Observations and Remote Sensing (J-STARS)*, **7**, 3225–3238, (2014).

100 U. Gessner, M. Machwitz, T. Esch, A. Tillack, V. Naeimi, C. Kuenzer and S. Dech, Multi-sensor mapping of West African land cover using MODIS, ASAR and TanDEM-X/TerraSAR-X data, *Remote Sensing of Environment*, (2015).

101 J. Karvonen, M. Simila and M. Makynen, Open water detection from Baltic Sea ice Radarsat-1 SAR imagery, *Geoscience and Remote Sensing Letters, IEEE*, **2**, 275–279, (2005).

102 B. Brisco, N. Short, J.V.D. Sanden, R. Landry and D. Raymond, A semi-automated tool for surface water mapping with RADARSAT-1, *Canadian Journal of Remote Sensing*, **35**, 336–344, (2009).

103 S. Martinis, A. Twele and S. Voigt, Towards operational near real-time flood detection using a split-based automatic thresholding procedure on high resolution TerraSAR-X data, *Natural Hazards and Earth System Science*, **9**, 303–314, (2009).

104 T. Hahmann, S. Martinis, A. Twele and M. Buchroithner, Strategies for the automatic mapping of flooded areas and other water bodies from high resolution TerraSAR-X data, *Cartography and Geoinformatics for Early Warning and Emergency Management: Towards Better Solutions*, Masaryk University, Brno, Czech Republic, 2009.

105 S. Martinis, A. Twele, C. Strobl, J. Kersten and E. Stein, A Multi-Scale Flood Monitoring System Based on Fully Automatic MODIS and TerraSAR-X Processing Chains, *Remote Sensing*, **5**, 5598–5619, (2013).

106 A. Wendleder, B. Wessel, A. Roth, M. Breunig, K. Martin and S. Wagenbrenner, TanDEM-X water indication mask: Generation and first evaluation results, *Journal of Selected Topics in Applied Earth Observations and Remote Sensing*, **6**, 171–179, (2013).

107 S. Schlaffer, P. Matgen, M. Hollaus and W. Wagner, Flood detection from multi-temporal SAR data using harmonic analysis and change detection, *International Journal of Applied Earth Observation and Geoinformation*, **38**, 15–24, (2015).

108 C. Kuenzer, J. Huth, S. Martinis, L. Lu and S. Dech, SAR time series for the analysis of inundation patterns in the Yellow River Delta, China, in *Remote Sensing Time Series*, C. Kuenzer, S. Dech and W. Wagner (Eds), Springer, 2015.

109 E. Kasischke and L. Bourgeau-Chavez, Monitoring South Florida wetlands using ERS-1 SAR imagery, *Photogrammetric Engineering & Remote Sensing*, **63**, 281–291, (1997).

110 R. Touzi, A. Deschamps and G. Rother, Wetland characterization using polarimetric RADARSAT-2 capability, *Canadian Journal of Remote Sensing*, **33**, 56–67, (2007).

111 F.M. Henderson and A.J. Lewis, Radar detection of wetland ecosystems: a review, *International Journal of Remote Sensing*, **29**, 5809–5835, (2008).

112 P. Patel, H.S. Srivastava and R.R. Navalgund, Use of synthetic aperture radar polarimetry to characterize wetland targets of Keoladeo National Park, Bharatpur, India, *Current Science*, **97**, 529–537, (2009).

113 A.K. Milne and I.J. Tapley, *Wetland monitoring of flood-extent, inundation patterns and vegetation, Mekong River Basin, Southeast Asia, and Murray-Darling Basin, Australia*, ALOS Kyoto & Carbon Initiative, Science Team Reports, Phase 1, Japanese Aerospace Exploration Agency, Japan, (2010).

114 B. Brisco, A. Schmitt, K. Murnaghan, S. Kaya and A. Roth, SAR polarimetric change detection for flooded vegetation, *International Journal of Digital Earth*, **6**, 103–114, (2011).

115 A. Schmitt and B. Brisco, Wetland monitoring using the curvelet-based change detection method on polarimetric SAR imagery, *Water*, **5**, 1036–1051, (2013).

116 A.L. Gallant, S.G. Kaya, L. White, B. Brisco, M.F. Roth, W. Sadinski and J. Rover, Detecting emergence, growth, and senescence of wetland vegetation with polarimetric Synthetic Aperture Radar (SAR) data, *Water*, **6**, 694–722, (2014).

117 J. Betbeder, V. Gond, F. Frappart, N.N. Baghdadi, G. Briant and E. Bartholome, Mapping of Central Africa forested wetlands using remote sensing, *Journal of Selected Topics in Applied Earth Observations and Remote Sensing (J-STARS)*, **7**, (2014).

118 L. White, B. Brisco, M. Dabboor, A. Schmitt and A. Pratt, A Collection of SAR Methodologies for Monitoring Wetlands, *Remote Sensing*, **7**, 7615–7645, (2015).

119 S. Wdowinski, S.-W. Kim, F. Amelung, T.H. Dixon, F. Miralles-Wilhelm and R. Sonenshein, Space-based detection of wetlands' surface water level changes from L-band SAR interferometry, *Remote Sensing of Environment*, **112**, 681–696, (2008).

120 S.-H. Hong, S. Wdowinski and S.-W. Kim, Evaluation of TerraSAR-X observations for wetland InSAR application, *IEEE Transactions on Geoscience and Remote Sensing*, **48**, 864–873, (2010).

121 X. Chou, S. Yun and W. Zi, InSAR analysis over Yellow River Delta for mapping water-level changes over wetland, in *18th International Conference on Geoinformatics*, IEEE, 1–5, (2010).

122 X. Chou, S. Yun, X. Ji, W. Zi and F. Liang, Analysis of ALOS PALSAR InSAR data for mapping water level changes in Yellow River Delta wetlands, *International Journal of Remote Sensing*, **34**, 2047–2056, (2013).

123 L.L. Bourgeau-Chavez, K. Riordan, R.B. Powell, N. Miller and M. Nowels, Improving wetland characterization with multi-sensor, multi-temporal SAR and optical/infrared data fusion, in *Advances in Geoscience and Remote Sensing*, G. Jedlovec (Eds) InTech, 2009.

124 C. Kuenzer, A. Bluemel, S. Gebhardt, T.V. Quoc and S. Dech, Remote sensing of mangrove ecosystems: A review, *Remote Sensing*, **3**, 878–928, (2011).

125 A. Schmitt, A. Wendleder and S. Hinz, The Kennaugh element framework for multi-scale, multi-polarized, multi-temporal and multi-frequency SAR image preparation, *ISPRS Journal of Photogrammetry and Remote Sensing*, **102**, 122–139, (2015).

126 R. Ouedraogo, *Fish and fisheries prospective in arid inland waters of Burkina Faso, West Africa*, PhD, University of Natural Resources and Life Sciences, 2010.

127 T. Krauss, P. D'Angelo, M. Schneider and V. Gstaiger, The fully automatic optical processing system CATENA at DLR, *International Archives of the Photogrammetry, Remote Sensing and Spatial Information Sciences*, **1**, W1, (2013).

128 R. Richter, A spatially adaptive fast atmospheric correction algorithm, *International Journal of Remote Sensing*, **17**, 1201–1214, (1996).

129 R. Richter and D. Schläpfer, *Atmospheric/Topographic Correction for Satellite Imagery (ATCOR-2/3 User Guide, Version 8.3.1, February 2014)*, ReSe Applications Schläpfer, Langeggweg 3, CH-9500 Wil, Switzerland, 2014.

130 R. Müller, T. Krauss, M. Schneider and P. Reinartz, Automated georeferencing of optical satellite data with integrated sensor model improvement, *Photogrammetric Engineering & Remote Sensing*, **78**, 61–74, (2012).

131 E. Vermote, S. Kotchenova and J. Ray, *MODIS Surface Reflectance User's Guide*, MODIS Land Surface Reflectance Science Computing Facility, 2011.

132 M. Berger, J. Moreno, J.A. Johannessen, P.F. Levelt and R.F. Hanssen, ESA's sentinel missions in support of Earth system science, *Remote Sensing of Environment*, **120**, 84–90, (2012).

133 M.E. Brown, J.E. Pinzón, K. Didan, J.T. Morisette and C.J. Tucker, Evaluation of the consistency of long-term NDVI time series derived from AVHRR, SPOT-Vegetation, SeaWiFS, MODIS, and Landsat ETM+ sensors, *IEEE Transactions on Geoscience and Remote Sensing*, **44**, 1787–1793, (2006).

134 J. Pedelty, S. Devadiga, E. Masuoka, M. Brown, J. Pinzon, C. Tucker, D. Roy, J. Ju, E. Vermote and S. Prince, Generating a long-term land data record from the AVHRR and MODIS instruments, in *IEEE International Geoscience and Remote Sensing Symposium 2007 (IGARSS)*, IEEE, 1021–1025, Barcelona, Spain, 2007.

135 T.F. Lee, S.D. Miller, C. Schueler and S. Miller, NASA MODIS Previews NPOESS VIIRS Capabilities, *Weather and forecasting*, **21**, 649–655, (2006).

136 C.J. Tucker and K.A. Yager, Ten years of MODIS in space: Lessons learned and future perspectives, *Italian Journal of Remote Sensing*, **43**, 7–18, (2011).

137 E. Vermote, C. Justice and I. Csiszar, Early evaluation of the VIIRS calibration, cloud mask and surface reflectance Earth data records, *Remote Sensing of Environment*, **148**, 134–145, (2014).

138 C. Huang, Y. Chen, J. Wu, L. Li and R. Liu, An evaluation of Suomi NPP-VIIRS data for surface water detection, *Remote Sensing Letters*, **6**, 155–164, (2015).

139 C. Donlon, B. Berruti, A. Buongiorno, M.-H. Ferreira, P. Féménias, J. Frerick, P. Goryl, U. Klein, H. Laur and C. Mavrocordatos, The global monitoring for environment and security (GMES) sentinel-3 mission, *Remote Sensing of Environment*, **120**, 37–57, (2012).

140 European Space Agency (ESA), *ESA Earthnet Online, ESA Earth Observation Missions – Overview, Copernicus – Sentinels*, Available: https://earth.esa.int/web/guest/missions/esa-eo-missions; Accessed: January 2017.

141 W. Dierckx, S. Sterckx, I. Benhadj, S. Livens, G. Duhoux, T. Van Achteren, M. Francois, K. Mellab and G. Saint, PROBA-V mission for global vegetation monitoring: standard products and image quality, *International Journal of Remote Sensing*, **35**, 2589–2614, (2014).

142 S. Sterckx, I. Benhadj, G. Duhoux, S. Livens, W. Dierckx, E. Goor, S. Adriaensen, W. Heyns, K. Van Hoof and G. Strackx, The PROBA-V mission: image processing and calibration, *International Journal of Remote Sensing*, **35**, 2565–2588, (2014).

143 M. Drusch, U. Del Bello, S. Carlier, O. Colin, V. Fernandez, F. Gascon, B. Hoersch, C. Isola, P. Laberinti and P. Martimort, Sentinel-2: ESA's optical high-resolution mission for GMES operational services, *Remote Sensing of Environment*, **120**, 25–36, (2012).

144 USGS, *Landsat 8*, Available: https://landsat.usgs.gov/landsat-8; Accessed: January 2017.

145 R. Torres, P. Snoeij, D. Geudtner, D. Bibby, M. Davidson, E. Attema, P. Potin, B. Rommen, N. Floury and M. Brown, GMES Sentinel-1 mission, *Remote Sensing of Environment*, **120**, 9–24, (2012).

146 L. Moser, A. Schmitt, A. Wendleder and A. Roth, Monitoring of the Lac Bam Wetland Extent Using Dual-Polarized X-Band SAR Data, *Remote Sensing*, **8**(4), 302, (2016).

147 L. Moser, A. Schmitt and A. Wendleder, Automated Wetland delineation from multi-frequency and multi-polarized SAR images in high temporal and spatial resolution, ISPRS Annals of the Photogrammetry, Remote Sensing and Spatial Information Sciences, **III**-8, 57–64, (2016).

148 M. Srinivasan, A. Andral, M. Dejus, F. Hossain, C. Peterson, E. Beighley, T. Pavelsky, Y. Chao, B. Doorn and E. Bronner, Engaging the applications community of the future surface water and ocean topography (SWOT) mission, *International Archives of the Photogrammetry, Remote Sensing & Spatial Information Sciences*, (2015).

149 C. Mbow, M. Brandt, I. Ouedraogo, J. De Leeuw and M. Marshall, What Four Decades of Earth Observation Tell Us about Land Degradation in the Sahel? *Remote Sensing*, 7, 4048–4067, (2015).

150 A.K. Mishra and V.P. Singh, A review of drought concepts, *Journal of Hydrology*, **391**, 202–216, (2010).

151 R.P. Singh, S. Roy and F. Kogan, Vegetation and temperature condition indices from NOAA AVHRR data for drought monitoring over India, *International Journal of Remote Sensing*, **24**, 4393–4402, (2003).

152 Z. Wan, P. Wang and X. Li, Using MODIS land surface temperature and normalized difference vegetation index products for monitoring drought in the southern Great Plains, USA, *International Journal of Remote Sensing*, **25**, 61–72, (2004).

153 Z. Balint, F. Mutua and P. Muchiri, *Drought Monitoring with the Combined Drought Index*, FAO-SWALIM, Nairobi, Kenya, 2011.

15

Satellite Derived Information for Drought Detection and Estimation of Water Balance

M. Ofwono[1,2], K. Dabrowska-Zielinska[1], J. Kaduk[2] and V. Nicolás-Perea[2]

[1] Institute of Geodesy and Cartography, Warsaw, Poland
[2] University of Leicester, Centre for Landscape and Climate Research, Department of Geography, Leicester, UK

15.1 Introduction

Drought is a complex natural disaster that causes serious environmental, social, and economic consequences worldwide. It impacts both surface and groundwater resources and can lead to reduced water supply, deteriorated water quality, crop failure, reduced range productivity, diminished power generation, disturbed riparian habitats, and suspended recreation activities, as well as affecting a host of economic and social activities [1]. Droughts occur in virtually all climatic zones and are mostly related to the reduction in the amount of precipitation received over an extended period of time. However, semi-arid regions are especially susceptible to drought conditions because of their low annual precipitation and sensitivity to climate changes.

In recent years, large-scale intensive droughts have been observed on all continents, affecting large areas in Europe, Africa, Asia, Australia, South America, Central America, and North America [2,3]. Demuth and Stahl [3] observed that the drought situation in many European regions has already become more severe and that northern to north-eastern Europe is most prone to a rise in flood frequencies while southern and south-eastern Europe shows significant increases in drought frequencies. Average precipitation is expected to increase and its variability is expected for northern European regions, suggesting higher flood risks, while less rainfall, prolonged dry spells and increased evapotranspiration may increase the frequency of droughts in Southern Europe [4,5].

Over the past 30 years, Europe has been affected by a number of major drought events, most notably in 1976 (northern and western Europe), 1989 (most of Europe), 1991 (most of Europe), and more recently, the prolonged drought over large parts of Europe associated with the summer heat wave in 2003 [6]. The most serious drought in the Iberian Peninsula in 60 years occurred in 2005, reducing overall European Union (EU) cereal yields by an estimated ten percent [7].

Earth Observation for Land and Emergency Monitoring, First Edition. Edited by Heiko Balzter.
© 2017 John Wiley & Sons Ltd. Published 2017 by John Wiley & Sons Ltd.

15.2   Literature Review

In an attempt to define drought, Wilhite and Glantz [8] classified drought into four different categories: meteorological, agricultural, hydrological, and socioeconomic droughts. All four categories have varying descriptions of drought impacts, but revolve around a central theme that drought is a "period of abnormally dry weather sufficiently prolonged for the lack of precipitation to cause a serious hydrological imbalance".

Meteorological drought is defined as a lack of precipitation over a region for a given period of time. The complex meteorological drought phenomenon can be simplified into a drought index, such as the Standardized Precipitation Index (SPI), which is a single number assimilating a large amount of water supply data. The index allows the quantification of climate anomalies in terms of intensity, duration, and spatial extent [9]. Several studies have analysed drought using monthly precipitation data with respect to average values, while other studies have analysed drought duration and intensity in relation to cumulative precipitation deficits [10–12]. However, calculation of meteorological drought indices requires a long record of climatic data, which may not be available because of the inaccessibility of a region and or a lack of human activity. This is where long-term satellite remote sensing of climatic and land cover variables are instrumental in determining drought characteristics.

This study focuses on agricultural drought, which is defined as a combination of short-term precipitation shortages and increased evapotranspiration (ET) demands from high-temperature anomalies that lead to adverse agricultural impacts [8]. Agricultural drought usually refers to a period with declining soil moisture and consequently crop failure without any reference to surface water resources. A decline of soil moisture depends on several factors, which affect meteorological and hydrological droughts along with differences between actual evapotranspiration and potential evapotranspiration. Plant water demand depends on prevailing weather conditions, biological characteristics of the specific plant and stage of growth, and the physical and biological properties of soil. Water stress develops in crops when the evaporative demand exceeds the supply of water from the soil. As a result, plant water status declines and that affects physiological processes, such as leaf expansion and other growth processes. Most crops are very sensitive to water deficits, and their yield is negatively affected even by short-term water deficits [9,13,14]. Several drought indices, based on a combination of precipitation, temperature and soil moisture, have been derived to study agricultural droughts [15].

A drought index is a prime variable integrating various hydrological and meteorological parameters such as rainfall, evapotranspiration (ET), runoff and other water supply indicators into a single number for assessing the effect of drought and defining different drought parameters, i.e. duration, intensity and severity, which provides a comprehensive picture for decision-making [15]. A drought index should be able to quantify the drought for different time scales, of which a long time series is essential. Annual analysis is the widely used time scale for drought, followed by a monthly basis. Although the yearly time scale is long, it can also be used to abstract information on the regional behaviour of droughts. However, the monthly time scale is more appropriate for agricultural drought monitoring, water supply and ground water abstractions [16].

Several drought indices have been derived in recent decades to quantify drought. Among the various drought indices are: the Palmer Drought Severity Index (PDSI) [17]; Crop Moisture Index (CMI) [18]; Standardized Precipitation Index (SPI) [19]; Keetch–Byram Drought Index [20]; Soil moisture drought index (SMDI) [21]; Surface Water Supply Index (SWSI) [22]; Crop Specific Drought Index (CSDI) [23]; and Vegetation Condition Index (VCI) [24]. These indices are used extensively for water resources management, agricultural drought monitoring and forecasting [25–27]. Each of these drought indices have their strengths and limitations, some of which are briefly explained below.

15.2.1   Palmer Drought Severity Index (PDSI) and Crop Moisture Index (CMI)

One of the most widely used drought index is the PDSI [17]. PDSI is primarily a meteorological drought index formulated to evaluate prolonged periods of both abnormally wet and abnormally dry weather conditions. The PDSI has been used to investigate spatial and temporal characteristics of drought e.g. [28–30], crop forecasting, assessing potential fire severity and drought forecasting [31,32]. Guttman *et al.* [33] reported that PDSI depicted spatially does not identify areas of equal hydrologic drought intensity owing to the variability of precipitation and groundwater levels over large areas. Hayes *et al.* [34] state that PDSI responds slowly to detect short-term dry spells which could be deteriorating during critical stages of crop growth.

Palmer [18] developed the Crop Moisture Index (CMI) as an index for short-term agricultural drought from procedures within the calculation of the PDSI using weekly precipitation and temperature to derive a simple moisture budget. PDSI is calculated from precipitation deficits for monitoring long-term drought conditions, whereas CMI is calculated from evapotranspiration deficits for monitoring short-term agricultural drought conditions that affect crop growth.

PDSI and CMI have similar limitations in that both models assume that parameters such as land use/land cover, and soil properties are uniform over the entire climatic zone ($7000–100,000 \, \text{km}^2$); it assumes that all precipitation is rain. Hence snow is not accounted for. It underestimates runoff by assuming that runoff occurs when the soil is saturated. However, in reality, parameters such as land use/land cover and soil properties vary widely. In addition to the large spatial lumping of physical parameters, several studies have highlighted various limitations of PDSI and CMI [35]. For example in PDSI, potential evapotranspiration (ET_p) is calculated using Thornthwaite's method. Thornthwaite's equation for estimating ET is based on an empirical relationship between ET and temperature [36].

Jensen *et al.*[37] evaluated and ranked different methods of estimating ET under various climatic conditions and concluded that the poorest performing method overall was the Thornthwaite equation. Leading to uncertainties in the PDSI and hence, the need for a robust method for deriving ET. Moreover, the water balance model used by Palmer [17] is a two-layer lumped parameter model. Palmer (1965) assumed an average water holding capacity of the top two soil layers for the entire region in a climatic division ($7000–100,000 \, \text{km}^2$). However, in reality, soil properties vary widely on a much smaller scale. The rainfall is also highly variable spatially in arid and semi-arid climatic zones. This often makes it difficult to spatially delineate the areas affected by localized drought events. Further, PDSI and CMI do not account for the effect of land use/land cover on the water balance. Palmer [17] assumed runoff occurs when the top two soil layers become completely saturated. However, runoff depends on soil type, land use, and management practices which are not accounted for in the PDSI of Palmer [17].

15.2.2 Standardized Precipitation Index (SPI)

McKee *et al.* [19] state that Standardized Precipitation Index (SPI) is primarily a meteorological drought index based on long term precipitation records for a given period of time. In calculating the SPI, the long-term precipitation record is fitted to the Gamma probability distribution. The Gamma distribution is then transformed to a Gaussian (normal) distribution (standard normal distribution with mean zero and variance of one), which gives the value of the SPI for the time scale used. Unlike PDSI and CMI, SPI takes into account the stochastic nature of the drought. In addition, the SPI can be calculated for a variety of time scales ranging from 3, 6, 9, 12, 24 to 48 months [15]. This versatility therefore allows SPI to monitor short- and long-term meteorological drought. However, SPI does not account for the effect of soil, land use characteristic, crop growth, and temperature anomalies which are critical for agricultural drought monitoring, even though it has the capacity to monitor short-term meteorological drought.

15.2.3 Surface Water Supply Index (SWSI)

The Surface Water Supply Index (SWSI) developed by Shafer and Dezman [22] was primarily developed as a hydrological drought index with an intention to replace PDSI for areas where local precipitation is not the sole or primary source of water. The SWSI is calculated based on monthly non-exceedance probability from available historical records of reservoir storage, stream flow, snow pack, and precipitation. The purpose of SWSI is primarily to monitor the abnormalities in surface water supply sources. Hence, it is a good measure to monitor the impact of hydrologic drought on urban and industrial water supplies, irrigation and hydroelectric power generation. However, there is a time lag before precipitation deficiencies are detected in surface and subsurface water sources. As a result, the hydrological drought is out of phase from agricultural drought. Because of this phase difference, SWSI is not a suitable indicator for agricultural drought.

15.2.4 Crop Specific Drought Index (CSDI) and Vegetation Condition Index (VCI)

A comprehensive crop-specific drought index (CSDI) for corn [23,38] was developed by taking into account the water use during specific periods of crop growth, using a simple water balance model, at the spatial scale of a crop reporting district. Wilhelmi *et al.* [39] developed a GIS based agricultural drought vulnerability assessment method for the state of Nebraska and found the seasonal crop moisture deficiency to be a useful measure for spatial characterization of the state's agroclimatology. Satellite remote sensing has proved to be a valuable source of timely and spatially continuous data, which has improved information on the monitoring of vegetation dynamics over large areas [40,41]. The Vegetation Condition Index (VCI) derived from AVHRR visible and near infrared bands adjusted for land, climate, ecology and weather conditions has shown promise in drought detection and monitoring [42,43]. The VCI enables the quantification of drought in terms of onset, intensity and duration. Unfortunately, the VCI is not reliable in winter seasons when the vegetation is largely dormant [44].

Agricultural crops are sensitive to soil moisture. The soil moisture deficit in the root zone during various stages of the crop growth cycle will have a profound impact on crop yield. For example, a 10% water deficit during the tasselling, pollination stage of corn could reduce the yield by as much as 25% [45]. Hence, the development of a reliable drought index for agriculture requires proper consideration of vegetation type, crop growth and root development, soil properties, antecedent soil moisture condition, evapotranspiration, and temperature. The existing drought indices currently used for drought monitoring do not give proper consideration to the aforementioned variables. Hence, a better tool for agricultural drought monitoring is essential for the farming community and the decision-makers.

As precipitation and soil properties have a high degree of spatial variability, an approach that is fundamentally linked to soil moisture and its measurement in the vadose zone, can only serve to enhance our ability to quantify local drought, especially during the crop growing season, for the obvious linkages with evapotranspiration, precipitation, and deep soil moisture [46,48].

15.3 Data

The soil moisture used in this study was derived from the **S**pecial **S**ensor **M**icrowave **I**mager (SSMI). The SSMI has been measuring brightness temperature since 1987. It is on board the **D**efense **M**eteorological **S**atellite **P**rogram (DMSP) and it has seven-channels, four-frequencies, and it is an orthogonally polarized passive microwave radiometer. The DMSP has a series of satellites which include; the F8 Special Sensor Microwave Imager (9 July 1987, through 18 December 1991), the F11 SSMI (3 December 1991, through 12 January 1998), and the F13 SSMI (3 May 1995, through 31 December 2007) [49].

Figure 15.1 shows global soil moisture retrieved from the SSMI brightness temperature using an iterative algorithm, which minimizes the differences between the satellites measured brightness temperature and the modelled brightness temperature obtained through the inversion of the tau-omega model, which accounts for the influence of vegetation on emitted microwave radiation. The empirical model of Wang and Choudhury [50] was used to correct land surface emissivity for surface roughness, because it accounts for depolarization caused by scattering of radiation between different surfaces.

15.4 Methodology

Given that the potential of crops to extract water from deeper soil layers varies during different growth stages and by crop type, and given that satellite derived soil moisture is approximated for the top 5 cm of the soil, the soil moisture deficit index (SMDI) is computed at the root zone which for this study is assumed to be 30 cm. The root zone soil moisture is derived using the recursive exponential filter for time sensitive data proposed by Stroud [51] and adopted by Albergel *et al.* [52]. The recursive equation for the redistribution of surface soil moisture to deeper soil layers is written as:

$$SWI_m = SWI_{m(n-1)} + K_n \left(ms(t_n) - SWI_{m(n-1)} \right) \tag{15.1}$$

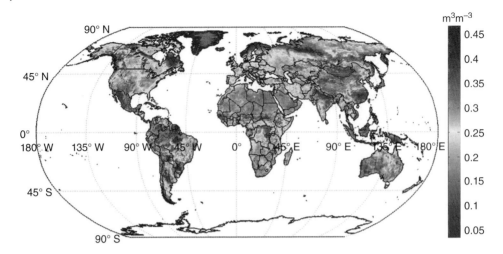

Figure 15.1 SSMI derived Global soil moisture averaged over the first 3 days of May 2010.

Where SWI_m is the soil water index, ms is the satellite derived soil moisture at time t_n and the gain K_n at a time t_n is given by:

$$K_n = \frac{1}{1 + \sum_i^{n-1} e^{-\frac{(t_n - t_i)}{T}}}$$
(15.2)

In recursive format, the gain is written as:

$$k_n = \frac{k_{n-1}}{k_{n-1} + e^{-\frac{(t_n - t_{n-1})}{T}}}$$
(15.3)

The parameter T represents a characteristic time length. This parameter can be considered as a surrogate parameter for all the processes affecting the temporal dynamics of soil moisture, such as the thickness of the soil layer, soil hydraulic properties, evaporation, run-off and vertical gradient of soil properties (texture, density). T represents the time scale of soil moisture variation, in units of days. The SWI_m at time t_n is calculated if there is at least one measurement in the time interval $[t_n - T, t_n]$ and at least four measurements in the interval $[t_n - 3T, t_n]$ [53].

The range of the gain K is [0, 1]. In the presence of extensive temporal data gaps (relative to the filter time scale), Equation 15.3 tends toward unity. In that particular case, the previous estimates are disregarded when new observations are obtained and the new estimate takes on the value of the new observation. For the initialization of the filter, $K_1 = 1$ and $SWIm(1) = ms(t_1)$. This recursive formulation is preferred over the original exponential filter developed by Wagner, [54] as the only requirement for an update of the SWI_m is (apart from the previous SWIm and K values) the availability of a new $ms(t_n)$ observation and the time interval since the last observation $(t_n - t_{n-1})$. The resultant soil moisture is shown in Figure 15.2.

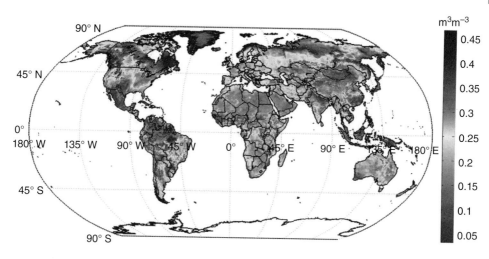

Figure 15.2 SSMI derived global soil moisture, averaged over the first 3 days of May 2010 and scaled to 30 cm soil depth.

The soil moisture deficit index (SMDI) as a measure of drought characteristic is based on the long-term median, long-term maximum and long-term minimum soil moisture values of a given pixel. The SSM/I daily soil moisture output is averaged over a 10-day period to get decadal soil water for each of the 37 decades in a year for each pixel. The long-term soil moisture statistics for each pixel is obtained by stacking the 25 years of daily images (1987–2012) and then calculating the desired statistics. The median is chosen over the mean as a measure of "normal" soil moisture because median is more stable and is not significantly influenced by few outliers. Using this long-term median, maximum and minimum soil moisture values, decadal percentage soil moisture deficit or excess soil water over the 25 years is calculated as:

$$SD_{i,j} = \begin{cases} \dfrac{SW_{i,j} - MSW_j}{MSW_j - \min SW_j} *100 & if \quad SW_{i,j} < MSW_j \\[4mm] \dfrac{SW_{i,j} - MSW_j}{\max SW_j - MSW_j} *100 & if \quad SW_{i,j} > MSW_j \end{cases} \tag{15.4}$$

Where $SD_{i,j}$ is the soil water deficit (%), $SW_{i,j}$ the mean decadal soil water available in the soil profile (mm), MSW_j the long-term median available soil water in the soil profile (mm), max.SW_j the long-term maximum available soil water in the soil profile (mm), and min.SW_j is the long-term minimum available soil water in the soil profile (mm) ($i = 1987–2012$ and $j = 1–37$ decades). By using Equation 15.4, the seasonality inherent in soil moisture was removed. Hence, the deficit values can be compared across seasons. The SD values during a decade range from −100 to +100 indicating very dry to very wet conditions. As the SD values for all the pixels are scaled between −100 and +100 they are also spatially comparable across different climatic zones (humid or arid). The SD value during any decade gives the dryness/wetness during that decade when compared to long-term historical data. Drought occurs only when the dryness continues for a prolonged period of time that can affect crop growth. As the limits of SD values are between -100 and $+100$, the worst drought can be represented by a straight line with the equation:

$$\sum_{t=1}^{j} z_t = -100t - 100 \qquad (15.5)$$

Where t is the time in decades. If this line defines the worst drought (i.e., -3 for the drought index to be comparable with PDSI), then SMDI for any given decade can be calculated by:

$$SMDI_j = \frac{\sum_{t=1}^{j} SD_t}{25t + 25} \qquad (15.6)$$

Now the task is to choose the time period (decades) over which the dryness values need to be accumulated to determine drought severity. In order to overcome this and take the time period into account indirectly, the drought index is calculated on an incremental basis as suggested by Palmer [17]:

$$SMDI_j = SMDI_{j-1} + \Delta SMDI_j \qquad (15.7)$$

In order to evaluate the contribution of each month to drought severity, we can set $i = 1$ and $t = 1$ in Equation 15.6 and we have:

$$SMDI_1 = \frac{SD_1}{50} \qquad (15.8)$$

Since this is the initial month:

$$SMDI_1 - SMDI_0 = \Delta SMDI_1 = \frac{SD_1}{50} \qquad (15.9)$$

A drought will not continue in the extreme category if subsequent months are normal or near normal. Therefore, the rate at which SD must increase in order to maintain a constant value of SMDI depends on the value of SMDI to be maintained. For this reason, an additional term must be added to Equation 15.9 for all months following an initial dry month:

$$\Delta SMDI_j = \frac{SD_j}{50} + cSMDI_{j-1} \qquad (15.10)$$

Where $\Delta SMDI_j = SMDI_j - SMDI_{j-1}$
Equation 15.10 can now be solved for c. By assuming SMDI is -3 during subsequent time steps, then SDi should be -100:

$$\Delta SMDI_j = \frac{-100}{50} + c(-3.0), \qquad (15.11)$$

$$0 = -2 - 3c$$

Therefore, drought severity in any given decade is given by:

$$SMDI_j = SMDI_{j-1} + \frac{SD_j}{50} - 0.5SMDI_{j-1} \tag{15.12}$$

SMDI during any decade will range from −3 to +3 representing dry to wet conditions.

15.5 Results

Figure 15.3 shows the SSMI derived soil moisture deficit index for the first 10 days of May 2000 through 2005 with −3 indicating severe drought while +3 indicates very moist conditions. The prolonged drought that occurred over large parts of Europe associated

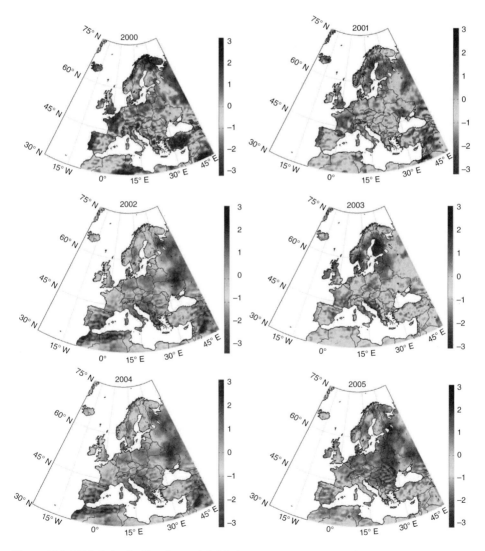

Figure 15.3 SSMI derived soil moisture deficit index.

with the summer heat wave of 2003 [6] is obvious in the image of 2003 with France and mainly southern European states bearing the brunt of the 2003 summer drought. The summer drought of 2003 did not affect Poland, while the drought significantly affected Germany. Likewise, the most serious drought in the Iberian Peninsula in 60 years, which occurred in 2005, reducing overall European Union (EU) cereal yields by an estimated ten percent [7] is seen in the 2005 image. However, the rest of Europe was not affected by the 2005 drought. Results also show that there was severe drought in central Europe in the year 2000 and severe drought in Eastern Europe in the year 2002. The drought in 2002 also affected France and the Iberian peninsula albeit moderately. Scandinavia is usually associated with high moisture content. However, it can be seen that in 2004, there was a moderate drought in the whole Scandinavia with the exception of Denmark. The 2004 drought also affected the northern parts of the United Kingdom and parts of France while the rest of Europe remained moist.

15.6 Discussion

The drought index derived from the SSMI soil moisture is in agreement with data from the Polish Central Statistics Office, best represented by the extreme event of 2000 [55]. In 2000 the thermal-humidity conditions were highly varied and severe weather anomalies occurred. It was the warmest year in Poland from 1961–2005 and warmer than an average of 1961–1990 [55,56]. Yearly average air temperature was about 9.6 °C and was higher than the long-term period (1961–1990) average. Compared to the long-term averages, the warmest months were: February (by about 4.1 °C), April (by about 4.3 °C, in the third decade +8.5 °C), October (by about 3.3 °C) and November (by about 3.2 °C), while the coolest were July and September with the average air temperature below the normal of about 1.1 °C and 1.0 °C respectively. Paradoxically the average annual sum of atmospheric precipitation was about 644 mm and exceeded the long-term period (1971–2000) value by about 7%. The highest excess of precipitation compared to normal value was observed in March – above 130% (in the first decade of March – 285%) and in July – above 80%, while the highest deficit was noticed in the third decade of September – about 90%, in the second decade of October – above 80% and in the third decade of October – about 60% [55]. The prolonged drought was therefore a result of poor temporal distribution of annual precipitation, with nearly all annual precipitation being received in the month of March to April.

The growing season during the year 2000 was extremely long – it lasted about 275 days (from 266 in Podlaskie Voivodship in north-east Poland to 297 days in Zachodniopomorskie Voivodship in north-west Poland) and was longer than previous one by about five weeks, and from the long-term period (1981–95) by an average of nearly four weeks [55]. This seasonal anomaly is attributed to the prolonged drought of 2000 as shown by the SMDI outlined in Figure 15.3. High variability of weather conditions caused high diversity of vegetation growing conditions, however, the level of this year's production (yields and harvests) of main agricultural and horticultural crops was, above all, determined by the agro-meteorological conditions during spring (drought and frost in May) and autumn (extremely warm, sunny and long) [55].

Despite the weather variations in January and February, winter conditions didn't generate larger risks for winter crops. However, recurring thawing and freezing processes of soil and strong drying wind were unfavourable for winter crops. Warm weather in the third decade of February and first decade of March caused an early start of the growing season in the Western and Central Poland. In the first decade of March 2000 abundant rainfalls, which constituted 200–300% of the long-term period (1971–2000) entailed excessive amount of soil moisture, resulting in inundations of many rivers (especially in southern Poland), causing localized flooding. These entailed significant losses in crops in fields with stagnant water.

Due to a sudden drop in air temperature that occurred again in the second decade of March, the growing processes were slowed down. Locally, in the western regions of Poland farmers started spring field work. Definite increase of air temperature in the third decade of March caused the start of vegetative growth of winter crops and permanent grasslands in the whole country (as in the previous year). Generally, spring field work started and in many regions – sowing of spring cereals. However, due to frequent and excessive rainfall amounts field work was locally (in some regions) delayed. Also at the beginning of April, rather cold weather and excessive soil moisture delayed field work. Therefore at the end of the first decade of April progress of these works was lower compared to the previous year. From 5–7 April there was intense precipitation in southeast Voivodships with rainfall totals considerably exceeded ten-day long-term periodic averages (by 300–400%) [55]. Locally, such heavy rainfalls caused floods and inundation of fields and grasslands. Very warm and sunny weather in the second decade of April contributed to drying up fields and significantly influenced the tempo of vegetation growth and development. Sufficient soil moisture values in these regions of the country also ensured good crop vegetative conditions. Favourable agro-meteorological conditions were observed also in the third decade of April, however, these favourable agro-meteorological conditions were short-lived as extremely high air temperature, intense insolation and lack of precipitation caused in many regions excessive over drying of soil surface, especially in northern and central Poland and by the middle of May 2000 drought covered almost the entire country as shown by SMDI in Figure 15.3. The state of crops drastically degraded, particularly in light soils (sandy, sandy-loam). Prolonged drought was extremely adverse for cereals, which were in a phase of increased water demand. The drought lasted up to the end of June and in some regions such as Podlaskie, Mazowieckie, Lodzkie, Wielkopolskie, Opolskie, Lubuskie, Kujawsko-Pomorskie, Warminsko-Mazurskie, lack of soil water content even intensified in summer.

15.7 Conclusion

Drought is a complex natural disaster that causes serious environmental, social, and economic consequences worldwide. It impacts both surface and groundwater resources and can lead to reduced water supply, deteriorated water quality and crop failure. Soil moisture deficit in the root zone during various stages of the crop growth cycle will have a profound impact on crop yield. As precipitation and soil properties have a high degree of spatial variability, determination of a drought index using an approach that is fundamentally linked to soil moisture and its measurement in the vadose zone, can only serve to enhance our ability to quantify local drought, especially during the crop-growing season. However, drought indices currently used for drought monitoring do not give proper consideration to the use of satellite derived soil moisture in the determination of a drought index. In this study soil moisture data derived from the special sensor microwave imager (SSMI) was first scaled to vadose zone soil moisture using the recursive exponential filter for time sensitive data. The resultant data was then used to determine the drought index over continental Europe over the past 25 years taking into account the long-term minimum, long-term maximum and long-term median soil moisture values for each pixel. This procedure captured the prolonged drought that occurred over large parts of Europe associated with the summer heat wave of 2003 with France and mainly southern European states bearing the brunt of the 2003 summer drought. The summer drought of 2003 did not affect Poland while the drought significantly affected Germany. Likewise, the result also shows the severe drought that hit the Iberian Peninsula in 2005. The drought index derived from the SSMI soil moisture is in agreement with data from the Polish Central Statistics Office, best represented by the extreme event of 2000. In 2000, the thermal-humidity conditions were highly varied and severe weather anomalies occurred. It was the warmest year in Poland from 1961–2005 and warmer than an average of 1961–90.

References

1 W.E. Riebsame, S.A. Changnon and T.R. Karl, *Drought and Natural Resource Management in the United States: Impacts and Implications of the 1987–89 Drought*, Westview Press, p. 174, (1990).

2 D. Le Comte, Weather highlights around the world. *Weatherwise*, **48**, (1995).

3 S. Demuth and K. Stahl (eds.), Assessment of the Regional Impact of Droughts in Europe. *Final Report to the European Union, ENV-CT97-0553:* Institute of Hydrology, University of Freiburg, Germany, (2001).

4 R. Voss, W. May and E. Roeckner, Enhanced resolution modeling study on anthropogenic climate change: changes in extremes of the hydrological cycle. *International Journal of Climatology*, **22**, 755–777, (2002).

5 R.T. Watson, M.C. Zinyowera and R.H. Moss (eds.), *The Regional Impacts of Climate Change – An Assessment of Vulnerability, IPCC Special Report, Summary for Policymakers*: Intergovernmental Panel of Climate Change, ISBN: 92-9169-110-0, (1997).

6 L. Feyen and R. Dankers, Impact of global warming on streamflow drought in Europe. *J. Geophys. Res.*, **114**, D17116, (2009).

7 United Nations Environment Programme, *Geo Year Book 2006: An Overview of Our Changing Environment, Nairobi*, (2006).

8 D.A. Wilhite and M.H. Glantz, Understanding: the Drought Phenomenon: The Role of Definitions. *Water International*, **10**, 111–120, (1985).

9 L. Ji and A.J. Peters, Assessing vegetation response to drought in the northern Great Plains using vegetation and drought indices. *Remote Sensing of Environment*, **87**, 85–98, (2003).

10 E.A.B. Eltahir, Drought frequency analysis in Central and Western Sudan. *Hydrological Sciences Journal*, **37**, 185–199, (1992).

11 T.J. Chang and X.A. Kleopa, A proposed method for drought monitoring. *Water Resources Bulletin*, **27**, 275–281, (1991).

12 M.A. Santos, Regional droughts: A stochastic characterization. *Journal of Hydrology*, **66**, 183–211, (1983).

13 H. Latifi and B. Galos, Remote sensing-supported vegetation parameters for regional climate models: a brief review. *iForest*, **3**, 98–101, (2010).

14 P.J. Zarco-Tejada and G. Sepulcre-Cantó, Remote sensing of vegetation biophysical parameters for detecting stress condition and land cover changes. *Studies of the Unsaturated Zone of the Soil*, **VIII**, 37–44, (2007).

15 A.K. Mishra and V.P. Singh, A review of drought concepts. *Journal of Hydrology*, **391**, 202–216, (2010).

16 U.S. Panu and T.C. Sharma, Challenges in drought research: some perspectives and future directions. *Hydrological Sciences Journal*, **47**, S19–S30, (2002).

17 W.C. Palmer, *Meteorological drought, Research Paper 45*. U.S. Department of Commerce, Weather Bureau, Washington, DC, (1965).

18 W.C. Palmer, Keeping track of crop moisture conditions, nationwide: the new crop moisture index. *Weatherwise*, **21**, 156–161, (1968).

19 T.B. McKee, N.J. Doesken and J. Kleist, The relationship of drought frequency and duration to time scales. In: *Proceedings of the 8th Conference on Applied Climatology*, American Meteorological Society, Anaheim, CA, Boston, MA, 17–22 January, pp.179–184, (1993).

20 J.J. Keetch and G.M. Byram, *A Drought Index for Forest Fire Control*. Southeastern Forest Experiment Station, Asheville, NC, (1968).

21 S.E. Hollinger, S.A. Isard and M.R. Welford, A new soil moisture drought index for predicting crop yields. In: *Preprints, Eighth Conference on Applied Climatology*, Anaheim, CA, American Meteorological Society, pp. 187–190, (1993).

22 B.A. Shafer and L.E. Dezman, Development of surface water supply index (SWSI) to assess the severity of drought conditions in snow pack runoff areas. In: *Proceedings of the 50th Western Snow Conference, Colorado State University Press*, Reno, NV/FortCollins, CO, pp. 164–175, (1982).

23 S.J. Meyer, K.G. Hubbard and D.A. Wilhite, A crop-specific drought index for Corn. I. Model development and validation. *Agronomy Journal*, **85**, 388–395, (1993).

24 W.T. Liu and F.N. Kogan, Monitoring regional drought using the Vegetation Condition Index. *International Journal of Remote Sensing*, **17**, 2761–2782, (1996).

25 B. Narasimhan and R. Srinivasan, Development and evaluation of Soil Moisture Deficit Index (SMDI) and Evapotranspiration Deficit Index (ETDI) for agricultural drought monitoring. *Agricultural and Forest Meteorology*, **133**, 69–88, (2005).

26 E.D. Hunt, K.G. Hubbard, D.A. Wilhite, T.J. Arkebauer and A.L. Dutcher, The development and evaluation of a soil moisture index. *International Journal of Climatology*, **29**, 747–759, (2009).

27 V. Sridhar, K.G. Hubbard, J. You and E.D. Hunt, Development of the Soil Moisture Index to Quantify Agricultural Drought and Its "User Friendliness" in Severity-Area-Duration Assessment. *Journal of Hydrometeorology*, **9**, 660–676, (2008).

28 H.F. Diaz, Some aspects of major dry and wet periods in the contiguous United States. *Journal of Climate and Applied Meteorology*, **22**, 1895–1981, (1983).

29 P.T. Soulé, Spatial patterns of drought frequency and duration in the contiguous USA based on multiple drought event definitions. *International Journal of Climatology*, **12**, 11–24, (1992).

30 P.D. Jones, M. Hulme, K.R. Brifta, C.G. Jones, J.F.B. Mitchell and J.B. Murphy, Summer moisture accumulation over Europe in the Hadley center general circulation model based on the Palmer drought severity index. *International Journal of Climatology*, **16**, 155–172, (1996).

31 T.B. Heddinghaus and P. Sahol, A Review of the Palmer Drought Severity Index and Where Do We Go From Here? In: *Proc. 7th Conf. on Applied Climatology*, September 10–13, 1991, American Meteorological Society, Boston, Massachusetts, pp. 242–246, 1991.

32 T. Kim and J.B. Valdes, Nonlinear model for drought forecasting based on a conjunction of wavelet transforms and neural networks. *Journal of Hydrologic Engineering, ASCE* **8**, 319–328, (2003).

33 N.B. Guttman, J.R. Wallis and J.R.M. Hosking, Spatial comparability of the Palmer drought severity index. *JAWRA Journal of the American Water Resources Association*, **28**, 1111–1119, (1992).

34 M.J. Hayes, M.D. Svoboda, D.A. Wilhite and O.V. Vanyarkho, Monitoring the 1996 drought using the standardized precipitation index. *Bulletin of American Meteorological Society*, **80**, 429–438, (1999).

35 N.B. Guttman, Comparing the Palmer drought index and the standardized precipitation index. *Journal of American Water Resources Association*, **34**, 113–121, (1997).

36 C.W. Thornthwaite, An approach toward a rational classification of climate. *Geographical Review*, **38**, 55–94, (1948).

37 M.E. Jensen, R.D. Burman and R.G. Allen, *Evapotranspiration and irrigation water requirements, ASCE Manuals and Reports on Engineering Practice No. 70*. American Society of Civil Engineers, New York, NY, (1990).

38 S.J. Meyer, K.G. Hubbard and D.A. Wilhite, A crop-specific drought index for Corn. II. Application in drought monitoring and assessment. *Agronomy Journal*, **85**, 396–399, (1993).

39 O.V. Wilhelmi, K.G. Hubbard and D.A. Wilhite, Spatial representation of agroclimatology in a study of agricultural drought. *International Journal of Climatology*, **22**, 1399–1414, (2002).

40 Y. Gu, J.F. Brown, J.P. Verdin and B. Wardlow, A five-year analysis of MODIS NDVI and NDWI for grassland drought assessment over the central Great Plains of the United States. *Geophysical Research Letters*, **34**, L06407, (2007).

41 A.J. Peters, E.A. Walter-Shea, Lei Ji., A. Vina, M. Hayes and M.D. Svoboda, *Drought monitoring with NDVI-based Standardized Vegetation Index*, Bethesda, MD, ETATS-UNIS, American Society for Photogrammetry and Remote Sensing, (2002).

42 F.N. Kogan, Droughts of the late 1980s in the United States as derived from NOAA polar-orbiting satellite data. *Bulletin of American Meteorological Society*, **76**, 655–668, (1995).

43 K. Dabrowska-Zielinska, F. Kogan, A. Ciolkosz, M. Gruszczynska and W. Kowalik, Modelling of crop growth conditions and crop yield in Poland using AVHRR-based indices. *International Journal of Remote Sensing*, **23**, 1109–1123, (2002).

44 R. Heim, A review of twentieth-century drought indices used in the United States. *Bulletin of American Meteorological Society*, **83**, 1149–1165, (2002).

45 D.C. Hane and F.V. Pumphrey, Crop water use curves for irrigation scheduling. In: *Agricultural Experiment, Station, Oregon State University*, Corvallis, OR, (1984).

46 R.M. Sandvig and F.M. Phillips, Ecohydrological controls on soil moisture fluxes in arid to semiarid vadose zones. *Water Resources Research*, **42**, W08422, (2006).

47 J. Sheffield, G. Goteti, F. Wen and E.F. Wood, A simulated soil moisture based drought analysis for the United States. *Journal of Geophysical Research-Atmospheres*, **109**, D24108, (2004).

48 R.D. Koster, P.A. Dirmeyer, Z. Guo, G. Bonan *et al.*, Regions of strong coupling between soil moisture and precipitation. *Science*, **305**, 1138–1140, (2004).

49 D.J. Cavalieri, C.L. Parkinson, N. Digirolamo and A. Ivanoff, Intersensor Calibration Between F13 SSMI and F17 SSMIS for Global Sea Ice Data Records. *IEEE Geoscience and Remote Sensing Letters*, **9**, 233–236, (2012).

50 J.R. Wang and B.J. Choudhury, Remote sensing of soil moisture content over bare field at 1.4 GHz frequency. *Journal of Geophysical Research-Oceans*, **86**, 5277–5282, (1981).

51 P.D. Stroud, *A recursive exponential filter for time-sensitive data*, Los Alamos National Laboratory, LAUR-99-5573, available at: public.lanl.gov/stroud/ExpFilter/ExpFilter995573.pdf. (1999).

52 C. Albergel, C. Rüdiger, T. Pellarin, J.-C. Calvet, N. Fritz, F. Froissard, D. Suquia, A. Petitpa, B. Piguet and E. Martin, From near-surface to root-zone soil moisture using an exponential filter: an assessment of the method based on in-situ observations and model simulation. *Hydrology and Earth System Science*, **5**, 1323–1337, (2008).

53 T. Pellarin, J.-C. Calvet and W. Wagner, Evaluation of ERS scatterometer soil moisture products over a half-degree region in southwestern France. *Geophysical Research Letters*, **33**, (2006).

54 W. Wagner, *Soil moisture retrieval from ERS scatterometer data*, Ph.D. thesis, University of Technology, Vienna, 101 pp., (1998).

55 A. Bogumił, B. Dawidek, M. Lewandowska, M. Pawelec-Potapska, M. Wojciechowski and A. Wróblewska, *An evaluation of situation in agriculture with respect to change in production and economic conditions and prodaction results (harvest level, trends in livestock growth and level of output and market supply of animal production) against a background of use of production factors*, Central Statistical Office of Poland, al. Niepodległości 208, 00-925 Warsaw, (2001).

56 A. Bogumił, B. Dawidek, M. Lewandowska, M. Pawelec-Potapska, M. Wojciechowski and A. Wróblewska, *An evaluation of situation in agriculture with respect to change in production and economic conditions and prodaction results (harvest level, trends in livestock growth and level of output and market supply of animal production) against a background of use of production factors*, Central Statistical Office of Poland, al. Niepodległości 208, 00-925 Warsaw, (2010).

16

Conclusions

Heiko Balzter

National Centre for Earth Observation, University of Leicester, Centre for Landscape and Climate Research, Department of Geography, Leicester, UK

The contributing chapters in this book have highlighted how satellites and other remote sensing platforms can be used in better ways to monitor the environment we live in. Ecosystem services provide vital life support to the human population on this planet. We have to use them wisely and in ways that do not threaten their sustainability.

Besides the JAXA global forest/non-forest maps from JERS-1 and ALOS-1 and 2 [1], there is currently no operational forest monitoring service using Synthetic Aperture Radar (SAR) data, despite successful demonstration projects that have mapped other biophysical parameters such as total growing stock (e.g. the SIBERIA project). Forest monitoring methods using Synthetic Aperture Radar (SAR) are still in the research and development stage, with the exception of forest cover mapping where global products are emerging from Japan and the US. Very few operational services using SAR are available to support sustainable forestry today. Forest landowners, forest managers and policy makers require regular access to additional information, including forest canopy height and forest cover change. Forest canopy height is an important economic indicator of the estimated timber yield from managed forests, indicates areas affected by disturbances and shows the general environmental state of the forest. Forest cover change needs to be monitored much more frequently than is provided by the CORINE and HRL data products, which are produced every ~6 years.

Part I of this book has presented three approaches to forest monitoring involving SAR sensors. In Chapter 2, current approaches to the mapping and monitoring of aboveground forest biomass from optical, radar and LiDAR sensors are contrasted. A maximum entropy approach for the estimation of forest aboveground biomass from a combination of geospatial data is presented that improves forest area maps as well. The validation with forest inventory data shows that multi-sensor monitoring concepts achieve higher accuracy than single satellites alone, and can be used for a Global Biomass Information System. This concept has informed the European Space Agency's current GLOBBIOMASS initiative. The new satellites from the ESA Copernicus Programme (the Sentinels), JAXA (ALOS PALSAR-2) and NASA (i.e. Landsat 8, IceSAT-2) will continue delivering such global maps. By the end of this decade, the BIOMASS mission by ESA will achieve even better accuracies of forest biomass from the first P-band SAR in space.

Chapter 3 has shown how radar remote sensing can be used for aboveground biomass estimation in boreal forests. It shows how radar backscatter is related to increasing biomass up to a saturation limit that can physically be explained by the water cloud model. The findings show that the radar signal saturates at a certain biomass level depending on the wavelength, polarisation, forest structure and wet/dry conditions. Interestingly, the integration of a large number of repeated SAR acquisitions can increase saturation limits of the radar signal. It can also be increased by using several polarizations and wavelengths. Interferometric SAR data can provide canopy height and coherence is lower for high biomass forests due to volume decorrelation and can therefore both be used for biomass retrieval. Biomass estimation errors reported in the literature varied from 7.4% (semi-empirical models of multi-temporal C-band coherence and backscatter) to over 100%.

Chapter 4 describes how SAR data can produce more accurate maps of forest/non-forest over the Congo basin, when compared to JAXA's Kyoto and Carbon Initiative forest/non-forest product. A supervised regional algorithm produces better results, particularly in forested forests, which can be misclassified as non-forest in the JAXA map. The Copernicus Sentinel 1 mission, the first of satellite of which was launched in April 2014, can provide six-day repeat acquisitions, which could become a useful information service for rapid deforestation monitoring. In particular if Sentinel-1 is analysed in conjunction with ALOS-2, also launched in 2014, SAR data have the potential to become the primary data source for monitoring in frequently cloudy tropical forest environments.

Emergency and disaster monitoring services are currently mainly provided by the International Charter on Space and Natural Disasters, which when activated provides frequent coverage with all available satellites. The main SAR data supply of the future of the Copernicus services will be from the Sentinel missions, which will start with C-band SAR of Sentinel-1 and later with the optical satellites Sentinel-2 and 3. For the monitoring of natural disasters and emergencies, rapid land cover mapping is often required. Because this can be difficult to achieve with optical sensors under cloudy conditions, SAR sensors are needed to provide cloud-independent imaging capability. Chapter 5 has analysed multi-frequency SAR data over selected African test sites in order to assess how well they can discriminate between basic land cover types. The frequent revisit time of Sentinel-1 (six days when both satellites are launched) is providing multi-temporal C-band SAR imagery that can achieve high accuracies when used for land cover mapping [2]. The chapter shows the additional value of using multi-frequency SAR. Even adding a second frequency, particularly when L-band and X-band are combined, the classification accuracy can be much improved, as was shown for both the Tanzania and Congo sites. The choice of imagery during the wet and dry season must be carefully considered because SAR is sensitive to moisture.

Chapter 6 presents a technique based on the combination of pixel-based and object-based approaches for an unsupervised analysis of high-resolution images. It uses multi-date image segmentation to generate objects that can be further improved using conditional image object fusion. The results presented in this chapter confirm the effectiveness of the approach. It has very high operational potential in situations where rapid maps of change/no change are required.

In Chapter 7, the Earth Observation Land Data Assimilation System (EO-LDAS) is investigated over the Barrax agricultural area in Spain. Different constraints were tested, including prior information on the canopy parameters, time regularization, spatial regularization and multi-sensor information. State variables such as leaf area index and chlorophyll content were estimated and hyperspectral signatures of the canopy were simulated. The chapter shows that EO-LDAS is able to estimate leaf area index and chlorophyll concentration with reasonable accuracy (correlation from 0.63 to 0.95). The feasibility of the EO-LDAS time regularization was demonstrated using four spectral bands of MISR at 275 m. The results are very encouraging and show that the leaf area index has lower uncertainties of 0.015 rather than 42. The results in this chapter show that EO-LDAS has huge potential for innovative applications.

In Chapter 8, airborne multi-frequency polarimetric SAR is applied to a salt marsh site in Wales. The SAR data can differentiate between vegetation habitats that differ in three-dimensional appearance. S-band and X-band respond differently to these vegetation type and topographic changes. X-band performed worse than S-band in differentiating salt marsh vegetation types but was better at detecting subtle differences in surface roughness in the pioneer zone and the channel networks of the salt marsh. In anticipation of the launch of NovaSAR-S, an S-band SAR mission being built in the UK, innovative applications of S-band SAR in forestry, agriculture and urban planning are required.

Chapter 9 has shown how reed plants along the shore of Lake Balaton in Hungary can be mapped using field spectroscopy and hyperspectral remote sensing. The results show the separability between reed species at leaf and canopy level and for different physiological states. The reed die-back syndrome is affecting many lakeshore areas around the world, and if a suitable mapping method can be demonstrated based on the work in this chapter, then the costs of surveys could be reduced. The classification of airborne hyperspectral imagery to map emergent macrophyte species resulted in a high accuracy. New sensors such as the Copernicus Sentinel-2 mission carry superspectral instruments that are superior to Landsat-type sensors.

Chapter 10 describes the retrieval, validation and mapping of chlorophyll-a (chl-*a*) concentrations in the optically complex lake waters of Lake Balaton, which contain a variable mixture of coloured dissolved organic matter (CDOM), chl-*a* and suspended particulate matter (SPM). Concentrations of chl-*a* are indicative of phytoplankton biomass and algal blooms, which underlie water productivity and quality, and have been accurately mapped in Lake Balaton with the fluorescence line height algorithm applied to MERIS satellite imagery. The processing chain to retrieve chl-*a* from MERIS can be transferred to Sentinel-3 OLCI and also adapted to Sentinel-2 MSI data.

In Chapter 11, two interferometric SAR (InSAR) methodologies are applied, the differential InSAR (DInSAR) and Persistent Scatterer Interferometry (PSI). A key limitation of these methods is the poor spatial coverage of the deformation maps when both methods are applied separately. The chapter demonstrates how both methods can be combined in order to improve the spatial coverage and provide better spatial information services to civil security agencies and public authorities.

Chapter 12 has examined the surface velocity of several Alpine rock glaciers in Switzerland from interferometric SAR satellite data and ground-based radar data. The large data archives of satellite SAR data can help in studying the historical behaviour of geomorphological processes and the kinematics of rock glaciers, while the ground-based radar can be deployed within a couple of hours after a mass movement if required. SAR data such as TerraSAR-X can be used for the compilation of an inventory of unstable slopes in the Alps.

In Chapter 13, sample supervised search-centric approaches in geographic object-based image analysis are reviewed. All mapping applications can benefit from such metaheuristic strategies, but particularly time-critical mapping tasks such as emergency response/rapid mapping approaches can benefit substantially. In such rapid emergency applications, image analysis solutions (supervised classifiers, rule sets) are not readily implemented beforehand and an operator has to interpret the satellite images. Concepts from machine learning, optimization, computer vision, image processing, remote sensing and geographic information science can all be applied to optimise the processing challenges posed by the demands of rapid mapping.

Chapter 14 provides a review on wetland mapping, classification, and monitoring applications using optical and SAR data of different spatial and temporal scales. The case study of Lac Bam in Burkina Faso has shown how remote sensing can be used to map and monitor wetland areas as indicators of water availability in semi-arid Africa. Synergistic combinations of multiple sensors can improve the detection of water bodies and reservoirs, flooded areas and floating vegetation on water. A novel aspect of this chapter is that it proposed time-series analysis of long-term data archives to analyse spatio-temporal trends of water bodies, wetlands and irrigation. The seasonality of the water cycle in some parts of Africa is absolutely critical to secure food supplies.

Chapter 15 has analysed the SSMI derived soil moisture deficit index and its ability to detect severe drought and extremely wet conditions in Poland, where summer droughts can significantly reduce harvest. The prolonged drought in Europe in 2003 was evident in the imagery, but the heat wave did not affect Poland at the time. A severe drought occurred in Eastern Europe in 2002, also affecting France and the Iberian Peninsula. The ability of the remote sensing indicators to detect the extent and severity of such droughts will improve our ability to cope with the impacts of such extreme weather, which is likely to increase under scenarios of climate change.

In conclusion, the breadth and depth of the studies shown in the chapter contributions in this book provide ample material for the improvement of existing monitoring services and the development of new innovative ways of using satellite and other remote sensing data for the benefits of citizens. Some of the applications described in this book have a very high potential to lead to either commercial or public information services in support of the European environmental monitoring programme from Earth Observation.

References

1 Masanobu Shimada, Takuya Itoh, Takeshi Motooka, Manabu Watanabe, Shiraishi Tomohiro, Rajesh Thapa and Richard Lucas, New Global Forest/Non-forest Maps from ALOS PALSAR Data (2007–2010), *Remote Sensing of Environment*, **155**, 13–31, (2014).
2 Heiko Balzter, Beth Cole, Christian Thiel and Christiane Schmullius, Mapping CORINE Land Cover from Sentinel-1 SAR and SRTM Digital Elevation Model using Random Forests. *Remote Sensing* 7, 14876–14898, (2015).

Index